T5-AGA-406

Wastewater Reuse—Risk Assessment, Decision-Making and Environmental Security

NATO Science for Peace and Security Series

This Series presents the results of scientific meetings supported under the NATO Programme: Science for Peace and Security (SPS).

The NATO SPS Programme supports meetings in the following Key Priority areas: (1) Defence Against Terrorism; (2) Countering other Threats to Security and (3) NATO, Partner and Mediterranean Dialogue Country Priorities. The types of meeting supported are generally "Advanced Study Institutes" and "Advanced Research Workshops". The NATO SPS Series collects together the results of these meetings. The meetings are co-organized by scientists from NATO countries and scientists from NATO's "Partner" or "Mediterranean Dialogue" countries. The observations and recommendations made at the meetings, as well as the contents of the volumes in the Series, reflect those of participants and contributors only; they should not necessarily be regarded as reflecting NATO views or policy.

Advanced Study Institutes (ASI) are high-level tutorial courses intended to convey the latest developments in a subject to an advanced-level audience

Advanced Research Workshops (ARW) are expert meetings where an intense but informal exchange of views at the frontiers of a subject aims at identifying directions for future action

Following a transformation of the programme in 2006 the Series has been re-named and re-organised. Recent volumes on topics not related to security, which result from meetings supported under the programme earlier, may be found in the NATO Science Series.

The Series is published by IOS Press, Amsterdam, and Springer, Dordrecht, in conjunction with the NATO Public Diplomacy Division.

Sub-Series

A.	Chemistry and Biology	Springer
B.	Physics and Biophysics	Springer
C.	Environmental Security	Springer
D.	Information and Communication Security	IOS Press
E.	Human and Societal Dynamics	IOS Press

http://www.nato.int/science
http://www.springer.com
http://www.iospress.nl

Series C: Environmental Security

Wastewater Reuse–Risk Assessment, Decision-Making and Environmental Security

Edited by

Mohammed K. Zaidi
Idaho State University,
College of Engineering,
Pocatello, ID, U.S.A.

Published in cooperation with NATO Public Diplomacy Division

Proceedings of the NATO Advanced Research Workshop on
Wastewater Reuse – Risk Assessment, Decision-Making
and Environmental Security
Istanbul, Turkey
12–16 October 2006

A C.I.P. Catalogue record for this book is available from the Library of Congress.

ISBN 978-1-4020-6026-7 (PB)
ISBN 978-1-4020-6025-0 (HB)
ISBN 978-1-4020-6027-4 (e-book)

Published by Springer,
P.O. Box 17, 3300 AA Dordrecht, The Netherlands.

www.springer.com

Printed on acid-free paper

TABLE OF CONTENTS

NATO ARW 2006 ISTANBUL—LIST OF PARTICIPANTS

DIRECTORS:

NATO ARW DIRECTOR: Mohammed K. Zaidi, M.S., Idaho State University, College of Engineering, Pocatello, ID. 83209-8060, USA. zaidmoha@isu.edu

NATO ARW CO DIRECTOR: Nava Haruvy, Ph.D., Prof., Netanya Academic College, 1 University Rd, Netanya 42100, Israel. navaharu@netvision.net.il

CO-EDITORS:

Nava Haruvy, Netanya Academic College, Netanya, Israel.

Eugene Levner, Holon Institute of Technology, Holon, Israel.

LIST OF SPEAKERS

AZERBAIJAN:

Azad A. Bayramov, D.Sc., Prof., Baku State University, Baku, Azerbaijan. Bayramov_azad@mail.ru

Islam Mustafaev, Ph.D., Institute of Radiation Problems of National Academy of Sciences, Baku, Azerbaijan. Imustafaev@Iatp.Az

CANADA:

Yehuda Kleiner, Ph.D., Institute for Research in Construction, National Research Council, Ottawa, Ontario, Canada. Yehuda. Kleiner@nrc-cnrc.gc.ca

M. Tahir Rashid, Ph.D., Guelph University, Guelph, Ontario, Canada. trashid@ uoguelph.ca

CZECH REPUBLIC:

Jiri Kubik, Ph.D., student, Brno University of Technology, Faculty of Civil Engineering, Institute of Municipal Water Management, Zizkova 17, 602 00 Brno, Czech Republic. hlavinek.p@fce.vutbr.cz

Zdenek Filip, Ph.D., Department of Biochemistry and Microbiology, Institute of Chemical Technology, Prague, Czech Republic. Zdenek.Filip@vscht.cz

EGYPT:

Hussein I. Abdel-Shafy, Ph.D., Department of Water Research & Pollution Control, National Research Centre, Dokki, Cairo, Egypt. husseinshafy@yahoo.com
Mohamed Tawfic Ahmed, Ph.D., Prof., Suez Canal University, Ismailia, Egypt. motawfic@tedata.net.eg

GREECE:

Phoebe Koundouri, Ph.D., Assistant Professor in Economics (Epicouros), Athens University of Economics and Business, Department of International and European Economic Studies, 76, Patission Street, Athens 104 34, Greece. pkoundouri@aueb.gr

ISRAEL:

Saul Arlosoroff, MBA, Director, and Chairman of its Finance/Economic Committee, "Mekorot"—The National Water Corporation of Israel, Tel Aviv, Israel. sarlo@inter.net.il

Gilad Axelrad, Ph.D., student, Department of Agricultural Economics and Management, The Hebrew University of Jerusalem, Faculty of Agricultural, P.O. Box 12, Rehovot 76100, Israel. axelrad@agri.huji.ac.il

Yosef Dreizin, Ph.D., Head, Water Desalination Administration, Water Commission, Tel Aviv, Israel. jossefdr10@water.gov.il

Yossi Inbar, Ph.D., Senior Deputy Director General, Industries, Ministry of the Environment, Jerusalem, Israel. yossii@sviva.gov.il

Yoav Kislov, Ph.D., Prof. Emeritus, Department of Agricultural Economics, Hebrew University, Rehovot 76100, Israel. kislev@agri.huji.ac.il

Eugene Levner, Ph.D., Prof., Holon Academic Institute of Technology, 52 Golomb St. Holon 58102, Israel. levner@hait.ac.il

Yossi Manor, Head, Insurance Organization, Hamasger 14, Tel Aviv, Israel. yossi@manor.co.il

Reuven Pedatzur, Ph.D., Natanya Academic College, School of Communication, Netanya Academic College, 1 University Rd, Netanya 42100, Israel. pedat@smile.net.il

Ezra Sadan, Ph.D., Prof., Manager, Sadan and Lovental, Hebrew University of Jerusalem, Israel. sezra@netvision.net.il

Naomi Tsur, M.A., Director, Jerusalem Branch of Society of Protection of Nature in Israel (SPNI), Jerusalem, Israel. naomit@actcom.net.il

ITALY:

Sureyya Meric, Ph.D., University of Salerno, Civil Engineering Department, Solimena, Italy. msureyya@unisa.it

JORDAN:

Abujamous Mahmoud, Project Manager, Ministry of Agriculture, Jabel El-Hussain, Amman, Jordan. mah12446@yahoo.com

Nizar Halasah, M.Sc., Environmental Research Centre, Royal Scientific Society P.O. Box 1438 Al-Jubiaha, Amman, Jordan. halasah@rss.gov.jo

Zein Nsheiwat, M.A., student, Albert Katz International School of Desert Studies, P.O. Box 201, Amman 11821, Jordan. zein_nsheiwat@hotmail.com

PORTUGAL:

Maria Helena Ferreira Marecos do Monte, Ph.D., Instituto Superior de Engenharia de Lisboa, Rua Conselheiro Emídio Navarro, 1950-062 Lisbon, Portugal. hmarecos@dec.isel.ipl.pt

SPAIN:

David Alcaide, Ph.D., Departamento de Estadística, Investigación Operativa y Computación, University of La Laguna, Campus de Anchieta, 38204 La Laguna, Tenerife, Spain. dalcaide@ull.es

The NETHERLANDS:

Joop de Schutter, M.Sc., UNESCO-IHE, Institute of Water Education, Westvest 7, Delft, The Netherlands 2601 DA. j.deschutter@unesco-ihe.org

TUNISIA:

Hamadi Kallali, Wastewater Treatment and Recycling Laboratory, Water Research and Technologies Centre Route touristique de Soliman, P.O. Box 273, 8020 Soliman, Tunisia. Kallali.Hamadi@inrst.rnrt.tn

TURKEY:

Alper Baba, Ph.D., Canakkale Onsekiz Mart University, Engineering and Architecture Faculty, Department of Geological Engineering, 17020 Canakkale, Turkey. alperbaba@comu.edu.tr

Orhan Gündüz, Ph.D., Dokuz Eylül University, Engineering Faculty, Department of Environmental Engineering, Kaynaklar Campus, 35160-Buca, İzmir, Turkey. orhan.gunduz@deu.edu.tr

Gurdal Kanat, Ph.D., Yildiz Technical University, Environmental Engineering Department Besiktas 34349 Istanbul, Turkey.

Hasan Göksel Özdilek, Ph.D., Çanakkale Onsekiz Mart University, Faculty of Engineering and Architecture, Department of Environmental Engineering, Terzioğlu Campus 17020 Çanakkale, Turkey. G-Ozdilek@yahoo.com

UKRAINE:

Atoyev Konstantyn, Ph.D., Glushkov Institute of Cybernetics, 40, Prospekt Glushkova, Kiev 03022, Ukraine. atoe@isofts.kiev.ua

Namik M. Rashydov, Ph.D., Department of Biophysics and Radiobiology, Institute Cell Biology and Genetic Engineering, NAS Ukraine, Kiev 03022, Ukraine. nrashydov@yahoo.com

UNITED STATES OF AMERICA:

Jobaid Kabir, Ph.D., P.E., Lyndon B. Johnson School of Public Affairs, University of Texas at Austin, Austin, TX, USA. jkabir@lcra.org

John Letey, Ph.D., Prof., Emeritus, Department of Environmental Sciences, University of California, Riverside, CA 92512, USA. John.letey@ucr.edu

Sarit Shalhevet, M.B.A., Senior Researcher, Shalhevet Consulting, 126 Thorndike Street, Brookline, MA 02446, USA. sarit.miki@gmail.com

Paul West, Ph.D., Department of Systems Engineering, United States Military Academy, West Point, New York, NY 10996, USA. Paul.west@usma.edu

Baqar R. Zaidi, Ph.D., Department of Marine Sciences, P.O. Box 9013, University of Puerto Rico, Mayaguez, PR 00681-9013, USA. Brzaidi@uprm.edu

UZBEKISTAN:

Renat Khaydarov, Ph.D., Institute of Nuclear Physics, Ulugbek, Tashkent, 702132 Uzbekistan. physicist@sarkor.uz

LIST OF OTHER CONTRIBUTERS/PARTICIPANTS—NATO ARW, ISTANBUL TURKEY.

EGYPT:
Al Shaarty Abeer, Ph.D., National Institute of Oceanography & Fisheries, Al-Anfoshy, Alexandria, Alexandria, Egypt. husseinshafy@yahoo.com

Aly, Raouf Okasha, Ph.D., Prof., National Centre for Radiation Research and Technology (NCRRT), Cairo, Egypt. raufokasha@yahoo.co.uk

Mohamed A. Amasha, Ph.D., Faculty of Specific Education-Mansoura University-34512 Egypt. Mw_amasha@yahoo.com

Tarek Mohammed Abou Elmaaty, Ph.D., Mansoura University, Damietta, Egypt. tasaid@mans.edu.eg

Mohamed Abdel Geleel Mohamed, Ph.D., Assistant professor, Physical and Environmental Chemistry, Ain Shams University, Nasr City and National Center for Nuclear Safety and Radiation Control, Cairo, Egypt. Mageleel2000@yahoo.com

Narmine Saleh Mahmoud, Ph.D., National Centre for Radiation Research and Technology, P.O. Box 29, Nasr City, Cairo, Egypt. mnarmine2@yahoo.com

Amir Abbas Fahmy, Ph.D., National Centre for Radiation Research and Technology, P.O. Box 29, Nasr City, Cairo, Egypt.

JORDAN:
Mu'taz Al-Alawi, M.Sc., Jordan Environment Society (JES), P.O. Box 16 – Al-Karak 61110, Jordan. alawi1979@yahoo.com

KYRGYZSTAN:
Azamat Tynybekov, Ph.D., Kyrgyz Russian Slavonic University, International Science Center, Bishkek, Kyrgyzstan. azamattynybekov@mail.ru

Charsky Vjacheslav, Ph.D., NGO "Club AGAT", Toktogula st. 173-18, Bishkek 720001, Kyrgyzstan. root@agat.freenet.kg

LITHUANIA:
Gaudenta Sakalauskiene, Ph.D., Director, Joint-stock Company "Daugela" UAB "Daugela", Balsiu str. 7, LT 14263 Vilnius, Lithuania. gaudenta@daugela.lt

ROMANIA:

Mihaela Lazarescu, Ph.D., Air Quality Control, Fluid Mechanics and Pollutant Dispersion Department in National Research and Development Institute for Environmental Protection – Icim Bucharest, Romania. mihaelalazarescu@yahoo.com

TURKMENISTAN:

Kurbangeldi Balliyev, Ph.D., Turkmenistan, Ashgabat 15, Bitarap Turkmenistan. kballyev@online.tm

Ishankuliev Dovletyar, Ph.D., Center for Physical and Mathematical Research, Turkmen State University, Ashgabat, Turkmenistan. geldy_ishankuliev@yahoo.com

TURKEY:

Zahide ACAR, Graduate student, Dokuz Eylul Univeristy, Engineering Faculty, Department of Geological Engineering, Kaynaklar BUCA- Izmir, Turkey. zahideacar@hotmail.com

Gul Asiye AYCIK, Ph.D., Mugla University, Chemistry Department, 48000, Mugla, Turkey. gulasiye@mu.edu.tr

Ozan, Deniz Ph.D., student, Dokuz Eylul Univeristy, Engineering Faculty, Department of Geological Engineering, Tinaz Tepe Kampus-Kaynaklar BUCA- Izmir, Turkey. ozandeniz@comu.edu.tr

Gülden Gökçen, Ph.D., Prof., Dokuz Eylul Univeristy, Engineering Faculty, Department of Geological Engineering, Tinaz Tepe Kampus-Kaynaklar BUCA- Izmir, Turkey. guldengokcen@iyte.edu.tr

Gülbin GÜRDAL, Ph.D., Dokuz Eylul Univeristy, Engineering Faculty, Department of Geological Engineering, Tinaz Tepe Kampus-Kaynaklar BUCA- Izmir, Turkey. ggurdal@comu.edu.tr

Ayhan Sirkeci, Ph.D., Istanbul Technical University, Maslak 34469 Istanbul, Turkey. sirkecia@itu.edu.tr

Dundar Renda, Ph.D., Istiklal Cad. Tunca Apt. No 471/1-1, Tunel, Istanbul, Turkey. drenda@superonline.com

PREFACE

This North Atlantic Treaty Organization (NATO) Advanced Research Workshop (ARW) was devoted to Wastewater Reuse – Risk Assessment, Decision-Making and Environmental Security held in Istanbul, Turkey, at the Hotel Villa Suites, Taksim during October 12–16, 2006. More than 100 scientists had requested to participate but only 63 could attend the meeting representing 20 countries—Azerbaijan, Canada, Czech Republic, Egypt, Greece, Italy, Israel, Jordan, Kyrgyzstan, Lithuania, Portugal, Romania, Spain, The Netherlands, Tunisia, Turkey, Turkmenistan, Ukraine, United States of America, and Uzbekistan.

48 papers and 10 posters were presented; only 45 research papers were selected and put in this proceedings manual. Other papers, although they had good information, could not be included due to poor data, not related to the topic or failed to meet the deadline. You may feel some difficulty in understanding some of the papers due to the fact that they were initially written in presenter's home language and then translated into English by nonscientific people, who have very little knowledge or interest in putting correct scientific terms.

Financial support came from the NATO Program for Security through Science, Public Diplomacy Division and sponsored by the Society of Risk Analysis (SRA). Twelve (12) individual participants contributed towards their travel and two (2) for their living expenses.

The ARW participants, the organizing committee, the host committee, and specially the codirector, Professor Nava Haruvy have to be thanked for their services. I am personally thankful to all the participants for taking their time to come and participate. Furthermore, I want to express my gratitude to those who had submitted their papers for publication.

I am thankful to RESL Director, Dr. R.D. Carlson for his help. Dr. Deniz Betan, Director, NATO Scientific Programs and staff specially Lynne Campbell/Nolan were of great help during the proposal submission, revision of the proposal, its award, the final reports, financial report, and the submission of this manuscript. I am thankful to my wife for helping me and giving me moral support during this painful job of editing this proceedings manual.

SESSION 1. RISK ASSESSMENT METHODS: ENVIRONMENTAL HAZARD IMPACT

EFFECTS OF MINING ACTIVITIES ON WATER AROUND THE ÇANAKKALE PLAIN, TURKEY

Alper Baba[1], O. Deniz[1], and O. Gülen[2]
[1]Çanakkale Onsekiz Mart University, Engineering-Architecture Faculty, Geological Engineering Department, Terzioğlu Campus, 17100, [2]JEOTEK-Hydrogeological Investigation Company, 17100, Çanakkale, Turkey

Corresponding author: ozandeniz@comu.edu.tr

Abstract:

Çan region is rich in clay mines and lignite deposits. Results of major anion-cation and some trace elements in groundwater and surface water around the Çan Plain, showed that groundwater is very reach with calcium-magnesium-sulfate (Ca-Mg-SO_4). Its aluminum (Al) concentrations are more than the acceptable maximum standard value. The results show that mining activity has contaminated water sources. The heavily polluted water is currently under the international standard value around the plain. However, if precautions are not taken, these contaminants may spread in surface water and groundwater.

Keywords: Çan (Çanakkale), water contamination, anion–cation, trace element, mining.

1. INTRODUCTION

Çan district, is located in the Biga is richer than the other parts in point of some natural sources such as mines, fossil and renewable energy (wind, geothermal etc.) types. Besides the Peninsula is situated on an active fault zone named as "North Anatolian Fault Zone". This fault zone passes through the Turkey from the east to the west and many natural sources in the region had occurred due to movements of this fault zone. Thermal water sources concerned with active fault zones are widespread around the Çan region. Çan is so important place in point of mining sector of the country due to its natural sources. Rich clay mines and lignite (coal) deposits of the Neogene aged sediments is found Çan and its surroundings. Many coal and clay mines have been already worked in the region. Activities of these mines have increased day to day [1].

Çanakkale region has many rivers which are arisen from Kaz dağı (Mount Ida) (Table 1). Five main streams are found in the Biga peninsula. These streams have been shaped the peninsula from past to the present and have been caused many

M.K. Zaidi (ed.), Wastewater Reuse–Risk Assessment, Decision-Making and Environmental Security, 3–10.

productive plain near the seashore around the peninsula. Kocaçay (Kocabaş stream) is the second biggest river in point of lengthiness in the province and pass through the study area from southwest to the northeast. It has a dendritic type drainage network and streams are collected in one main stream such as a network. Annual average flow of this steam is 7.79 m^3/s.

Marmara type of the Mediterranean climate dominates in the region. Temperature changing between summer and winter is considerably high. According to the meteorological data annual average precipitation is 713.6 mm in the region. This value becomes least on August (10.6 mm) and at the most on December (106.4 mm) [3].

1.1. Geology

Many geological investigations were carried out in the study area from MTA (General Directorate of Mineral Research and Exploration), TKI (General Directorate of Turkish Coal) and countless private mining company before. Investigations which are conducted by private mining company have increased recently due to existence of valuable natural raw material resources in the region.

Table 1. Characteristics of streams in the Çanakkale Province [2]

Name	Length (km)	Minimum flow (m^3)	Maximum flow (m^3)	Region	Source	River mouth
Menderes stream	110	60–70	1530	Biga peninsula	Kaz dağı	Çanakkale strait
Kocabaş stream	80	15–20	1345	Biga peninsula	Kaz dağı	Marmara Sea
Tuzla stream	52	10–15	1400	Biga peninsula	Kaz dağı	Aegean Sea
Sarıçay stream	40	15–20	1300	Biga peninsula	Kaz dağı	Çanakkale strait
Mıhilı stream	12	–	75	Biga peninsula	Kaz dağı	Aegean Sea

The study area is located in the Biga peninsula which is an active region in point of tectonic. Basement rocks of the peninsula are composed of Paleozoic aged metamorphic rocks named as Kaz dağı group. Rocks of the Karakaya complex were settled on this group with tectonic boundary. Volcanic and sedimentary rock series cover these rocks in the peninsula. Volcanic rocks are dominant rock types in the region but many type of volcanic, sedimentary and metamorphic rocks are observed. Most of these rocks had altered and fractured due to effects of active faults. Basement rocks are composed of Oligocene aged volcanic rocks such as andesite, dacite, rhyodacite, basalt, tuff, and agglomerate. Several mineral deposits that can be mined profitable and precious metals may have been found in the alteration zones or fractured parts of these volcanic rocks. Neogene-aged sedimentary rocks overlie these rocks. These sediments consist of mostly fine grained components like sand, mill, clay and may include thick coal veins in different layers. All of the rocks are covered by Quaternary-aged alluvium which has a heterogeneous loose structure and is composed of gravel, sand, silt and clay. Most part of Çan city centre (Çan plain) is settled on the alluvium and volcanic rocks.

1.2. Hydrogeochemistry

Mining and ceramic industry have already developed Çan region. Kaolen and coal quarries, a thermal power plant and ceramic factories are founded in the region. Because of these developing activities a great deal of groundwater and surface water has consumed today. This study was conducted to determine effects of mining activities on surface and groundwater around the Çan (Çanakkale) plain. For this reason, water samples were taken two times (January 2005 and May 2005) from 12 locations at different part of the Çan plain.

pH, *T* (temperature) and EC (electrical conductivity) properties of these samples were measured from well in the field (Figure 1). Mean pH values are between 7.1and 8.2, mean EC values are between 627 and 2025 μS/cm and mean *T* values are between 10.6°C and 19.9°C. pH values of drinking water does not effect of human health directly. However, pH is affected by many factors in the environment. These factors can be harmful to human health indirectly. pH value of water is the measurement of acidity and alkalinity degree. Especially water hardness causes to change pH. If the water samples taken from study area are examined it is seen that pH value of sample 11 is high (pH = 8.2). This sample was taken from Kocaçay stream and this stream can have been polluted by several quarries and ceramic factory. pH values of coal ashes are between 9 and 10 in the study area.

Major anion–cation (Na^+, K^+, Ca^{++}, Mg^{++}, SO_4^-, HCO_3^-, CO_3^-, and Cl^-) and some trace elements (Al, Ba, Cd, Co, Cr, Cu, Fe, Mn, Pb, Rb, Sr, and Zn) were measured in these samples (Tables 2 and 3). Major anions–cations and trace elements were analyzed by Central Chemistry Laboratory of Çanakkale Onsekiz Mart University. Chemical analysis results of water samples were evaluated with two softwares; AquaChem 3.7 [4] and Water Quality Analyzer [5].

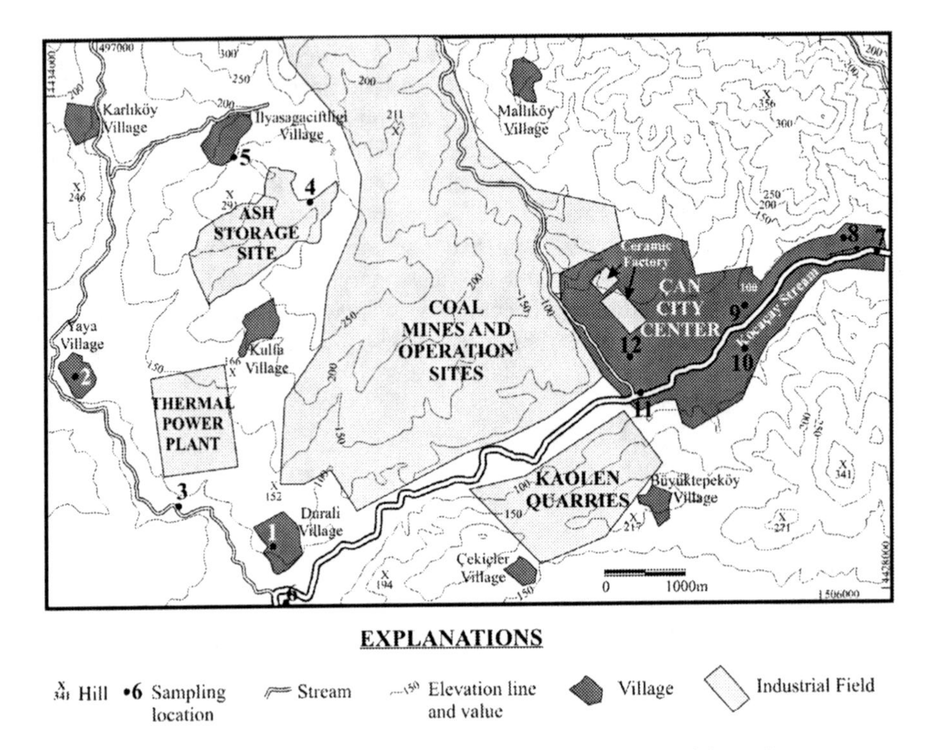

Figure 1. Water sample locations and industrial activities around the study area

Major anion–cation chemical analysis results of water samples are given below in Table 2. It is shown that water is reach with Ca-Mg-SO_4 ions in the study area. Particularly sample 1 is the richest sample in point of ion concentrations. This sample was taken from a spring in Durali village. Generally greenish, blackish, and gray colored basalts which have large hornblende phenocrystallines are found typically in Durali village and its surroundings. Hornblende which is existed in these basalts is the commonest mineral of the amphibole group: $(Ca, Na)_{2\text{–}3}$ $(MgFe^{+2}, Fe^{+3}, Al)_5$ $(AlSi)_8$ $O_{22}(OH)_2$. It has a variable composition, and may contain potassium and appreciable fluorine. Hornblende is commonly black, dark green or brown, and occurs in distinct monoclinic crystals or in columnar, fibrous, or granular forms. It is a constituent of many acid and intermediate igneous rocks (granite, syenite, diorite, and andesite) and less commonly of basic igneous rocks [6]. Altered (kaolen) tuffs are located commonly between these basalts.

Table 2. Major anion–cation analysis results of groundwater in the Çan plain

Sample number	Mean pH	EC Mean (μS/cm)	*T* Mean (°C)	Na (mg/l)	K (mg/l)	Ca (mg/l)	Mg (mg/l)	Cl (mg/l)	HCO_3 (mg/l)	SO_4 (mg/l)	CO_3 (mg/l)
1	7.1	2025	17.2	29.4	1.62	395.8	85.1	70.8	48.8	1196.0	6.0
2	7.8	1367	18.1	34.7	8.84	166.3	71.7	155.9	21.3	524.9	6.0
3	7.7	1054	17.2	37.7	3.79	137.3	51.0	99.3	284.4	192.3	13.5
4	7.7	864	10.6	10.1	7.99	78.2	60.7	47.9	27.4	358.3	10.5
5	7.5	627	13.4	18.6	0.68	76.1	51.6	60.3	33.5	312.9	6.0
6	7.7	727	14.5	15.9	1.53	88.2	50.4	60.3	61.0	317.2	3.0
7	7.9	1023	19.2	49.9	3.48	73.2	40.1	65.6	190.2	162.3	3.0
8	7.3	1115	17.2	29.9	0.45	125.9	58.3	92.2	42.7	472.8	0
9	7.4	1154	15.5	9.5	0.12	162.3	37.0	93.9	33.5	423.6	0
10	7.4	1157	15.4	48.9	0.33	129.3	47.4	101.0	212.5	255.4	1.5
11	8.2	1143	19.9	50.8	2.95	68.1	48.6	60.3	6.1	371.7	4.5
12	7.3	1711	16.9	90.5	0.18	23.5	59.6	145.4	51.8	753.3	0

The rocks which are composed of volcanics such as andesite, dacite, rhyodacite, basalt, tuff, and agglomerate cover many part of the study area. Potassium and sodium feldspars are common this kind of rocks. Generally, minerals which are existed in this kind of rocks are durable to decomposition and because of this reason dissolved salts are rare in the water deriving from these rocks [7–9]. But, it is seen that very common alteration zones are existed in volcanic rocks in the study area. Water samples which were taken from groundwater and surface water in the study area are plotted on the Piper diagram in Figure 2. In this diagram, some anions and cations are located corners of the small two triangles and their concentrations are shown as percent at the border lines. The quadrangle shows facies of water in this diagram. According to the diagram, water samples are taken part of Ca-Mg-SO_4 water facies and May 2005 water samples are richer than January 2005 water samples.

Chemical analysis results of water samples in Table 3 were compared with Turkish Drinking Water Standards [10] and with 2004 Edition of the Drinking Water Standards and Health Advisories of United States Environmental Protection Agency [11]. According to these standards groundwater and surface water have high contents of Al (aluminum). Excessive concentration of aluminum in drinking water caused some health problems in human body [12]. Other trace elements which were investigated in this study such as Barium (Ba), Cadmium (Cd), Chromium (Cr), Copper (Cu), Iron (Fe), Manganese (Mn), Lead (Pb), and Zinc (Zn) are remained under the acceptable contaminant level sector of drinking water standards aforementioned.

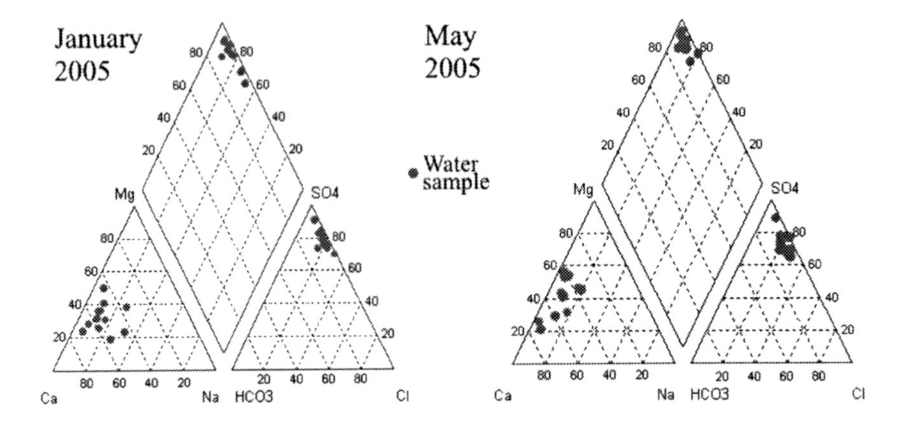

Figure 2. Projection in the Piper diagram of major anion–cation chemical analysis results of water samples

Table 3. Major anion–cation analysis results of the Çan Plain groundwater

Sample number	Al (mg/l)	Ba (mg/l)	Cd (mg/l)	Co (mg/l)	Cr (mg/l)	Cu (mg/l)	Fe (mg/l)	Mn (mg/l)	Ni (mg/l)	Pb (mg/l)	Zn (mg/l)
1	2.335	0.005	0.004	0.011	0.052	0.035	0.327	0.005	0.017	0.038	0.089
2	2.250	0.012	0.005	0.006	0.024	0.030	0.123	0.005	0.002	0.003	0.072
3	2.290	0.010	0.002	0.003	0.001	0.030	0.118	0.005	0.014	0.027	0.085
4	2.160	0.016	0.001	0.005	0.002	0.042	0.172	0.005	0.001	0.071	0.085
5	2.260	0.010	0.004	0.003	0.006	0.037	0.140	0.005	0.004	0.011	0.078
6	2.370	0.013	0.003	0.001	0.007	0.035	0.155	0.005	0.011	0.033	0.083
7	2.290	0.009	0.004	0.008	0.005	0.030	0.284	0.033	0.002	0.019	0.273
8	2.328	0.006	0.003	0.006	0.005	0.037	0.265	0.047	0.001	0.047	0.139
9	2.445	0.004	0.003	0.006	0.005	0.028	0.229	0.013	0.007	0.045	0.136
10	2.286	0.016	0.003	0.001	0.007	0.034	2.067	0.019	0.003	0.012	0.100
11	2.187	0.009	0.003	0.010	0.008	0.034	0.082	0.014	0.001	0.050	0.138
12	2.214	0.003	0.002	0.011	0.001	0.037	0.110	0.007	0.006	0.047	0.094
MACL (TS266, 1997)	0.200	0.300	0.005		0.050	0.300	0.200	0.050		0.050	5.000
MCL (USEPA, 2004)	0.200	2.000	0.005		0.100	1.000	0.300	0.050		0.015	5.000

MCL: Maximum contaminant level (mg/l), MAC: Maximum acceptable contaminant level (mg/l)

A coal combustion thermal power plant was commissioned in the western site of the Çan plain and it has started power production last year. Coal production in the Çan region has increased day to day due to activities of the thermal power plant. In the future, the Çan city will have more ash storage sites and mine wastes than today.

2. RESULTS AND DISCUSSIONS

Current data display that water environments in the area have not yet contaminated. The concentration of some trace elements such as aluminum (Al) is high in the water samples. The analysis results of Al are above "acceptable maximum standard value". The study area has much altered rocks mostly higher parts. These rocks contain more clay minerals such as kaoline and illite. Also, coal in the sedimentary sequence and ash from coal combustion in the power plant contain more Al_2O_3. The results show that mining activity have contaminated water source. Mining activity is going to increase around Çan plain for many more years. The operation of the coal combustion thermal power plant will produce ash and will spread via wind to the environment. Currently the heavily polluted water is limited around the plain. However, if precautions are not taken, these contaminants may spread in surface and groundwater nearby the Çan plain.

3. CONCLUSION

Çan region is rich in clay mines and lignite deposits and the groundwater is very reach with calcium-magnesium-sulfate (Ca-Mg-SO_4). Its aluminum (Al) concentrations are more than the acceptable maximum standard value. The results show that mining activity has contaminated water sources and if precautions are not taken, these contaminants may spread in surface water and groundwater.

Acknowledgment: The authors of this article thank the Scientific Research Fund and Geology Department of Engineering Faculty of Çanakkale Onsekiz Mart University and NATO ARW Organizing Committee for their financial supports.

4. REFERENCES

1. Deniz O. 2005. Investigation of Groundwater Quality of the Çanakkale City. MSc thesis, Çanakkale Onsekiz Mart University, Institute of Applied Sciences, p. 100. Çanakkale/Turkey.
2. Önder Ü, Çınar A, Serter G. 2001. Çanakkale Province Environment Condition Report. The Province Environment Directorate of Çanakkale Governorship of the Turkish Republic. p. 358, Çanakkale/Turkey.
3. Gülen Ö. 2005. Effects of Mining to the Groundwater in the Çan(Çanakkale) District. MSc thesis, Çanakkale Onsekiz Mart University, Institute of Applied Sciences, p. 119. Çanakkale/Turkey.
4. Calmbach L. 1997. AquaChem Computer Code-Version 3.7.42. Waterloo Hydrogeologic, Waterloo, ON.

5. Zorlu D. 1999. Water Quality Analyzer Software. Dokuz Eylül Üniversity Engineering Faculty, Department of Geology. İzmir/Turkey.
6. Bates RL, Jackson JA. 1987. Glossary of Geology, Third Edition. American Geological Institute, Alexandria, Virginia, USA.
7. Castany C. 1969. Practical Applications of Groundwater. Translation: Kazım Karacadağ and T. Adem ŞEBER. DSİ Printing house, Ankara/Turkey.
8. Schoeller H. 1973. Groundwater; Investigating, Operating and Evaluating of Resources. Translation: Kazım Karacadağ. Dizekonca Print. Ankara/Turkey.
9. Şahinci A. 1986. Geochemistry of Groundwater. Dokuz Eylül Üniversity Engineering Faculty Printing house. İzmir/Turkey.
10. TSE266, 1997. Turkish Drinking Water Standards. Turk. Std. Institute.
11. USEPA, 2004. 2004 Edition of the Drinking Water Standards and Health Advisories. EPH 822-R04-005, Office of Water U.S. Environmental Protection Agency, USA. http://www.epa.gov/safewater/mcl.html
12. Balkaya N, Açıkgöz A. 2004. Drinking Water Quality and Turkish Drinking Water Standards, J. Standard, Publ. Turk. Stand. Inst. n. 505, pp. 29–37.

DEVELOPMENT OF A DECISION SUPPORT SYSTEM FOR WASTEWATER REUSE IN THE MIDDLE EAST

Joop L.G. de Schutter
UNESCO-IHE Institute for Water Education
2601 DA, Delft, The Netherlands

Corresponding author: j.deschutter@unesco-ihe.org

Abstract:

The Regional Water Databanks (WDB) project is a project implemented within the framework of the Middle East peace process and coordinated by Executive Action Team (EXACT) and recommended by Middle East Multilateral Working Group on Water Resources (WWG) with the aim to promote regional cooperation among Israel, Jordan and the Palestinian Authority. The main aim of the regional water data bank project is to improve the joint monitoring, data availability and information exchange among water managers of the three core parties in the Middle East. The EU is financing it since 1995. A decision support system (DSS) for wastewater reuse is presented.

Keywords: Wastewater, regional water databank, UNESCO, exact-me, agriculture.

1. INTRODUCTION

1.1. Regional Water Data Banks Project

The Regional Water Databanks Project is a project implemented within the framework of the Middle-East Peace process. The program is coordinated by Executive Action Team (EXACT) on the basis of recommendations of the Middle-East Multilateral Working Group on Water Resources (WWG). Five donor parties support the WWG: the European Union, the United States, Canada, France, and the Netherlands. One of the recommendations of this WWG has been to establish regional water databank to improve the joint monitoring, data availability, and information exchange among water managers of the three core parties in the Middle-East. These recommendations were reaffirmed during the fifth WWG meeting in Muscat, Oman during April 1994, and were further elaborated into an implementation plan [1].

The implementation plan for this project aims to enable proper assessment of the state of the region's water resources through development of accurate and up-to-date databank for hydro-meteorological, hydrological, hydro-geological, and water

M.K. Zaidi (ed.), Wastewater Reuse–Risk Assessment, Decision-Making and Environmental Security, 11–21.

quality data, which are mutually comparable and exchangeable. The plan proposes approaches to develop a new Palestinian hydrometric database facility, and to upgrade and strengthen the Israeli and Jordanian existing water data programs, as well as introducing internationally recognized minimum quality standards for all. The implementation plan focuses furthermore on issues as: training for water managers and field technicians; communication and information; network review and evaluation; field data collection; laboratory analysis; databank enhancement; quality assurance, and control.

Since 1995, EU is financing the WDB program. From 1996 until 1998 phases 1 and 2 were implemented and water databank in the three core parties were upgraded and reinforced and their water monitoring networks were reviewed. Furthermore, a hydrometric database facility was developed for the Palestinian Water Authority. Technical assistance was furthermore provided in terms of database management and the setting up of a water related news bulletin for the Palestinian and Jordanian core parties. Also, during phase 1 and 2 a common ground was reached in terms of the type of information that could be shared amongst the core parties, which is to be considered of major importance with relevance to the overall objectives of the project.

Since early 1998 nonconventional water resources came into focus within the WWG, in particular the possibilities for reusing treated wastewater and production of desalinated water. Assessments were made about current flows of produced, treated and reused wastewater in each of the core party. Also existing treatment plants and standards were inventoried, and the three institutional arrangements in connection to wastewater treatment and reuse were described [2–3].

Phase 3 of the WDB was to elaborated on the outstanding issues on treatment and reuse of wastewater. It addressed issues such as design of monitoring networks and implementation of baseline surveys on the emission, treatment and reuse of wastewater; procedures and equipment for data collection and monitoring; data storage and interpretation; and development of decision support tools for small scale wastewater treatment technologies and agricultural reuse practices. The wastewater monitoring networks were fragmented along different institutions and authorities, standardisation, wastewater data collection, basically no wastewater flow measurements, and many different organisations have a stake in wastewater treatment and reuse, meanwhile lacking a platform for proper communication and coordination.

Phase 3 was concluded with a number of recommendations including the development of a statistically sound monitoring network programme; training on wastewater sampling and development of standards; extensive baselines surveys especially in Jordan and the Palestinian areas; procurement of field measurement equipment; improved data management; development of decision support systems for wastewater reuse practices, groundwater recharge and agricultural water management.

The presently ongoing WDB Phase IV is dealing with most of these recommendations and to concentrate on the development of a decision support system (DSS) for wastewater reuse that is intended to be shared among the core parties.

1.2. Wastewater Reuse in Israel, Jordan, and the Palestinian Areas

The surface area of the three core parties and their major river basins. This paragraph describes the main characteristics with respect to the current status of wastewater treatment and reuse within the areas of responsibility of the three core parties in particular.

1.2.1. Israel

Israel has a surface area of more then 20,000 km^2, and currently inhabits a population of nearly 7 million, which is growing rapidly. About 60% of its population is concentrated in the urban areas along the coast from Haifa through Tel Aviv to Ashdod. Its rainfall distribution is variable and gradually decreases from the north in Galilee (more then 1000 mm/year) to the south in the Negev (less than 250 mm/year), most of which falls in the period from November-March. Israel's main water sources are:

(1) Lake Kineret (Sea of Galilee), which accounts for about 30% of Israel's water supply. Its water is mainly conveyed through the National Water Carrier.
(2) Groundwater from the limestone
(3) Aquifer in the mountainous eastern areas, the sand and sandstone aquifer along the coast, and natural springs in the north
(4) Nonconventional water resources, including desalinated seawater and treated wastewater

Israel's water market is currently under stress as demand is exceeding supply. The aggregated water consumption for the year 2001 has been 1,760 million cubic meter (MCM) and was the lowest since 1994. This decline reflects the drastic cut in water quota for the agricultural market segment due to water shortages. The recent years of drought have deteriorated the country's water reserves to an all-time low. The National Water Commissioner decided to lower the red line levels, which determine the extent to which these reservoirs may be used. Nevertheless, the water withdrawal from these strategic reservoirs continues in order to meet the demand.

The main consumers in Israel's water market are domestic water supply, agricultural and industry (including power plants). The average per capita consumption of the entire population is around 110 CM/year (approximately 128 CM/year/person for Jewish population, and 47 CM/year/person in non-Jewish settlements). The agricultural sector remains the largest water consumer requiring over a BCM/year of water. The demand of the industrial segment has been constant over the years requiring some 130 MCM/year of water. The growth in water usage by the domestic segment is expected to continue as a result of the natural growth of the population and the rising standard of living.

Realizing the level to which the water market has deteriorated and the need for effective measures, the government, by way of the Water Commission and water-related legislation, has issued an aggressive master plan attempting to revolutionize the water market in Israel. Included in this plan is the reuse of treated wastewater that needs to increase from the current 300 MCM/year to about 500 MCM in the year 2020. This massive reuse of treated wastewater is foreseen either by upgrading the wastewater treatment plants and tunnelling the treated water to increased agricultural use, or by insertion of treated water into the ground (groundwater recharge) in order to preserve the level of the country's underground reservoirs.

This treatment and reuse of wastewater still is a governmental goal. In 2002 only 65% of its potential was reused for agricultural irrigation or groundwater recharge. Consequently, treatment and reuse of wastewater is a growing market in Israel during the next decade. Most of the growth is expected in the building new advanced treatment plants and in the upgrading of existing ones in order to enable the use of treated wastewater for a much wider industrial and agriculture reuse market. The second segment is the construction of plants and infrastructure, designed for the incorporation of treated wastewater into existing water systems where the use of such water would be allowed.

1.2.2. Jordan

Jordan has a surface area of close to 92,000 km^2. It currently inhabits a population of 6.5 million, which is growing about 3% per year. About 90% of the population is living in the northern highlands and in the Jordan valley in the west. Rainfall distribution in Jordan is variable. In the North East, including the Jerash area it is more then 650 mm/year, but decreasing sharply towards the south and east where semi-arid and desert conditions occur (less than 50 mm/year). Most of the rain falls in the period from November–March.

Jordan is a typically water-poor country with an average water availability of less then 240 CM/year/person. It is using approximately 780 MCM/year of water including approximately 275 MCM generated from groundwater. Major surface water sources are the Yarmouk and Zarqa rivers with the King Talal Dam, Wadi Shueib and Wadi Arab in the Jordan valley, which contain water in the winter period. Currently water consumption exceeds safe yields by an average of 20%, leading to a continuous depletion of the renewable water resources. Another acute problem is the severe water losses through its municipal water supply systems (up to 50%) and irrigation systems (up to 60%).

The Jordanian Water Strategy (Ministry of Water and Irrigation) addresses these challenges through a number of initiatives, including the development of reclaimed wastewater and desalination of brackish (including brackish water wells) water and seawater. Other initiatives are to enhance monitoring and promote conservation and reuse practices. Currently, Jordan is treating approximately 50% of its municipal wastewater through approximately 20 wastewater treatment plants that generate more than 75 MCM/year. According to national planning the amount of treated wastewater effluent should reach 240 MCM by the year 2020. However, effluent pollution parameters still vary considerably and especially discharge of wastewater from the olive industry is still causing very high concentrations of biochemical oxygen demand (BOD) loads.

Monitoring of water quality is one of the crucial elements in optimising wastewater treatment and reuse schemes. Currently, the Laboratory and Monitoring Department at the Ministry of Water and Irrigation is responsible for field monitoring and analysis of wastewater flows and ambient groundwater quality. Also, the Department of Environmental Health of the Ministry of Health is involved in water quality monitoring, as well as the Ministry of Environment and the Royal Scientific Society (RSS). The combination of the monitoring database of the Ministry of Water and Irrigation (the Water Information System – WIS) and the Department of Health (LIMS database) will contribute to a more effective monitoring and management of renewable water resources in the future.

Meanwhile large reuse schemes are already in use in Jordan such as near the Khirbet As-Samra wastewater treatment plant near Amman, which is the largest in Jordan and treating more than 180,000 CM per day, serving a population of more than 1.8 million. It provides treated water to large tree plantations in the vicinity of Amman. Altogether about 1500 ha are irrigated with treated wastewater, while another 9000 ha are provided with a mix of conventional surface water and treated wastewater. The WDB Project IV plan to develop a common framework for the structuring of these reuse schemes and provide a planning tool in order to be able to optimize existing schemes or develop new ones on the basis newly established combinations of wastewater treatment plants and reuser options.

1.2.3. Palestinian areas

The West Bank and Gaza have a total surface area of around 6,000 km^2 and currently consists a population of approximately 3 million people. About 55% of the Palestinians are living in the West Bank, 35% in Gaza and 10% in the eastern part of Jerusalem. Rainfall distribution varies from around 1000 mm/year in the northern parts of the West Bank to less than 250 mm/year in the southern region of Gaza. Most of the rain falls in the period from November–March. Groundwater is the largest source of water for the Palestinians in the West Bank and Gaza.

Other minor sources are springs, surface water occurrences and collected rain-water. Depletion of these fresh groundwater resources is already a very severe problem in Gaza and some parts of the West Bank. The main source of water for the 1.2 million people in Gaza is the coastal aquifer. The total Palestinian abstraction in Gaza is currently around 125 MCM/year and may need to increase up to 230 MCM/year in the year 2020. This is excluded of the Israeli utilization of wells around Gaza. Meanwhile the total renewable amount from rainfall is only about 85 MCM/year. This leads to serious depletion and infiltration of salt seawater into the coastal aquifer system that is causing rapid depletion of the freshwater volumes available to Gaza.

The annual Palestinian water demand in the West Bank is probably close to 200 MCM/year. Here also groundwater is the main source of water. These resources are divided over the western, eastern and north eastern aquifer system. The annual recharge of the Western system is close to 400 MCM/year, of which some 350 MCM/year is abstracted for Israeli use and 50 MCM for Palestinian use. The northeastern aquifer system has a recharge of about 160 MCM/year, of which about 130 MCM is abstracted for Israeli use and 30 MCM for Palestinian use. The eastern aquifer system, which drains into the Jordan valley, has an annual recharge of approximately 200 MCM/year and is largely available for further Palestinian development. All in all, the Palestinian water supply in the West Bank faces a structural deficit of more then 50 MCM/year, which will grow substantially during the coming decades.

Alarming signals have been reported from some places in the West Bank and Gaza about groundwater pollution with concentrations of chloride up to 400 mg/l, sodium up to 200 mg/l and nitrate up to 250 mg/l. Whereas chloride is linked to siltation of the water resources, nitrate directly links to infiltration of raw wastewater. About 65% of the Palestinian wastewater flow is currently collected in cesspits or directly discharged in the wadi's. Existing sewerage collection and treatment facilities in the Palestinian areas are facing problems and currently only the WWPT of the Al-Bireh municipality near Ramalla is functioning properly on the West Bank. Another problem is that average water consumption per capita is low (about 82 l pcpd), as a result of which the wastewater contains higher concentration of organic matter which is relatively expensive to remove. Moreover, the wastewater contains high levels of chloride, specifically in Gaza, which cannot be removed through conventional WWT techniques. As a consequence, many existing WWTPs are currently not properly dimensioned and are overloaded such as is the case Tulkarm, Ramallah, and Hebron.

At the same time there is a lot of recent experience with development of small-scale wastewater treatment plants in rural areas, which provide valuable perspectives for the future. These systems have been prepared and constructed by various authorities and NGOs, such as the Palestinian Agricultural Relief Committee (PARC) and the Palestinian Hydrology Group (PHG) with support from the Palestinian Water Authority. PHG and PARC have for instance constructed about 300 household treatment plants and six communal treatment plants receiving each up to 20 CM per day (or up to 400 persons equivalent). These plants are often relatively simple, consisting of a septic tank followed by upflow gravel filters, duckweed ponds or trickling filter techniques. Also, experience with more advanced but still cheap techniques has been obtained, such as with the upflow anaerobic sludge blanket reactor (UASB reactor) near the village of Artas, and at the Birzeit University [4]. These techniques have removal efficiency of 80% or more, and have proven to be suitable system components for integrated agricultural reuse schemes.

2. THE WDB IV DECISION SUPPORT SYSTEM (DS)

2.1. DSS Principle and Model Structure

The overall objective of the WDB IV project is to enhance the ability of the core parties (Israel, Jordan, Palestinian areas) to quantify problems related to wastewater reuse and identify measures to be taken to improve the situation. An added objective of the WDB project is to facilitate a common water resources monitoring framework that will allow transfer and exchange of data and that will facilitate to identify, plan and analyse different reuser options. The key issue of the project is to improve the availability of water treatment and reuse related information for Israeli, Jordanian, and Palestinian water managers, planners, and operators in order to promote regional cooperation between the core parties. A main component of the project therefore is the development of a DSS that can match a wide range of wastewater quality resources to an ever increasing number of reuser options, mainly agriculture. In connection to these objectives the project is contributing to increased capacity for field monitoring (water quality and water quantity) and development of the database.

The outline of the DSS has started from a combination of a systems analysis (model), the available database and model and the decision framework according to the global structure. A decision support system is helpful in situations where a decision depends on, or is influenced by, a large number of factors, rendering the decision procedure complex.

A properly designed DSS should provide an easy-to-use, usually graphics enhanced, working environment for the development, processing, and analysis of decision alternatives on the basis of a policy analysis framework. In the case of the WDB IV project, the DSS is a tool to enable the core parties to evaluate different options for improving their decision making about the optimization of possible combinations of water treatment on the one side and water use and reuse options on the other.

A wastewater treatment plant (WWTP-1) will discharge effluent to a wadi system and to direct reuser for agriculture or industry. Further downstream there is new area where agriculture development is planned in combination with a new WWTP-2. Water supply to this new area will take place using a new reservoir that is constructed by building a new dam in the Wadi downstream of the wastewater treatment plants. The water quality and quantity that will become available to the new user area can be calculated as a combination of the base flow in the wadi and the discharge figures from the wastewater treatment plants.

The phase WDB IV DSS in principle consists of two sets of excel spreadsheets that are connected with a data exchange sheet. One set is dealing with the quantity and quality of the water resources (urban wastewater, storm water, industrial wastewater) available to the system. The data series are in one month time steps. With the influent data the total load to the wastewater treatment plant can be calculated. Based on quality requirements for a reuser the DSS allows to selected possible treatment options (depending on treatment efficiencies for certain pollutants) from a base list of over 40 different treatment technologies. In addition to these treatment technologies, there is selection of effluent polishing systems available that will allow to deal with removal of specific pollutants (coli mainly). The same set of sheets has cost information that will calculate the delivery price of the effluent at the location of the reuser.

The second set of spreadsheets is dealing with actual wastewater reuse options for agriculture. This module has a complete set of crop combinations and irrigation methods (over 180 combinations) that is now being completed with irrigation water requirement hydrographs per combination over the year. This module will allow match water quantities and water qualities with soil and crop data per farm area. The DSS will therefore support to investigate which possible crops can be grown on the basis of the water available and how much water must be made available on the basis of a given crop and irrigation method pattern in a certain farm area. The main decision criterion will be save money. This criterion will be linked to the selection of treatment method (price/m^3 effluent produced as a function of the treatment technology) and reuse option.

2.2. Quality Criteria and Data Structure

In view of the existing situation with reference to water scarcity all core parties have done a lot of work with reference to the required effluent quality and reuse criteria. Especially water reuse for the irrigated agriculture sector has been researched and translated into regulations and guidelines, although these may differ per core party. In addition to these reuse criteria groundwater pollution has been recognized as a main threat to overall water management in the region and water quality standards for groundwater recharge have now become allowed to vary between narrow limits only. The following list of parameters is now being developed for WWTP influent and effluent in the WDB IV DSS. Where no figures are given most parties agree that the standards such as those according to the "Fraunhofer List" that is used under the EU water framework directive should apply.

One very specific issue is the acceptability of reused wastewater as a resource, which is an important issue for development of the concept worldwide. Wastewater reuse is often negatively perceived by the public and it is therefore important to survey the public attitude and inform users through public campaigns and clear examples.

Table 1. Proposed water quality parameters for the WDB IV DSS (partially)

Influent	Effluent	Units	Israel	West Bank and Gaza	Jordan
			Average threshold values effluent (highest category)		
Quantity	Quantity	m^3/day			
BOD_5	BOD_5	mg/l	20	20	30
TSS	TSS	mg/l	10	30	50
Ammonia	Ammonia	mg/l	15	20	20
COD	COD	mg/l	70	100	100
Total N	Total N	mg/l	25	25	45
	Kjeldhal N	mg/l		n.a.	n.a.
Total P	Total P	mg/l	5	5	n.a.
Temperature	Temperature	°C			
	Chloride	mg/l	250	250	
	Boron	mg/l	0.4	0.4	n.a.
	Detergents	mg/l			
	Nematodes	Egg/l	<1	<1	<1
	Coliforms	#/100 ml	2	100	100
	pH	#	6.5–8.5	6.5–8.5	6–9
EC	EC	Ds/m	1.4	1.4	n.a.
	SAR variable - Ca^{2+}, Mg^{2+}, Na^+	#	5	5	5
	Heavy metals index	#	n.a.	n.a.	n.a.

The database used by the core parties are presently being renewed and will become available in combination with a spatial database. The database in general covers the following items addressing volume, quality, cost of production, systems for waste-water, surface water, groundwater, other freshwater resources and the water demand (Table 1).

The design of the DSS is presently being finalized on the basis of actual case studies made up by each of the core parties. The final version of the DSS should be operational soon. At that time the software should be connected to the database of the core parties as well and become available to the water managers and water planners of the participating organisations. Due to the very interactive (workshops, meetings, special training sessions) way of developing the model and DSS and there will eventually be a small core of experts who are familiar with the model and should be able to use it within their planning environment. A wider use on a sub-regional planning as foreseen in the terms of reference of the project will take more time and effort.

3. CONCLUSIONS

The EXACT project is continuing to playing an important role in the development of a common framework for water resources management and monitoring in the middle-east. The development of the WDB IV DSS in its present format derives for experience in the previous phases of the project mainly and will not produce a generic model and tool that will be widely available on the sub-regional planning level. A major achievement of this phase of the project will be that the DSS of the previous phase has been transferred in a working version capable of supporting case studies and completed with a good link to the new database that have just now been developed. It is equally important that the core parties have been working together on the model and participated in the supporting training.

Acknowledgment: I would like to express my deepest gratitude and sincere thanks to the NATO program for Security through Science for cooperation and help.

4. REFERENCES

1. http://exact-me.org and www.unesco-ihe.org
2. Exact WWG (June 2001). Regional Water Data Banks Phase 3, Decision Support System on Wastewater Treatment technologies for Small Communities, Technical Report. EU Contract no ME1/B7-4100/95/MEPP/003
3. Exact WWG (June 2001). Regional Water Data Banks Phase 3, Decision Support System on Water Reuse for Agriculture, Technical Report. EU Contract no ME1/B7-4100/95/MEPP/003
4. El-Khateeb MA, El-Goharary FA. (2003). Combining UASB Technology and Constructed Wetland for Domestic Wastewater, Water Supp. 3(40), 201–208

WASTEWATER TREATMENT IN THE MEDITERRANEAN COUNTRIES

Claudia Wendland*, Ismail Al Baz**, Göksel Akcin***,
Gürdal Kanat***, and Ralf Otterpohl*
* Institute of Wastewater Management, Hamburg University of Technology (TUHH)
** InWEnt Capacity Building International, Germany
*** Yildiz Technical University, Istanbul, Turkey

Corresponding author: c.wendland@tu-harburg.de

Abstract:

This paper describes a new wastewater treatment concept which was assessed to be technically and economically feasible for suburban and rural areas in Mediterranean countries. The treatment consists of a two-step anaerobic high-rate reactor like upflow anaerobic sludge blanket (UASB) reactor followed by vertical flow constructed wetland (CW) and UV radiation. It fulfills the defined criteria, as meeting the standards for water reuse, costs especially for energy are low, sewage sludge production is limited, and operation and maintenance are simple.

Keywords: Constructed wetland, water reuse, anaerobic treatment, high-rate anaerobic reactor, UASB.

1. INTRODUCTION

Although there are much experience and data available about design and operation of constructed wetland (CW) and ultraviolet (UV) radiation, the knowledge about anaerobic treatment of municipal wastewater in UASB reactor at moderate temperature is limited. Therefore, bench-scale experiments with presettled high-strength municipal wastewater were carried out in a one step UASB reactor of 55 l. The results show a satisfactory effluent quality at fluctuating hydraulic retention time (HRT). During nine months of continuous operation, total chemical oxygen demand (COD) removal was 46% and 60% with HRT of 8–16 h and more than 24 h, respectively. Below an HRT of 24 h, the removal of suspended solids was limited, sludge wash-out took place due to fluctuating HRT. The removal of colloidal COD was satisfying with around 60% for all HRT. These results indicate that combined with a second high-rate anaerobic reactor and/or a settler after the UASB reactor, constructed wetland, and UV radiation, an adequate tertiary effluent for water reuse in irrigation will be achieved.

M.K. Zaidi (ed.), Wastewater Reuse–Risk Assessment, Decision-Making and Environmental Security, 23–32.

In Mediterranean countries (MEDAC) with serious water shortage, the reuse of treated wastewater is increasingly demanded for different purposes like agriculture and tourism. In rural areas there is often a lack of energy and skilled personnel to run conventional activated sludge systems successfully. Appropriate wastewater treatment with low running costs, easy operation, and producing a safe effluent for water reuse purposes are urgently needed.

The Efficient Management of Wastewater, its treatment and reuse in the Mediterranean countries (EMWater) project aims at increasing efficiency and effectiveness in the wastewater management in the MEDAC by helping decision makers in choosing the most suitable and appropriate technology. The beneficiary MEDAC of this project are Jordan, Lebanon, Palestine, and Turkey. The project has been funded by the EC under the EU-MEDA "Regional Program for Local Water Management" and is cofunded by the German Ministry for Economic Cooperation and Development (BMZ).

Specific tasks being implemented consist in the elaboration of regional policy guidelines for wastewater treatment and reuse, the design and the construction of pilot plants applying appropriate and low-cost techniques for demonstration and training purposes, the definition and implementation of an adapted training and capacity building programs (local, regional, and web based) for technicians, engineers, and employees of authorities and nongovernment organizations in the field of wastewater treatment and reuse (local stakeholders and professionals). In this paper, a wastewater treatment cycle for municipal wastewater is presented that was assessed to be appropriate for suburban and rural areas in Mediterranean climate in terms of fulfilling the requirements for long-term sustainability and water reuse in irrigation. This treatment concept will be applied on the pilot plant in Turkey within the EMWater project. The feasibility of the primary anaerobic step was investigated at the Institute of Wastewater Management at Hamburg University of Technology (TUHH) in preliminary experiments. The main objective was to study the performance of a single UASB reactor for the treatment of municipal presettled sewage with an average COD concentration of 500–600 mg/l at temperatures of around 20°C.

1.1. The Wastewater Treatment System

In order to assess an appropriate treatment concept for the MEDAC and thus for the pilot plant in Turkey, an evaluation was performed as first step. The following criteria were found to be most important for long-term sustainability of water reuse concepts in suburban and rural areas of the MEDAC:

- Affordable; especially low operation costs
- Operable; operation must be easily possible with locally available staff and support

- Reliable; producing a safe effluent for water reuse
- Environmentally sound; little sludge production and low energy consumption
- Suitable in Mediterranean climate (average wastewater temperature in Istanbul 23°C in July and 15°C in January)

The treatment cycle as shown in Figure 1 was assessed to fulfil sound these criteria for the following reasons. The operation costs are very limited, consists mainly of pumping costs, no aeration is necessary which represents usually the major financial part. Maintenance can easily be done by educating the local staff.

As primary treatment, the anaerobic step provides physical as well as biological treatment with a very low sludge production, less than 10% of conventional systems with activated sludge. Sludge disposal which often causes further pollution can be better managed because the sludge is simultaneously stabilised within the high-rate reactor [1]. Its discharge is only necessary once or twice a year and can be applied on the fields according to agricultural needs. Depending on the ambient temperature, one or two high-rate reactors, e.g., UASB are required. Because of limited hydrolysis below 15°C, a first reactor is designed to enhance hydrolysis [2]. Elmitwalli (2000) reported that at a temperature of 13°C an anaerobic filter followed by a UASB reactor is a suitable system for the treatment of domestic sewage [3].

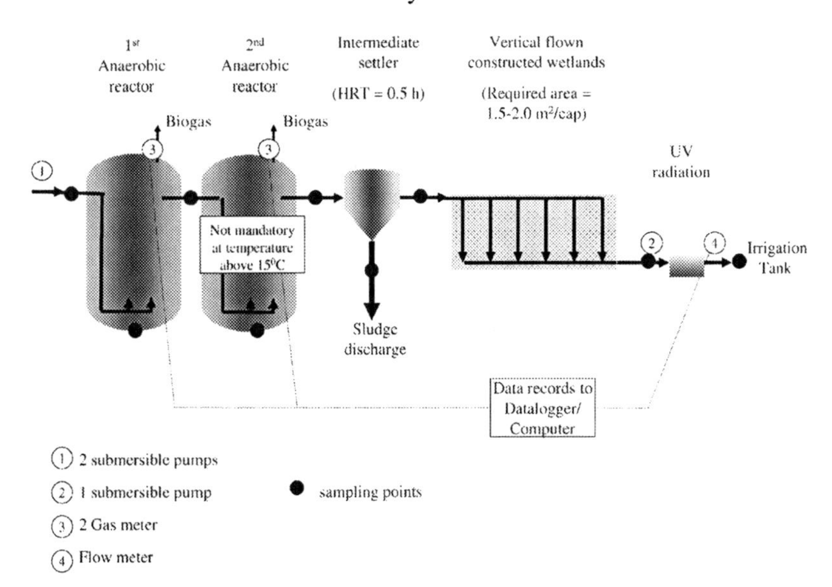

Figure 1. Treatment system for the pilot plant in Turkey

The small settler after the UASB reactor is designed for investigations about sludge wash-out. Lettinga proposed such a settler as complementary removal device for suspended solids after the UASB reactor as substantial further improvement in its performance [4].

As secondary treatment, CW was chosen to be the basic unit. CW act as biofilters combining physical, chemical, and biological treatment and are especially suitable for low diluted water flows [5–6] as many plants prove, e.g., CW realized within the SWAMP project (funded by the EU) [7]. Sousa, El-Khateeb, and El-Gohary showed that submerged as well as free water CW are suitable to treat anaerobic effluents to total COD of 60 and 70 mg/l respectively [8, 9]. In the last years, vertical flow CW were developed in Europe which provide aerobic conditions due to intermittent feeding and a better performance than submerged beds or free water CW. Moreover, the vertical flow CW require less space than the other types of CW and other natural systems [6]. Due to high HRT, the operation is reliable and can cope with fluctuating influent flows and temperature. CW need a start up phase of about three months to achieve a reliable effluent standard in terms of COD and total suspended solids (TSS) reduction.

As tertiary treatment, UV radiation as an environmentally sound technology is selected for disinfection because most of the reuse standards require a disinfection step for safety reasons like the WHO guidelines [10]. This treatment cycle does not remove nutrients. The nutrients like nitrogen, phosphorus, and potassium can additionally fertilize the plants when applying the treated water on the fields.

There are sufficient experiences and data available about the treatment performance of CW and UV radiation for municipal wastewater even in cold climate. But there is only limited research on the applicability of UASB reactors at moderate and cold temperature. Since anaerobic treatment efficiency is highly dependent on the temperature, further research at moderate temperature was performed in the following preliminary experiments.

2. EXPERIMENTS

2.1. Material and Methods

The experiments were carried out at TUHH to get preliminary results for the operation of the Turkish pilot plant in Istanbul, Turkey. A UASB reactor with a total volume of 55 l was built of polyvinyl chloride (PVC) with a diameter of 0.15 m. The scheme is shown in Figure 2.

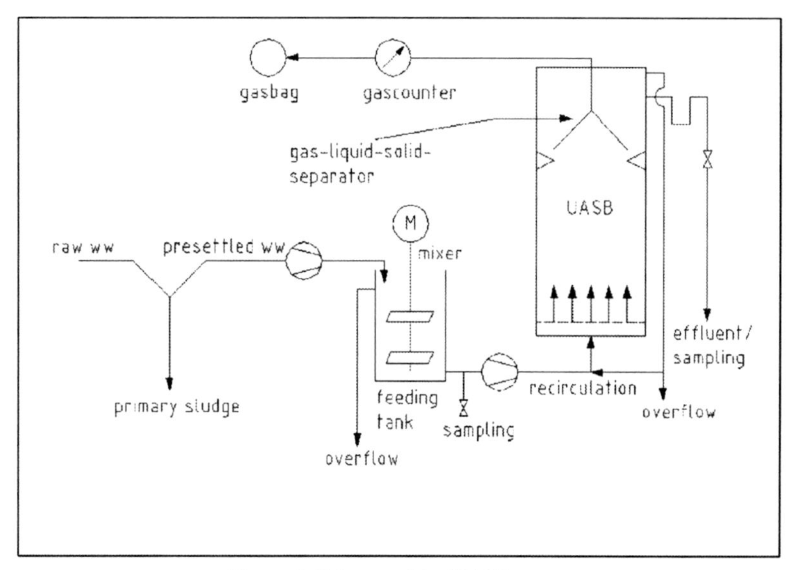

Figure 2. Scheme of the UASB reactor

The characteristics of the raw and municipal wastewater from the city of Hamburg are shown in Table 1. The UASB reactor was started to treat primary and secondary municipal sewage sludge and operated continuously for 4 months as start-up phase. For determination of anaerobic biodegradability a batch experiment was carried out. Therefore, 100 l of raw wastewater was recirculated for more than 10 days with an upflow velocity of 0.25 m/h.

Secondly, the UASB reactor was operated continuously for 9 months with presettled wastewater. During the last 4 months, influent and effluent were analysed twice a week (2 h composite samples). The COD was analyzed as COD_{total}, COD_{diss} (<0.45 μm), COD_{coll} (0.45 < COD < 25 μm), and COD_{ss} (>25 μm) with analytical cuvettes. Total suspended solids (TSS), NH_4, and TN were measured as described in Standard Methods [11].

During the investigation period, no sludge was discharged. The reactor temperature was operated at the ambient temperature, between 18–25°C. Due to the local conditions, it was not possible to control the upflow velocity so that the HRT was fluctuating between 3 h and more than 24 h during the investigation period.

3. RESULTS AND DISCUSSION

3.1. Influent Characteristics

The influent wastewater is a typical municipal wastewater with high-domestic influence and low diluted due to limited drinking water consumption which is in Hamburg Water Works (HWW) 113 l/(cap*d) [12]. These wastewater characteristics are also typical for rural areas of water limited regions like the MEDAC.

Table 1. Characteristics of raw wastewater, average values in mg/l with std.dev

	COD_{total}	COD_{diss}	COD_{coll}	COD_{ss}	TOC	Total N	NH_4-N	TSS
Raw wastewater	1005 ± 134	209 ± 59	264 ± 70	450 ± 134	271 ± 32	96 ± 24	62 ± 15	240 ± 83

3.2. Anaerobic Biodegradability

Within the batch recirculation experiment, the maximal anaerobic biodegradability can be calculated. In Table 2 the removal ratios are given. Colloidal, dissolved COD, NH_4-N and TSS were not measured explicitly. The total COD removal is 85% and in the typical range for municipal wastewater with high domestic influence. There is a very good correlation with total organic carbon (TOC). Removal of colloidal COD is often limited in single UASB reactors. However, $COD_{coll+diss}$-removal in this UASB reactor of 79% is a high value as compared to that reported [3, 13, 14]. This high removal of COD <25 µm might be due to difference in the definition of colloidal COD. They defined that COD_{coll} ranges between 0.45 µm and 4.4 µm and not 25 µm. Suspended COD is removed to a major part (86%), but still the effluent was not clear which is a draw back of anaerobic treatment of municipal wastewater. Even reaching the maximal anaerobic biodegradability, a posttreatment is required to produce a high quality clear effluent. The removal of total N is in the usual range of anaerobic treatment; N is needed for biomass production and is therefore found in the suspended particles.

Table 2. Anaerobic biodegradability

	COD_{total}	$COD_{coll+diss}$	COD_{ss}	TOC	Total N	NH_4-N	TSS
Maximum Removal	85%	79%	86%	82%	18%	—	—

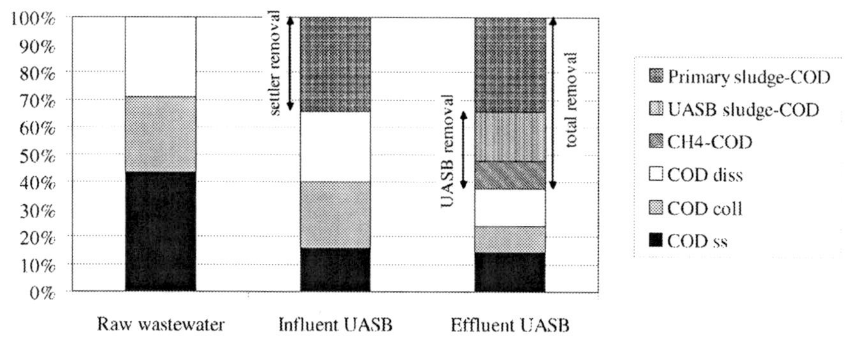

Figure 3. Mass balance for COD

3.3. COD and Total Organic Carbon (TOC) Removal

Figure 3 shows the COD mass balance in the raw wastewater, influent, and effluent of the UASB reactor. The sludge-COD was not measured but calculated based on the raw wastewater-COD. As expected there is a significant removal of COD_{ss} in the sedimentation step but only limited removal of COD_{diss} and COD_{coll}. In the UASB reactor there is almost no further removal of COD_{ss} but COD_{diss} and COD_{coll} is converted to CH_4 and anaerobic sludge as seen in Figure 3. When looking on the effluent characteristic (Table 1) there is a high variation in the all parameters which are caused by the significantly fluctuating HRT. Therefore the removal efficiency of the UASB reactor was analyzed (Table 3).

The high standard deviations show a high variation in the COD removal. However, the trend becomes obvious, the COD removal for HRT >16 h was very good with COD_{total} of 47% and 60% respectively. The total COD removal is comparable to that reported by Halalsheh [1] in the treatment of concentrated sewage in a one step UASB reactor at moderate temperature. Although it was already presettled wastewater, a very good removal of COD_{coll} of 40% and 66% was achieved. Even at HRT below 16 h, the removal of COD_{coll} of 61% and 62% was very high. This confirms the results of the anaerobic biodegradability test (Table 2). The difference compared to other research may be due to the different definition of colloidal COD. Due to the unstable conditions, there was a significant washout of COD >25 μm at HRT < 16 h.

As COD consists of settlable solids (>25 μm), addition of a settler after the UASB reactor, as proposed for the pilot plant in Turkey (Figure 1), will result in removal of the washed COD_{ss} from the UASB reactor. Accordingly, COD_{total} as high as 60% can be achieved at UASB-HRT of 8–16 h with a settler addition after the UASB reactor.

Table 3. COD and TOC removal with standard deviation

	HRT of UASB reactor in hours (h)			
	3 – 8 h	8 – 16 h	16 – 24 h	24 – 30 h
COD_{total}	27% ± 24%	46% ± 19%	47% ± 19%	60% ± 11%
COD_{diss}	40% ± 12%	43% ± 21%	40% ± 25%	40% ± 20%
COD_{coll}	61% ± 24%	62% ± 22%	40% ± 29%	66% ± 15%
COD_{ss}	sludge washout	55% ± 36%		64% ± 39%
TOC	29% ± 30%	42% ± 18%	39% ± 24%	39% ± 15%
Expected COD_{total} after addl settler	53% ± 24%	60% ± 16%	61% ± 13%	70% ± 6%

The UASB reactor showed a stable and efficient removal of COD_{diss} (40–43%) at different HRTs. Compared to the anaerobic biodegradability of $COD_{coll+diss}$ (Table 2) and values of 54% and 65% for domestic wastewater from Last and Lettinga and Elmitwalli [15, 16], there is still an anaerobic removal potential of COD_{diss} of more than 11%. Therefore, a two-step UASB reactor will result in a higher removal efficiency of COD_{diss}. TOC represents the organic substances similar to COD, but does not correlate always to the total COD values. For the COD analyse more volume of water (some ml) is needed than for determination of TOC (one drop). As influent and effluent contain considerable amounts of suspended solids, the COD measurement is therefore more reliable.

3.4. Nitrogen Removal

Nitrogen as main nutrient in domestic wastewater is not removed by sedimentation or anaerobic digestion (Table 1). This was expected as nutrient removal is not required for further reuse in irrigation. The increase in NH_4-N during the treatment is caused by the conversion of organic nitrogen to ammonium, due to protein hydrolysis.

3.5. Biogas composition

The biogas produced in the UASB reactor contains 80% methane, 16% nitrogen, and 4% carbon dioxide (CO_2) which are typical concentrations for a high, rate anaerobic process applied on municipal wastewater [17].

4. CONCLUSIONS

The evaluation indicates that high-rate anaerobic treatment like the UASB reactor followed by constructed wetlands and UV radiation is an appropriate and cost efficient wastewater treatment for suburban and rural areas of the MEDAC. If adapted to ambient temperature, it is an easy operable and sustainable treatment cycle. The preliminary experiments with presettled high strength wastewater show that at ambient temperature around 20°C and a fluctuating HRT between 3 and 30 h, the removal of COD_{total} is varying between 27% and 60%. To obtain a stable performance it is however recommended to keep the HRT stable. Improved with a second high-rate anaerobic reactor and/or an additional settler after the UASB reactor, this treatment concept with CW and UV radiation is expected to meet the standards for water reuse in irrigation. Nutrient removal is not an issue in this context because nutrients like N, P, and K can partly replace the use of chemical fertilizer. Further detailed research about the anaerobic treatment of municipal wastewater under Mediterranean conditions are recommended and will be carried out in the pilot plant in Istanbul, Turkey, within the EMWater project.

Acknowledgment: This research project is funded partly by the European Community under the MEDA WATER PROGRAMME initiative - ME8/AIDCO/ 2001/0515/59641-P044. NATO provided stage to present this paper in the Advanced Research Workshop in Istanbul, Turkey during October 12–16, 2006.

5. REFERENCES

1. Halalsheh MM. (2002). Anaerobic pre-treatment of strong sewage. A proper solution for Jordan. Ph.D. thesis, Wageningen University, the Netherlands.
2. Sanders WT, Geerink M, Zeeman G and Lettinga G. (2000). Anaerobic hydrolysis kinetics of particulate substrates. Water Science Tech. 41 (3), pp. 17–24.
3. Elmitwalli TA. (2000). Anaerobic treatment of domestic sewage at low temperature. Ph.D. thesis, Wageningen University, the Netherlands.
4. Lettinga G. (2005). A good life environment for all through conceptual, technological and social innovations. VIII Latin American Workshop and Syposium on Anaerobic Digestion. 2 – 5 October 2005. Uruguay.
5. US EPA (2000). Constructed Wetlands, Treatment of Municipal Wastewater. Manual. EPA/625/R-99/010.
6. SWAMP (2002). Sustainable Water Management and Wastewater Purification in Tourism Facilities. Website: www.swamp-eu.org
7. Sousa JT, Haandel AC and Guimarães AAV. (2001). Post treatment of anaerobic effluents in wetland systems. Water Science Tech. 44 (4), pp. 213–219.
8. El-Khateeb MA and El-Gohary FA. (2003). Combining UASB technology and constructed wetland for domestic wastewater reclamation and reuse. Water Supp. 3 (4), pp. 201–208.
9. Masi F. (2005). Sustainable Sanitation by CWs in Mediterranean Countries. Proc. First International Zero-m conference, Istanbul, Turkey, March 2005.
10. Mara D and Cairncross S. (1989). Guidelines for the safe use of wastewater and excreta in agriculture and aquaculture, WHO, Geneva.

11. APHA (1995). Standard Methods for the Examination of Water and Wastewater. 19th ed., American Public Health Association, Washington DC, USA.
12. HWW (2005). Annual report of Hamburg Water Works 2004 (in German). Website: www.hww-hamburg.de/hww_prod_engine.shtml?id=1110
13. Wang JY. (1994). Integrated anaerobic and aerobic treatmnet of sewage. Ph.D. thesis, Wageningen University, The Netherlands.
14. Mahmoud N. (2002) Anaerobic pre-treatment of sewage under low temperature (15°C) conditions in an integrated UASB-digester system. Ph.D. thesis, Wageningen University, The Netherlands.
15. Last ARM and Lettinga G. (1992). Anaerobic treatment of domestic sewage under moderate climatic (Dutch) conditions using upflow reactors at increased superficial velocities. Water Science Tech. 25 (7), pp. 167–178.
16. Elmitwalli TA, Soellner J, Keizer A, Bruning H, Zeeman G and Lettinga G. (2001). Biodegradability and change of physical characteristics of particles during anaerobic digestion of domestic sewage. Water Research 35 (5), pp. 1311–1317.
17. Elmitwalli TA, Zeeman G, Oahn KLT and Lettinga G. (2002). Treatment of domestic sewage in two step system anaerobic filter/anaerobic hybrid system at low temperature. Water Research 36 (9), pp. 2225–2232.

RISK/COST ANALYSIS OF SUSTAINABLE MANAGEMENT OF WASTEWATER FOR IRRIGATION: SUPPLY CHAIN APPROACH

E. Levner
Holon Institute of Technology, Holon, 58102, Israel

Corresponding author: levner@hait.ac.il

Abstract:

The problem considered in this paper is to coordinate the costs and ecological risks of all stakeholders in the environmental water/wastewater supply chain. Using the concept of the environmental supply chain we construct a corresponding decision tree and a risk matrix, which quantitatively estimate the risk level. We propose an economic–mathematical model permitting to mitigate the integrated risk to population and society under economic, technological and social constraints.

Keywords: Environmental supply chain, risk analysis.

1. INTRODUCTION

Wastewater reuse in irrigation is a substantial resource for agricultural production [1–4]. In Israel, about 300 million cubic meters being treated to varying degrees. Currently, more than 70% of this sewage is reused in agriculture for irrigation. Advantages of using wastewater in irrigation are evident:

- Abundant, low-cost, and highly reliable water supply throughout the year
- Preservation of freshwater sources
- Less use of chemical fertilizers and decreased costs
- Efficient way of disposal of the treated effluents
- Opportunities for land extension

To avoid the negative effects, the wastewater is treated. The higher is the wastewater treatment level, the higher are the treatment costs but, at the same time, the ecological risks for potential hazards are decreased [1–3]. A problem considered in this paper is devoted to the analysis and coordination of the costs and ecological risks. Using the concept of an "environmental supply chain" [4–6], a corresponding decision tree and the risk matrix are constructed that will estimate the integrated risk level. An economic–mathematical model was proposed to mitigate the integrated risks to population under social constraints. The model

M.K. Zaidi (ed.), Wastewater Reuse–Risk Assessment, Decision-Making and Environmental Security, 33–42.

Table 1. Wastewater ingredients and ecological risks

Ingredients	Negative effect and damage
Suspended solids (SS)	Production of sludge and anaerobic conditions in open water bodies and streams. Gradual increase in mineral contents in groundwater
Biodegradable organics (BOD, COD)	Produce dangerous conditions to aquatic species
Pathogenic organisms	A cause for infectious diseases
Heavy metals and dissolved inorganics	A cause for different diseases
Organic and inorganic pollutants	Toxic factors for different diseases

permits to simulate real-life situations and may serve as a basis for a master plan for sustainable wastewater use on a regional, national, and international level. Main negative ecological effects of wastewater ingredients (Table 1).

2. CONCEPT OF ENVIRONMENTAL SUPPLY CHAIN: AQUATIC LOGISTICS SUPPLY CHAIN

A concept of environmental supply chain, known also in the literature as a green, or closed-loop supply chain was independently suggested [4–7]. The main components of the conventional supply chain are demand forecasting and planning, material requisition and inventory, manufacturing and packaging, distribution and transportation, customer service, and wastewater treatment and reuse.

Graphically, the supply chain is represented as a graph of a network in which each agent or stakeholder in the system is presented as a node, and each chain link describes an interconnection transforming raw material into product and services. Thus, the supply chain can be viewed as a visual presentation of the technological activities of the agents (stakeholders) of the system during its life cycle. In its simplest form, a supply chain forms a linear supply chain [4] (Figure 1).

The definition of the environmental supply chain (ESC) integrates the above material, financial and information flows with flows of natural resources, throughout the product life cycle, and introduces new decisions for suppliers and

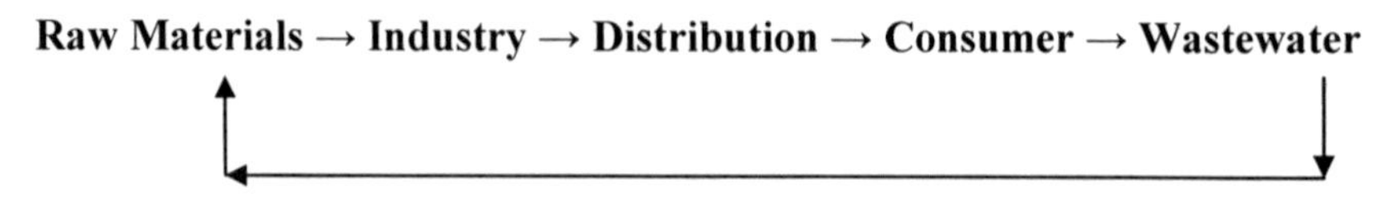

Figure 1. An example of a supply chain

manufacturers in the SC necessary to decrease waste flows and the environmental pollution, even beyond their sale and delivery interests [4–6]. As emphasized by Carter and Narasimhan [6], the environmental protection issues are critical in the ESC and incorporated into supply chain management strategies; the environmental dimension should be viewed as an inseparable part of business performance at all stages of the supply chain management. The environmental risk was explicitly incorporated into the ESC, along with the costs and benefits, as a main objective of the design, planning, and management. A special class of the environmental called the aquatic logistics supply chains. In this chain, water flows and constitute final products, by-products, and wastes occurring in the same supply chain.

The aquatic logistics supply chain (ALSC) is a management science paradigm to the concept of the hydrological water flow cycle in hydrological sciences. The aquatic logistics supply chain is a visual presentation of the technological activities of all the stakeholders of the ecosystem throughout the water life cycle. An example of a simple aquatic logistics supply chain is the following (Figure 2):

Water Sources → Water Preparation → Water Transportation → Customers → Wastewater Sources (Water Pollution Actions) → Wastewater Treatment → Treated Wastewater Usage (Irrigation)→Wastewater Disposal

Figure 2. A schematic aquatic logistics supply chain

A mathematical model will be formulated as an optimization problem on large-scale graphs representing the ecological supply chains.

3. RISK DEFINITION

According to EPA [9], the risk is a likelihood that a course of actions will result in an undesired event. Two main types of risk assessment are known in the literature:

- Discipline-oriented (engineering, biological, medical, ecological, etc.)
- Integrated (that integrates space, time, sources of risk, stressors, their pathways, results, and multiple endpoints)

Many researchers who studied the integrated risk management problems have noticed that it is comparatively easy to describe and formulate constraints of the problem but it is difficult and troublesome to formulate and quantify the objective function of the risk-management problem. As an illustration consider first the basic integrated decision-making model suggested [1], which has the following characteristics: Decision variables are various basic options of wastewater treatment and transportation and the corresponding composition of crops in a region. The decision variables are transparent and easily applicable. Here are some of them, among many others:

- Daily probability of infection through ingestion of pathogens [8].
- Annual probability of infection through ingestion of pathogens [3].
- Probability of nutrient loading to an adjacent water body [8].
- Costs of damage to aquifer, soil, and human health [1].
- Product of the probability to die due to wastewater contamination and value of life of a statistical individual [1, 10].
- Product of the probability and magnitude of the damage and the deficiency of the latter approach is discussed and motivated [8–12].
- Two-dimensional risk matrix.
- Probability and magnitude of damages [12–13].
- The risk matrix has the capability to evaluate the effectiveness of risk mitigation measures (Figure 3).

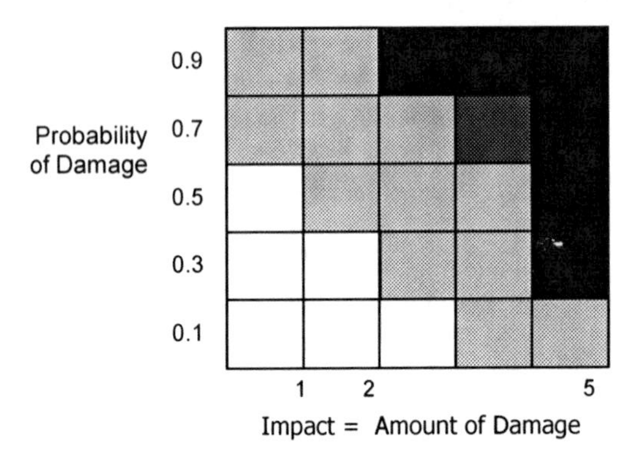

Figure 3. Two-dimensional characteristics of risk. The matrix serves to rank the risks: the first (white) tier denotes *low level*, the next one – acceptable and high, respectively

3.1. Risk Ranking

Example. As an illustration to the above concept, Table 2 reproduced from UN [14], ranks various wastewater treatment systems with respect to cost but also according to other multiple indices of effectiveness. The selection of the best alternative depends heavily on qualification and confidence of experts, local conditions, and practical requirements.

The first number in each square in Table 3 defines the rank according to the Expert Estimate EE*j*, *j*=1,…9. The second number in each square is the indivudial Borda count, equal to *N*-EE*j*, where *N*=5 is the number of ranks. The last column present the total (summary) Borda count. According to the Borda ranking method, the stabilization ponds are the best technology, next ones are the modified aeration activated sludge, aerated lagoons, and the conventional activated sludge. The oxidation ditches conclude the list. Details of the Borda method and its generalization for fuzzy, uncertain and missed data can be found in [14–16]. For simplicity, all the weights of individual estimates are accepted to be equal 1. Next section gives a method for defining nonunit weights.

Table 2. Ranking of five wastewater treatment systems

Rank (1 is best)	Initial costs (IC)	O&M costs (OMS)	Life Cycle cost (LCC)	Operability (Op)	Reliability (Re)	Land area (LA)	Sludge production (SP)	Power use (PU)	Effluent quality (EQ)
1	MA	SP	MA	SP	SP	MA	SP	SP	SP
2	AS	MA	AS	AL	AL	AS	AL	MA	AL
3	OD	AL	OD	OD	OD	OD	OD	AS	MA
4	AL	AS	AL	AS	MA	AL	AS	AL	OD
5	SP	OD	SP	MA	AS	SP	MA	OD	AS

SP stabilization ponds, *AL* aerated lagoons, *OD* oxidation ditches, *AS* conventional activated sludge, *MA* modified aeration activated sludge

Table 3. Calculations for the Borda ranking

Alternatives	EE1 IC	EE2 OMC	EE3 LCC	EE4 Op	EE5 Re	EE6 LA	EE7 SP	EE8 PU	EE9 EQ	Borda Count
SP	5/0	¼	5/0	¼	1/4	5/0	1/4	¼	¼	24
AL	4/1	3/2	4/1	2/3	2/3	4/1	2/3	4/1	2/3	18
OD	3/2	5/0	3/2	3/2	3/2	3/2	3/2	5/0	4/1	13
AS	2/3	4/1	2/3	4/1	5/0	2/3	4/1	3/2	5/0	14
MA	1/4	2/3	¼	5/0	4/1	1/4	5/0	2/3	3/2	21

3.2. Decision Trees

The crucial point in tuning and customization of the Borda ranking method is the correct determination of the weights applicable to all wastewater stakeholders. The decision tree is a reliable and effective tool for performing this mission. This section outlines the mathematical form of a risk minimization model in terms of multicriteria decision trees and fuzzy mathematical programming problem.

A decision tree is a flow chart or diagram representing a classification system or predictive model. The tree is structured as a sequence of simple questions, and the answers to these questions trace a path down the tree. The tree consists of nodes and arcs. The end node reached determines the final objective of classification or prediction made by the model, which can be a qualitative judgment (e.g., the linguistic varibles or a numerical forecast (e.g., cost, risk, benefit, etc.). A square node of the tree represents a point at which a decision must be made, and a circular node represents situations where the outcome is uncertain. Each arc leading from a square represents a possible decision; while arc leading from a circle represent a possible outcome and its probability assigned. The procedure for defining risk weights is based on using multiple decision trees. Each of different decision trees is related to a single specific objective function (such as, for instance, the probability of the fatal disease in the area, the impact of pollution on ecological losses, in particular the loss for farmers, the severity of ecological losses). The structure of each tree reflects the ecological supply chain like:

Pollution Sources → Stressors → Protection Targets, or Pollution Sources → Pollution-Mitigation Activities → Stressors Mitigated→ Protection Targets.

The first layer represents the main pollution sources, each of them being a basic object to which pollution prevention activities are applied. The second layer represents the main activities for pollution prevention and measures for risk aversion such as regular monitoring/prevention of all types of water/soil pollution; introduction and development of clean technologies; introduction of water/ wastewater filters and other wastewater treatment facilities; technical reconstruction and reequipment of the water treatment plants, preventive and remediation measures including ecological literacy education.

The third level represents different classes and subclasses of ecological stressors (physical, chemical, and biological sources of damages and environmental risks), whose impact is to be mitigated by using the activities enlisted in the second layer when imposed upon the hazard sources enlisted in the first layer. The arc leading from the node i in the second layer of the tree to a node j in the third layer may be divided into several possible strategies (e.g., adventurous', basic' cautious) with a set of corresponding probabilities p_{ijt} of the mitigation of a harmful stress and effects e_{ijt} of strategy i upon decreasing stress j ($t = 1, 2,\ldots, N$) assigned; N is the total number of possible strategies (outcomes). Each third-level node representing a stress is supplied with a weight v_j depicting the importance, or severity of the stress.

The fourth level represents different classes and subclasses of the endpoints (stakeholders, groups of human population, plants, bio-organisms, etc.) which are to be protected by mitigating the effect of harmful and toxic stressors in the wastewater enlisted in the third layer. The arc leading from the node j in the third layer to a node k in the fourth layer is supplied with (a) a set of probabilities q_{jks} of decreasing the harmful impact of stress j upon end user k under an outcome s (where s = 1, 2,..., S) and (b) the positive effect f_{jks} upon end-user k due to the decreasing of stress j (s = 1, 2,...;S); here S is the total number of possible outcomes. Each fourth-level node representing an end-user is supplied with a weight w_k reflecting its size and importance in the ecological system. The values are assigned to the fourth-level nodes representing experts' evaluations of expected results of decreasing the stressors due to the counter-pollution measures and strategies, that are expressed either numerically or in linguistic terms (very strong, strong, medium, weak, and negligible) with respect to a selected objective function. The results of the fifth layer are the values and/or linguistic evaluations of an objective function. The final stage of the multiple-tree procedure is an integrated (quantitative and/or qualitative) estimation of the ecological risk based on the convolution of linguistic and numerical results provided by all decision trees, which is finally expressed as a Pareto point in the risk matrix.

3.3. Mathematical Programming Model

In our model, the damage $\boldsymbol{D}_\mathrm{k}$ to the human health caused by several toxic stressors in the wastewater, is represented as follows:

$$\boldsymbol{D}_\mathrm{k} = \boldsymbol{v} \bullet \boldsymbol{W}^\mathrm{k} = \sum_{j=1,\dots,LxM} v_j(W_{jk}) W_{jk}, \tag{1}$$

where: L: the number of stressors; M: the number of pollution sources; $v_j = v_j(W_j)$: the weight for the jth stressor; W_{jk}: the amount of the harm to the kth endpoint caused by the jth stressor; $\boldsymbol{W}^\mathrm{k} = (W_{jk})$: the damage vector for the kth object caused by all stressors. The total damage to the supply chain (that is, an exploited ecological system) consisting of K objects (protected endpoints, end-users, targets) is defined as follows:

$$\boldsymbol{D} = \sum_{k=1,\dots,K} u_\mathrm{k} \boldsymbol{D}_\mathrm{k} = \sum_{k=1,\dots,K} \sum_{j=1,\dots,LxM} v_j(W_{jk})\, u_\mathrm{k} W_{jk} x_{jka}, \tag{2}$$

where u_k is the weight (importance) of the kth endpoints. The number M of sources, fons et origo, of water pollutants causing damage to the human health as well as to the health of other biological organisms can be defined by the experts in advance. In particular, for measuring the total damage we can employ a rating scale defined by the Borda method in Section 4 presented in the last column in Table 3 and indicating the degree of damage. Similar to damage D in Equation 2, we define the mitigation level for damage *MD* caused by the counter-pollution activities, $a = 1,\ldots, A$.

$$\boldsymbol{MD} = \sum_{a=1,\ldots,A} \boldsymbol{MD}_{\mathrm{ka}} = \sum_{a=1,\ldots,A}\sum_{k=1,\ldots,K}\sum_{j=1,\ldots,LxM} v_j(W_{jk})\, u_{\mathrm{k}} W_{jka} x_{jka}, \tag{3}$$

where: $\boldsymbol{MD}_{\mathrm{ka}}$ is the mitigation of damage to the *k*th target caused by the *a*th counter-pollution strategy; W_{jka} is the amount of the harm decrease to the *k*th object (= endpoint) caused by the *j*th stressor as a result of the *a*th counter-pollution activity; decision variables x_{jka} define the selection of proper strategies and activities: $x_{jka}=1$ if the *a*th counter-pollution strategy is chosen and 0 otherwise. The problem is to choose the set of the counter-pollution strategies and activities so that to minimize the integrated risk (the integrated damage degree, damage strength), or, equivalently, to maximize the total risk mitigation rating level ***MD*** under constraints described below. The first set of constraint requires that the component-wise probabilities for each type of risk do not exceed the allowed levels.

$$P\Big(\sum_{a=1,\ldots,A}\sum_{k=1,\ldots,K}\sum_{j=1,\ldots,LxM} v_j(W_{jks})\, x_{jka}\Big) \le p_s,\ s = 1,\ldots, S.$$

The second set of problem constraints state that the total amount of social, ecological, and technological benefits expected as a result of applying integrated risk mitigating strategies must satisfy the corresponding demand:

$$\sum_{a=1,\ldots,A}\sum_{k=1,\ldots,K}\sum_{j=1,\ldots,LxM} v_j(W_{jk})\, u_{\mathrm{k}} B_{jkan} x_{jka} \ge B_n, \text{ for all types } n \text{ of benefits.}$$

The third set establishes the water balance conditions for water flows (with losses) in all intersection nodes of the graph representing the supply chain. The fourth set of constraints imposes the bounds R_m on material, financial, human, and other resources required for the establishment of technological solutions and water management policies:

$$\sum_{a=1,\ldots,A}\sum_{k=1,\ldots,K}\sum_{j=1,\ldots,LxM} v_j(W_{jk})\, u_{\mathrm{k}} R_{jkam} x_{jka} \le R_m, \text{ for all types } m \text{ of resources.}$$

4. CONCLUSION

The authorities should agree on solutions and stronger safeguard wastewater laws will be assigned, it is necessary for the water stakeholders in aquatic logistics supply chain and researchers to enter into consultation with one another, paying reasonable regards to the risk and legitimate interest of each party, in order to arrive at a scientifically rigorous and equitable resolution minimizing the possible risks for all participating parties.

Acknowledgment: The research was partly supported by the Ministry of Education and Science of Spain, under Grant SAB2005-0161. The author acknowledges the NATO financial support to attend and present this paper.

5. REFERENCES

1. Haruvy N. (2004). Irrigation with treated wastewater in Israel, assessment of environmental aspects, Kluwer, Dordrecht, 375–384.
2. Shuval HI, Lampert Y. Fattal B. (1997) Development of a risk assessment approach. Water Science and Technology. 25, 15–20.
3. Oron G, Campos C, Gillerman L, Salgot M. (1999). Wastewater treatment, renovation and reuse. Ag. Water Management, 38, 223–224.
4. Barrio HD, Irusta R, Fatta D, Papadopoulos A, Mentzis A, Loizidou M. (2006), Analysis of best practices and success stories for sustainable urban wastewater treatment. The MEDA WATER International Conference on Sustainable Water Management, Marrakech, Marocco, June 2006 8–10.
5. Bloemhof J, Van BP, Hordijk L,Van Wassenhove LN. (1995). Interactions between Oper. Res. and Env. Management, Euro. J. Oper. Res. 85, 229–243.
6. Carter JR, Narasimhan R. (1998). Environmental supply chain management, Center for Advanced Purchasing Studies, Focus study, 1–5.
7. Govil M, Proth JM. (2002). Supply Chain Design and Management: Strategic and Tactical Perspectives, Academic Press, New York.
8. US EPA – USA Environmental Protection Agency (1998). Guidelines for Ecological Risk Assessment, EPA/630/R-96/0O2F.
9. Daniel SE, Diakoulaki DC, Pappis SP. (1997). Operations Research and environmental planning, Euro. J. Operational Research, 102, 248–263.
10. US EPA/USAID – United States Environmental Protection Agency/United States Agency for International Development (1992). Guidelines for Water Reuse US Environmental Protection Agency Technical Report no. 81, p. 252.
11. Jones DS, Armstrong AQ, Muhlheim MD, Sorensen BV. (2001). Integrated risk assessment/management as applied to decentralized wastewater treatment: Risk-Based Decision Making. May 19–20, 2000, St. Louis, Missouri.
12. Zaidi Mohammed (2007). Risk assessment of modes of terrorist attack, Proceedings of the NATO ARW on Risk Analysis, April 2007 (in press).
13. UN – United Nations, (2003). Wastewater Treatment Technologies: A General Review, Report E/ESCWA/SDPD/2003, United Nations, N.Y.
14. Levner E, Alcaide D, Benayahu Y. (2007). Environmental risk ranking. Proceedings NATO ARW on Risk Analysis (in press).

15. Fishburn PC, Gehrlein WV. (1976). Borda's rule, positional voting, and Condorcet's simple majority principle, Public Choice, 28, pp. 79–88.
16. Lansdowne ZF. (1996). Ordinal ranking methods for multicriterion decision- making, Naval Research Logistics, 43, 613–627.

DIRECT OSMOSIS TECHNIQUE: NEW APPROACH TO WASTEWATER REUSE IN UZBEKISTAN

Renat R. Khaydarov and Rashid A. Khaydarov
Institute of Nuclear Physics,
Ulugbek, 702132, Tashkent, Uzbekistan

Corresponding author: physicist@sarkor.uz

Abstract:

The paper deals with a novel advanced solar-powered wastewater treatment technique based on direct osmosis process. The separation is driven by natural osmosis, which does not require external pumping energy as in the reverse osmosis process. Test results of the constructed pilot device having capacity of 1 m^3/h with various wastewater samples taken in villages of Aral Sea region have been discussed. The treated water might be reused as a substitute for potable quality water in many applications.

Keywords: Direct osmosis, wastewater, disinfection, Aral Sea.

1. INTRODUCTION

Many years of poor water management during Soviet Union period now threaten agriculture, soil quality, and availability of water for all purposes in Republic of Uzbekistan. Runoff from irrigation is negatively affecting the quality of drinking water from the main water streams. The annual discharge of collector and drainage water into surface water amounts to 20–25 km^3. Industrial users of water (energy, mining, metallurgy, and chemical industries) discharge into surface water 2.2 km^3 of wastewater, of which 131 million m^3 is untreated polluted water. Many cities have installations that only partially treat domestic sewage [1]. All above mentioned is the reason why the water management and its interrelation with agriculture and energy are considered the highest priority in Uzbekistan. This is particularly reflected in the National Environmental Action Plan, the Framework on Water Supply Development for the period up to 2010, the Framework on Groundwater Protection and Use, for the period up to 2010, and the Programme on Maintaining Population by Drinking Water and Gas up to 2010 [2]. There are numerous processes that can be used to clean up wastewaters depending on the type and extent of contamination. In general wastewater treatment requires the application of one or more of the following types of processes [3]:

M.K. Zaidi (ed.), Wastewater Reuse–Risk Assessment, Decision-Making and Environmental Security, 43–52.

(a) Primary treatments using mainly mechanical filtration (removal of suspended solids etc.)
(b) Secondary treatments using biological treatment processes (activated sludge processes, trickling filter, stabilization pond system, and other aerobic and anaerobic processes)
(c) Tertiary treatments (disinfection using chemicals etc.)
(d) Advanced treatments (carbon adsorption, reverse osmosis [RO], multi-stage flash [MSF] and multieffect distillation [MED], etc.) [4].

Usage of wastewater treatment techniques in developing countries must always take into account local climatic conditions, capital and operation costs, and manpower. Therefore, finding new low-energy consuming techniques of wastewater treatment is very important for such regions. As some of remote rural areas of Uzbekistan do not have enough access to electricity; therefore usage of solar energy for powering wastewater treatment devices seems to be a very perspective trend. In our previous works [5–6] we described two wastewater treatment methods developed in our Department of Development of Devices of Institute of Nuclear Physics of Uzbekistan Academy of Sciences. These are

(a) Oligodynamic disinfection technique used mainly as a part of tertiary wastewater treatment [5]
(b) Ion-exchange fiber sorbents technique used during both primary and tertiary wastewater treatments [6]

Both techniques proved to be very useful in hot climatic conditions of Uzbekistan and have been applied in a number of industrial plants of Central Asia region in order to treat wastewater. The purpose of this work was to develop new advanced low-energy consuming wastewater treatment method based on direct osmosis process that can effectively use solar cells as an energy source for pumping various fluids (feed, brine, product, working solution) and solar energy/ambient temperature/waste heat/geothermal energy for recovery of working solution.

2. METHOD

The proposed method is the advanced wastewater treatment technique allowing to reuse wastewater for drinking water purposes and implies further refining of the treated wastewater effluent. To describe the technique based on direct osmosis process let us suppose that (Figure 1) there is a feed containing contaminants with molar concentration of dissociated ions C_a in section A. In section B there is a water solution of a certain substance with initial molar concentration C_{bi}. Let us call them as a working fluid (WF) with a working substance (WS) respectively. Sections A and B are divided from each other by a semipermeable membrane, which is permeable for water molecules and impermeable for substances dissolved in section A and for the working substance. Let us also suppose that the condition $C_a << C_{bi}$ is met. Then in order to equilibrate the difference in chemical potential, osmosis movement starts, i.e., water begins to diffuse through the membrane from section A to section B. In other words, in section B water solution of WS is diluted by clean water pulled out from section A. Water solution of WS with final molar concentration C_{bf} from section B is transported to section C where water and WS are separated or a part of water is removed from the solution.

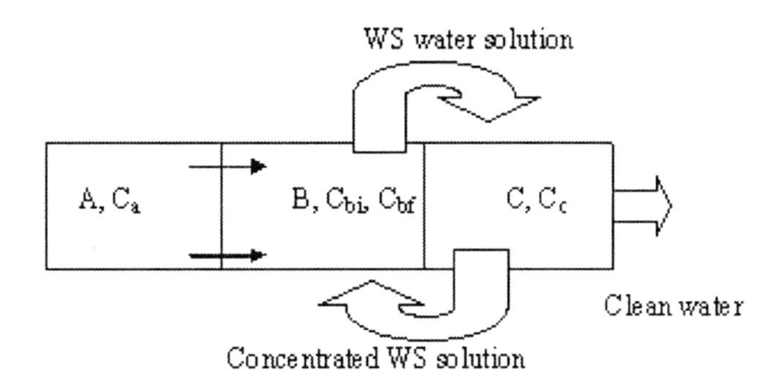

Figure 1. Principle of operation of direct osmosis method

The enthalpy diagram of the process is given in Figure 2. Let us suppose that ΔH_1 is the enthalpy of dilution of WF by the clean water passed through semipermeable membrane from section A to section B. Energy that is necessary to separate WS and clean water in section C is determined by enthalpy ΔH_2. It is clear that $|\Delta H_2|$ is less than $|\Delta H_4|$, where ΔH_4 is the enthalpy of the reverse osmosis (RO) process of removing organic and inorganic matters from the feed. In proposed direct osmosis water treatment technique, solar thermal energy or any other source of heat (ambient temperature/geothermal energy/waste heat) can be effectively used for the separation of clean water and WS. Therefore, total electrical energy consumption will be much less than that in case of RO technique. Moreover, as the coefficient of efficiency of solar cells usually is less than 20%, total solar energy consumption in the proposed method can be less than in the RO technique.

The main problem of the proposed water treatment method is in selection of WS. The simplest and the most accessible methods of separation of water from working substance in section C are evaporation, freezing, and forming crystalline hydrates. Definitely other methods of separation can be used as well. Some materials that might be utilized as WS (Table 1).

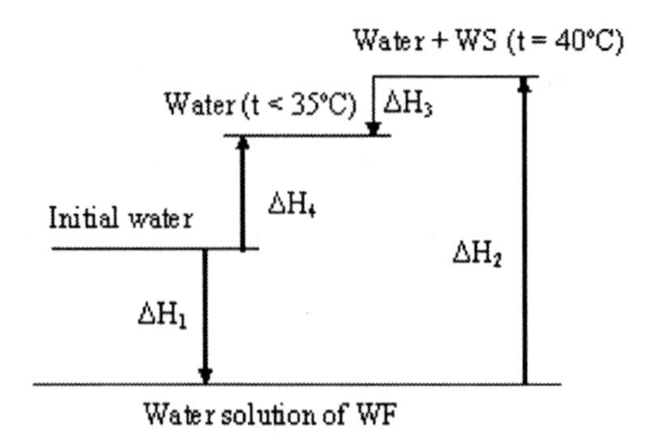

Figure 2. Enthalpy diagram of the treatment process

Table 1. Examples of materials can be used as a working substance (WS)

Type of WS	Name of WS	Specifications- t_{bp} - boiling point t_f - fusing point)	WS separation technique
Alcohol	Ethanol	t_{bp} = 78.39°C	WS evaporation
	Glycerin	t_f = 20°C	WS freezing
Ethers	Diethyl ether	t_{bp} = 35.6°C	Water freezing or WS evaporation
Amines	Ethylamine	t_{bp} = 16.6°C	WS evaporation
	Trimethylamine	t_{bp} = 3.5°C	WS evaporation
	Diethylamine	t_{bp} = 56.3°C	WS evaporation
	Methylamine	t_{bp} = –6.5°C	WS evaporation
Ketones	Acetone	t_{bp} = 56.24°C	WS evaporation

3. MATERIALS AND EQUIPMENT

In our experiments we used diethyl ether $(C_2H_5)_2O$ as the working substance and RE-1812-LP reverse osmotic membrane produced by SAEHAN Inc (South Korea) as the semipermeable membrane for direct osmosis process. These thin film composite (TFC) polyamide membranes (PA) consist of a porous support layer and a thin film dense layer which is a cross linked membrane skin and is formed in situ on the porous support layer made of polysulfone. The thin film dense layer is a crosslinked aromatic polyamide made from interfacial polymerization reaction of a polyfunctional amine such as *m*-phenylenediamine with a polyfunctional acid chloride such as trimesoyl chloride. The laboratory experimental water treatment device was constructed as shown in Figure 3. Vessel 1 with 80 l of NaCl solution was allocated at height of 0.5 m to exert a hydrostatic feed pressure of 5 kPa on a semipermeable membrane 2. The flow rate was regulated by valve 10 discharging concentrated water. On the other side of the membrane 2 there was a working liquid: water solution of diethyl ether $(C_2H_5)_2O$ having boiling point of 35.6°C and water solubility of 100 ml/l at 16°C.

The permeate (working liquid with clean water pulled out through the membrane 2) was divided into two parts by means of flow dividing valve 3. Then the most part of the permeate came back to the semipermeable membrane 2 by means of pump 4. The smaller part of permeate went in extending box 6 with an electrical heater (in the laboratory experimental device) holding the temperature at 38–40°C. As a result, clean water was accumulated in the collector 8 and vapor of diethyl ether went into condenser 7 cooled by water 9. Condensed diethyl ether was pulled in the flow of WF by the ejector 5 reconcentrating the diethyl ether water solution for the repeated use in the semipermeable membrane 2.

Deionized water was used to prepare the sodium chloride and diethyl ether solutions. Concentrations of Na and Cl were determined by neutron activation analysis [7]. Water samples were irradiated at the nuclear reactor of the Institute of Nuclear Physics (Tashkent, Uzbekistan). Ge(Li) detector with a resolution of about 1.9 keV at 1.33 MeV and a 6144-channel multichannel analyzer were used. The area under γ-peaks of radionuclides ^{24}Na (half-life is 15 h, energy of the γ-peaks are 1.368 MeV and 2.754 MeV) and ^{38}Cl (half-life is 37.7 min, energy of the γ-peaks are 1.642 and 2.167 MeV) were measured to calculate the concentrations of Na and Cl, respectively. Concentration of $(C_2H_5)_2O$ in water was determined by gas chromatograph "Svet-5" (Russia) using standard methods [8]. Flow rates of feed water and WF in the device were measured by flow meter "Zenner" (Germany) over a selected time period. Direct osmosis tests were conducted with the working liquid and the feed maintained at the same temperature 24±1.5°C. Energy E_h of water heating from $t_i = 24$°C to $t_f = 40$°C and energy E_s of separation of WS from clean water were calculated as follows: $E_h = cm_w(t_f - t_i)$ and $E_s = \lambda m_{ws}$, where $c = 4.2$ kJ/(kg·K) is specific heat of WF with concentration of WS equal to C_{bf}, $m_w = 25$ kg is mass of WF going for separation of WS from clean water, $\lambda = 384$ kJ/kg is specific heat of evaporation of diethyl ether, m_{ws} is mass of diethyl ether in WF going for separation of WS from clean water. Power of heater P_r that is necessary to obtain clean water output of 25 l/h was determined as $P_r = UI$, where U is electric heater voltage, V; I is electric current of heater, A.

In a pilot device, solar batteries of 12 V with a total capacity of 500 W produced by Physical-Technical Institute (Tashkent, Uzbekistan) were used as an energy source and ion-exchange fibroid sorbents developed in the Institute of Nuclear Physics of Academy of Sciences of Republic of Uzbekistan [6] were used to remove suspended solids, salts of hardness, and ions of iron. The specific surface of the fibroid sorbents is (2–3) × 10^4 m^2/kg and their exchange capacity is 3.5–4.0 meq/g. In the pilot device a water disinfection device based on oligodynamic method [5] developed in the Institute of Nuclear Physics was used to kill bacteria to prevent destruction of semipermeable membrane and prevent water contamination in emergency situations. The device provides killing all types of pathogens in water and has very low energy consumption of 0.1 W-h for water flow rate of 1 m^3/h. The above-mentioned pretreatment was necessary to prevent membrane fouling.

In the pilot device the vessel 1 with volume of 200 l was located at height of 3 m and filled by a water pump with power of 75 W. The water pretreatment unit containing standard commercially available mechanical filter, special water filter on the base of fibroid sorbents, and water disinfection device was installed between the vessel 1 and the semipermeable membrane 2. The pump 4 and pump for cooling water 9 had power of 75 W. Solar heater was used in the extending box 6 to heat water up to 38–40°C.

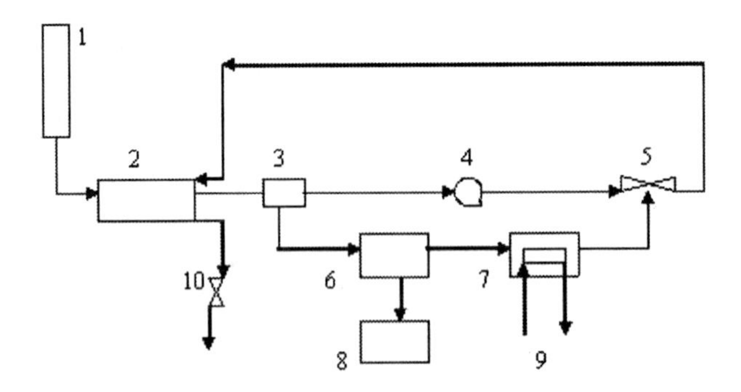

Figure 3. Scheme of the device: 1—Vessel with NaCl water solution, 2—semipermeable membrane, 3—flow dividing valve, 4—pump, 5—ejector, 6—extending box with heater, 7—condenser, 8—collector, 9—cooling water, 10—flow rate regulated valve

4. RESULTS AND DISCUSSION

Estimated value of osmotic pressure of our WF (water solution of $(C_2H_5)_2O$ with concentration of 100 ml/l) is 2.35 MPa. Meanwhile for instance, 4 g/l NaCl water solution (with degree of dissociation of 0.44) has osmotic pressure of 0.39 MPa. Thus the difference between the osmotic pressures was always large enough, i.e., there was self-sustaining osmotic movement without necessity of usage energy-consuming pumps in contrast to reverse osmosis water treatment method.

Some results of tests of the laboratory experimental model are given in Table 2. Three samples with various water salinities were chosen for the tests: 1 g/l as regulation limit for drinking water salinity in Uzbekistan, 4 g/l as sample of water salinity in the most regions of Uzbekistan, and 20 g/l as sample of water salinity in Aral Sea region of Uzbekistan. Mentioned in the Table 2 c_a, c_{bi}, c_{bf}, and c_c are mass concentrations corresponding to molar concentrations C_a, C_{bi}, C_{bf}, and C_c, appropriately. Flow rate of WF should be low to minimize the pump power consumption. Two values of the WF flow rate were tested: equal to about two values of productivity of the device (minimal allowable flow rate) and ten values of productivity. In our experimental device the energy was consumed mainly at the stage of WS separation from the water (the heater with maximum power of 600 W) and for WS circulation by the pump (power of 15 W) to obtain clean water output of 25 l/h. Temperature of influent water was 24°C. Comparison of calculated energy E_h of water heating up to 40°C and energy E_s of separation WS from the water shows that the main part of energy (from 85 to 95%) is consumed for water heating. Heater power of the device P_r that is necessary to obtain clean water output of 25 l/h is 500–600 W and practically does not depend on the initial concentrations of organic and inorganic matters in the range from 1 to 20 g/l. This power decreases when temperature of influent water increases and approaches 35–40°C because only 60–100 W is consumed for separation of clean water from WS.

Total electric energy consumption of the advanced treatment process can be essentially decreased by using a solar heater instead of the electric heater used in the laboratory experimental device. Thanks to the financial support of UNESCO in March, 2005, the pilot device with productivity of 1 m^3/h was installed in Turtkul village of Aral Sea region (Figure 4). Total concentration of organic and inorganic matters in initial wastewater (from saline containing drain water, sewage, pesticides, etc.) was about 17 g/l. Test results of the device are shown in Table 3. Due to replacing the electric heater 6 by the solar heater the total electric power of the device did not exceed 500 W, i.e., it was much less than energy consumption of RO method. The product salinity after the treatment was about 50 mg/l.

Table 2. Test results of the laboratory experimental model (clean water output is 25 l/h, initial temperature of water is 24°C)

c_a of NaCl (g/l)	Recovery-relation of permeate and feed flow rates	Flow rate of WF (l/h)	Part of WF for water separation	C_{bi} (g/l)	C_{bf} (g/l)	E_h (kJ)	E_h (kJ)
1	0.1	60	0.5	5	3.0	1,600	29
		300	0.1	5	4.6	1,600	44
	0.5	60	0.5	7	4.1	1,600	40
		300	0.1	7	6.4	1,600	62
4	0.1	60	0.5	20	12	1,600	112
		300	0.1	20	18	1,600	176
	0.5	60	0.5	25	15	1,600	140
		300	0.1	25	23	1,600	220
20	0.1	60	0.5	40	33	1,600	224
		300	0.1	40	37	1,600	352
	0.4	60	0.5	60	35	1,600	336
		300	0.1	60	55	1,600	528

Figure 4. Pilot device installed in Turtkul village of Aral Sea region

Our experiments have shown the following advantages of the proposed technique: (a) the method does not require a high pressure pump to push water through the membrane; (b) electric energy consumption of the method is very low (0.5 kWh/m^3 in comparison with 2–5 kWh/m^3 for RO and 2–4 kWh/m^3 for MSF and MED) when solar energy or heat of ambient air are used to heat the water up to 36–40°C; (c) surface area of the solar heat exchanger of the device is 3–30 times less than that of solar thermal desalination process because of lower solar energy consumption (80 MJ/m^3 in comparison with 2,500 MJ/m^3 for solar thermal desalination process without energy recovery and 250–500 MJ/m^3 with energy recovery); (d) the method can be used for wastewater treatment with organic and inorganic matters concentrations up to 40 g/l; (e) since the disinfecting device is used in water pretreatment process, the water keeps disinfecting properties for a long time (not less than 1 year) and water produced in hot time can be stored in water storage systems for cold seasons. Membrane scaling or fouling have been prevented by using ion exchange fibroid water filters to remove salts of hardness and iron ions and water disinfection device to kill bacteria.

Disadvantage of the method is a short lifetime of the membrane material (not more than 2–3 months) because of destroying by diethyl ether. We suppose that diethyl ether reacts with materials of membrane housing and agglutinative materials used in membrane assembling process. Thus, it is necessary to investigate other types of membranes, working substances, and process of separation of clean water from WS.

Table 3. Purification of water sample taken from Aral Sea region

Items	Influent water (mg/l)	Effluent water (mg/l)
pH (at 25°C)	6.7	6.7
Ca	560	1.2
Mg	400	1.6
Cl	4,590	7.1
NO_3	300	<0.02
SO_4	3,825	1.8
Na	3,300	2.3
K	1,200	1.6
Phenols	2.8	<0.02
Lindane	0.13	0.0001

5. CONCLUSION

Obtained test results have shown the efficacy of wastewater treatment device based on direct osmosis technique during advanced treatment stage. The devices utilizing solar batteries as an energy source can be used in remote regions for purification of pretreated wastewater effluent with concentrations of organic and inorganic matters up to 40 g/l. When solar energy or heat of ambient air are used to heat the working fluid up to 36–40°C, the electric energy consumption of the advanced treatment stage is very low: 0.5 kWh/m^3 in comparison with 2–5 kWh/m^3 for RO and 2–4 kWh/m^3 for MSF and MED when solar energy or heat of ambient air are used to heat the water up to 36–40°C. Due to simultaneous use of the oligodynamic disinfection method, the treated water might be reused as a substitute for potable quality water in many applications and keeps disinfecting properties for a long time (not less than 1 year).

Acknowledgment: We express our acknowledgment to NATO ARW organizing committee and the NATO Science Program for financial support for participation at the NATO ARW held in Istanbul, Turkey, during October 12–16, 2006.

6. REFERENCES

1. Environmental Performance Reviews. 2001. Uzbekistan. United Nations, New York and Geneva.
2. Environmental Protection. 2000. Collection of Laws and Regulations. *Tashkent.*
3. Kanbour F. 1994. Technology of wastewater treatment and reuse, Proceedings of the Conference on Water in the Islamic World: An Immient Crisis, Khartoum, Sudan.
4. Schippers J, Kennedy M. 2004. Intensive Course on Membrane Technology in Drinking & Industrial Water Treatment, Middle-East Desalination Res. Center, Muscat, Oman.
5. Khaydarov RA, Khaydarov RR, Olsen RL, Roger SE. 2004. Use of electrolytically generated silver, copper and gold for water disinfection, J. Water Supply RT-Aqua, 53, 567–572.
6. Khaydarov RA, Gapurova O, Khaydarov RR, Cho SY. 2005. Fibroid Sorbents For Water Purification, Modern Tools and Methods of Water Treatment for Improving Living Standards, NATO Science Series, IV. Earth and Environmental Sciences – Vol. 48, pp. 101–108
7. Alfassi ZB. 1990. Activation Analysis, Volumes I and II, CRC Press, Boca Raton, Florida.
8. MUK 4.1.655-96. 1996. Methodical instructions on gas chromatography determination of diethyl ether in water, Ministry of Health of Russian Federation.

SESSION 2. WASTEWATER REUSE—CASE STUDIES

WASTEWATER MANAGEMENT, TREATMENT, AND REUSE IN ISRAEL

Saul Arlosoroff
Mekorot-National Water Corporation, Tel-Aviv, Israel

Corresponding author: sarlo@inter.net.il

Abstract:

Israel, a water scarce country, had embarked on a national campaign to develop all its water resources and to integrate into water resources management strategy – recommends a comprehensive water conservation and efficient use of water by all users. This strategy stressed that treated wastewater is a valuable water resource and focused on the total wastewater treatment and reuse as a national objective, and the exchange of the treated effluent with the farmers' fresh water allocations. The Dan Region wastewater project is planned to help Israel face the water shortage issue.

Keywords: Wastewater, effluent, irrigation, microbiological standards, reuse policy.

1. INTRODUCTION

The unique characteristics of the water strategy of Israel is defined in times as the water demand management and conservation, a major shift of paradigm from the conventional supply management of water to the management of the demand side. This objective leads to the production of additional quantities of water for the needs of the people, through the creation of virtual quantities of water, whether by conservation policies, the treatment, reuse of human waste, use and desalination of sea and brackish sources, cloud seeding, and others. Legal and administrative measures, progressive block pricing, comprehensive water metering, and allocations are all part of the water demand management strategy [1].

The initiation of a comprehensive wastewater treatment and reuse program (WWT&RU) as a national sector policy dates back to the 1960s when realizing that the country will face serious water scarcity as well as pollution of its water resources, beaches and aquifers unless a total WWT&RU will be implemented, and will include the appropriate actions to maintain adequate risk assessment of environmental impacts.

M.K. Zaidi (ed.), Wastewater Reuse–Risk Assessment, Decision-Making and Environmental Security, 55–64.

The national program started with small lagoons type treatment plants (primary and secondary) functioning as oxidation ponds, using gravity or sprinkler irrigation gradually moving to the situation of 2005 when approximately 85% of all wastewater discharges are treated by activated sludge plants, as well as tertiary treatment of about 50% of the total discharges.

Most of the effluents are drip irrigated and/or use spray systems for crops and areas that minimize environmental risks [2]. Israel is already reusing approximately 350 mcm/year (millions cubic meters per year) of effluents, half of the total water allocation for irrigation in the country.

Regulations have been legislated in order to increase the quality levels of WWT plants and its effluents to maximize its reuse potential and minimize the health and environmental risks. These actions are enhancing the trading potential for its exchange for freshwater allocations, mainly for irrigation purposes. Since the 1990s, the allocation policy for irrigation concentrates on reduction of freshwater quantities to the farming community and replacing it with treated wastewater effluents. Total sewerage costs is borne by the city, while the reuse component by the water sector, when the links between discharges and use are funded by the State (Graph 1).

Graph 1. Forecasted reuse of treated human waste - effluent—Israel 1995

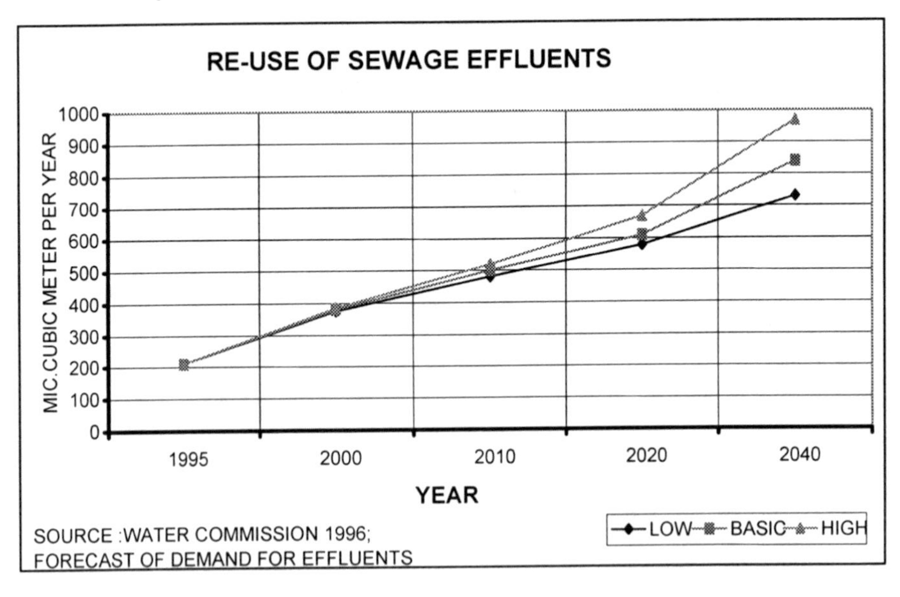

1.1. Background

Due to the severe shortages of water on one hand and the growing water demand by its intensive agriculture on the other hand, WWT&RU for crops irrigation has become a sector priority for the past four to five decades. As a good part of its agricultural produce is being supplied locally as well as exported, emphasis was given to environmental health and safety of the irrigated produce, the quality of the reclaimed waters, and the promoted efforts that have been invested in the development of appropriate technology, for wastewater irrigation. The authorities have been focusing and updating reclaimed wastewater quality standards and criteria, adapting them to international demands, the application of new irrigation techniques and to public concerns [3].

The total water demand by the agricultural sector in an average year, has been reduced by 30–40%, and is at present ≈ 1000 mcm/year of which approximately 350 mcm/year is from WWT&RU schemes. In the frequent dry years, the reclaimed wastewater can constitute 40–50% of all irrigation water demand. In fact, during drought years, reclaimed wastewater is the most reliable source of irrigation water to those farmers who are connected to the reclaimed wastewater distribution system, when freshwater allocations are decreased to match the natural recharge of the surface and groundwater resources.

Table 1. Water demand according to various uses in the period 2005–2020 (million cubic meters per year [mcm/year])

Consumptive use	Years			
	2005	2010	2015	2020
Domestic, freshwater (drinking quality)	720	840	960	1,080
Agriculture freshwater (drinking quality)	530	530	530	530
Agriculture saline waters	160	140	140	140
Agriculture reclaimed wastewaters	300	500	600	700
Industry freshwater (drinking quality)	100	315	500	650
Industry saline waters	30	50	80	80
Industry reclaimed wastewater	0	5	13	15
Natural impoundment	25	50	50	50
Aquifer restoration freshwater	100	200	0	0
Neighboring countries freshwater	100	110	130	150
Total demand	2,060	2,505	2,558	2,805

Source: Y. Dreisen, Israel Water Commissioner Office, 2005

Tables 1 and 2 illustrate respectively, the water demand according to the sectors of use and the sources of supply. It should be noted that seawater desalination is going to constitute an important part of Israel's water balance, in the future, when conservation and efficiency efforts would have been exhausted. The supply of water to Jordan and to the Palestinian authority, where water demand supersedes their dwindling water resources, is based on agreements signed during the 1990s.

Table 2. Sources of water supply in the period 2005–2020 mcm/year

Natural water resources	Years			
	2005	2010	2015	2020
Freshwater (drinking qual.)	1,470	1,470	1,470	1,470
Freshwater (drinking qual.)	160	140	140	140
Saline waters	160	140	140	140
Reclaimed wastewaters	300	450	520	600
Desalinated seawater	100	315	500	650
Desalinated saline waters	30	50	80	80
Total Supply	2,140	2,425	2,710	2,910

Source: Y. Dreisen, Israel Water Commissioner Office, 2005

1.2. Treated Wastewater Quality Issues

Israel's Ministry of Health appointed a state Committee to establish an up-to-date criteria based on modern risk assessment and risk management approach. These new standards have become effective in 1999. The new criteria for unrestricted irrigation with reclaimed wastewater of all crops, enforced basic requirements, as following:

- Fecal coliforms of average 10/100 ml
- Turbidity of 5 NTU (or 10 mg/l suspended matter)
- Effluent BOD_5 of less than 20 mg/l and suspended matter of less than 30 mg/l
- Granular filtration
- Residual chlorine concentration of at least 1.0 mg/l after at least 30 min contact time.

As a substitute to granular filtration, a 60 days average retention in a seasonal reservoir or 30 days in "sealed" (closed) retention should be provided. Once "risk management barriers" are provided, the above requirements may be relaxed. The barriers provide reduced public health risk, particularly risk of actual contamination of fruits or vegetables consumed by humans. The principal barriers are as follows:

- Distance of fruit or produce from the drip irrigation laterals
- Use of subsurface drip irrigation
- Plastic cover(mulch) of the drip-irrigation system
- Fruit or produce should be cooked
- Fruit or produce have a nonedible peel

In mid-2005, a draft of a new set of requirements by the Israel Ministry of Environmental protection, new regulations for effluents quality, was adopted after more than five years of deliberation by State Committee and has recommended more stringent requirements on the biochemical, chemical, and mineralogical aspects.

1.3. The WWT&RU Systems

There are over 70 municipalities in Israel which operate secondary or tertiary treatment and their effluents are being reused, totally or partially. The three major schemes are the Dan region, (Greater Tel-Aviv region), The Greater Jerusalem wastewater system, and the Greater Haifa Complex, (Kishon valley).

2. THE DAN REGION (GREATER TEL-AVIV) SCHEME

The system for treatment and reclamation of the greater Tel Aviv region, is the first, unique in it's design and size as a tertiary treatment and reuse project. It is one of the largest and most complex of its' kind in the world. It is treating and reusing the effluents discharged by population equivalent of over 2 million, (approximately 30–35% of the total population of the country), supplying over 125 mcm/year of effluents with adequate water quality. It constitutes the principle source for irrigation water supply in Israel. Following a secondary activated sludge with nitrification/denitrification treatment, the effluent is discharged on large infiltration basins and is recharged into sandy aquifer under the basin.

Following an average of 300 days of residence time in the aquifer a significant additional polishing occurs which is commonly defined as "soil-aquifer treatment" (SAT) [4]. The high-quality reclaimed water is pumped by a large number of recovery wells into a distribution system. The change in the major quality parameters along the treatment and SAT are shown in Table 3.

2.1. SAT Method—Description, Evaluation, Study, and Monitoring

The following is a brief description of the Dan region's SAT system:

The operation of the recharge basins is intermittent, i.e., flooding periods alternate with drying periods maintain high infiltration-rates through the upper soil layer and to allow oxygen penetrate into the soil, thus enhancing and diversifying the soil purification capacity. The recharged effluent gradually displaces the existing groundwater and flows towards a ring of recovery wells surrounding the recharge basin. The recovery well pumps the high-quality reclaimed water obtained after SAT to a separate, nonpotable conveyance system, which is used only for unrestricted irrigation of agricultural crops.

Table 3. Major parameters along the Dan region treatment and SAT stages

Parameter	Units	Raw sewage	T. Plant effluents	Reclaimed water*
Suspended solids	mg/l	378	11	0
BOD_5	mg/l	368	12	<0.5
COD	mg/l	851	49	8
NH4-N	mg/l	38.5	6.5	0.03
Total-N	mg/l	60.7	11.4	1.02
P	mg/l	15	3	0.13
Total bacteria	No./1 ml	2.8E+07	8.0E+05	456
coliforms	MPN/100 ml	2.3E+08	4.1E+05	0
Fecal coliforms	MPN/100 ml	3.2E+07	2.8E+04	0
Streptococus faecalis	MPN/100 ml	4.2E+06	9.0E+03	
Enteroviruses	PFU	2	0	

* Observation well No. 01—7 (*Source*: Idelovitch E and Michal M – 1985)

Environmental risks are prevented as the zone of the aquifer enclosed within the ring of recovery well is hydrologically separated from the rest of the aquifer, which is not affected by the effluent recharge operation and continues to supply potable water. A small portion of the potable aquifer is thus used for the effluent. A large number of observation wells, which are located between the recharge basin and the recovery well, provide the means for risk assessment and monitoring the SAT.

3. RISK ASSESSMENT STUDY

A major study of the Dan region's SAT system was carried out during 1999–2001 [5]. It consisted of a comprehensive analysis of the large amount of water quality data accumulated over a period of 25 years. The main objectives of the study were to

1. evaluate the recharge effluent quality prior to SAT;
2. evaluate the reclaimed water quality after SAT;
3. estimate the purification capacity of the soil aquifer system and its efficiency, after 25 years of operation;
4. analyze the performance of the physical, biological and chemical processes occurring during SAT; and predict their long-term behavior.

3.1. The Monitoring Procedure

The data for the study was based on an extensive monitoring program that accompanied the recharge–recovery operation since its initiation in 1977, thus providing an unusual wealth of reliable data on the effluent quality before and after SAT. The monitoring program has two major objectives: (1) to follow the movement of the recharged effluent in the aquifer (hydrological aspects); and (2) to detect the changes in the physical, chemical, and bacteriological characteristics of the groundwater in the recharge zone and its vicinity (water quality aspects). The monitoring program includes sampling points for the recharge effluent (RE) (before SAT), as well as some 50 observation and 100 recovery wells (after SAT).

The quality of the recharge effluent includes some 60 physicochemical and biological parameters which were analyzed with a frequency varying from once per week for common physicochemical parameters to twice per year for bacteriological indicators and some trace elements. In the observation wells (OW) and the recovery wells (RW), a "routine analysis" consisting of some 20 parameters is performed with a frequency varying from one to four times per year. In addition, a comprehensive analysis, consisting of some 50 physicochemical and biological parameters, were and are being carried out once or twice per year in a limited number of selected wells.

3.2. Monitoring and Study—Results and Discussion

In order to analyze, present, and interpret the large amount of data accumulated during the 25 years, the parameters included in the monitoring program were divided into seven groups as follows:

1. Chlorides, the main parameter that indicates the arrival of the recharged effluent at a certain well and permits estimating the percentage of effluent in the well water.
2. Parameters characterizing wastewater type and strength such as total and volatile suspended solids (SS and VSS), as well as total and filtered biochemical oxygen demand (BOD and BOD_f).

3. Parameters measuring organic substances such as: total and filtered chemical oxygen demand (COD and COD_f), total and dissolved organic carbon (TOC and DOC), ultraviolet absorbance (UV absorbance), and detergents.
4. Main nutrients: ammonia (NH_3/NH_4), Kjeldahl nitrogen (NK), nitrite and nitrate (NO_2 and NO_3), and phosphorus (P).
5. Parameters measuring the salinity of the effluent and its ionic composition.
6. Heavy metals and trace elements [6].
7. Microbiological parameters.

Table 4. Relative removal efficiency (RRE) of various contaminants by SAT (all concentrations are in mg/l, except for UV absorbance which is in $cm^{-1} \times 1{,}000$)

Parameter	Concentration		Average RRE
	before SAT	after SAT	
SS	10–80	0	100%
BOD	5–40	0.5	98%
COD	40–160	10–20	85%
COD_f	40–80	10–20	75%
DOC	15–20	3–6	74%
UV absorbance	150–400	30–80	80%
Detergents	0.4–1.0	0.05–0.2	82%
Total N	5–30	5–10	57%
Total P	3–10	0.01–0.03	99%

3.3. Long-Term Forecast

The long-term forecast with respect to the purification process occurring during SAT depends on the type of process. The physical process—filtration through the upper soil layer—which is responsible for the removal of suspended solids, will last for generations if the maintenance and cleaning of the recharge basins are done appropriately. The biological process—bacterial degradation and nitrification–denitrification, which are responsible for the removal of dissolved organic matter and nitrogen, respectively, will also last for very long periods, if the conditions required for bacterial activity and the delicate balance between aerobic and anaerobic conditions are preserved.

As for the chemical processes—chemical precipitation and adsorption, which are responsible for the removal of phosphorus as well as of heavy metals and trace elements, the long-term forecast is that they will last only a limited amount of time, until the precipitation and/or adsorption capacity are exhausted. However the study and monitoring maintain that, this is only a theoretical forecast. In reality, the above process can last for a long time as confirmed by the excellent and stable removal of phosphorus and trace elements obtained during 25 years of operation of the SAT in the Dan Region Project. This is due to the large soil volume participating in the process, as well as to the combined action of chemical and biological processes. Some of the contaminants adsorbed or precipitated by chemical process undergoes in time and thus liberate the soil sites for adsorption or precipitation of new contaminants.

It was not possible to separate between the organic matter removed by biological process that could last indefinitely, and that removed by adsorption that should last only during a limited period of time (until the soil adsorption capacity is exhausted). Similarly, it was not possible to separate ammonia by adsorption and is removed by nitrification–denitrification.

In the recent years, since the completion of the study, the infiltration capacity of the basin has increased and the ratio between the length of drying and flooding period has increased, too. Thus, aerobic conditions prevail again in a larger portion of the soil-aquifer system. As a result, the concentration of ammonia and UV absorbance in the observation wells have been decreasing. It is hoped that this tendency will continue in the future.

4. CONCLUSION

The SAT has proved to be a feasible program to meet the additional water needs of Israel while preserving the environmental constraints and risks, without a foreseeable danger to the long-term sustainability of the aquifers as well as the health of workers. The marginal costs, per unit of water, of the whole process of SAT, is much lower than the prevailing seawater desalination costs ($\approx$ 0.60–0.65 US\$/m^3) at the inlet to the conveyance system. Thus the economic feasibility which was doubted in the past as well, does not present any more questions, after 25 years of experience with the existing program.

It has been proved that wastewater is a valuable resource for irrigation in a water scarce country, and if properly treated and reused it will not cause an environmental hazard.

Acknowledgment: The paper was written with the support and material provided by the late Emanuel Idelovitch, an active consultant to Mekorot—The National Water Corporation of Israel and Ms. Nelly Icekson-Tal, manager of Mekorots' wastewater reclamation program. The paper is presented with the financial support of NATO Advanced Research Workshop grant.

5. REFERENCES

1. Idelovitch E, Icekson-Tal N, Avraham O, Michail M. (2002). The long-term performance of Soil Aquifer Treatment (SAT) for effluent reuse. Water Science and Technology, 3(4), 239–264.
2. Idelovitch E. (2001). Effluent Treatment by Soil-Aquifer-Treatment. Mekorot Water Co Ltd., Central District, Dan Region Unit, Israel.
3. Brill, E, Hochman E, Zilberman D. (1997). Allocation and pricing at the water district level. American J. Agricultural Economics, 79(3), 952–963.
4. Amiran DHK. (1978). Geographical aspects of national planning in israel: The management of limited resources. Transactions of the Institute of British Geographers, 3(1), 115–128.
5. Friedler E, Lahav O, Jizhaki H, Lahav T. (2006). Study of urban population attitudes towards various wastewater reuse options: Israel as a case study. J. Environmental Management, 81(4), 360–370.
6. Rattan RK, Datta SP, Chhonkar PK, Suribabu K, Singh AK. (2005). Long-term impact of irrigation with sewage effluents on heavy metal content in soils, crops and groundwater—a case study. Agriculture, Ecosystems & Environment, 109(3/4), 310–322.

SOIL AQUIFER TREATMENT AREAS IN TUNISIA: JERBA ISLAND

Makram Anane[1], Hamadi Kallali[1], Salah Jellali[1], and Mohamed Ouessar[2]
[1]Wastewater Treatment and Recycling Laboratory, Water Research and Technologies Centre, Soliman
[2]Arid Region Institute, Medenine, Tunisia

Corresponding author: Kallali.Hamadi@inrst.rnrt.tn

Abstract:

Soil aquifer treatment (SAT) has been practiced in various dry land and semiarid areas as means to reduce environmental risk by improving wastewater quality at disposal sites and to replenish aquifers enabling the use of reclaimed water for nonrestrictive crop irrigation. SAT success is preconditioned by an appropriate site selection, a correct design, and a continuous maintenance. The location of the SAT basins is determined by a multitude of criteria often less than inituitive. This study aims to identify potential sites for SAT in Jerba Island integrating geographic information systems (GIS) tools and the analytical hierarchy process (AHP).

Keywords: AHP, GIS, SAT, wastewater, Jerba Island.

1. INTRODUCTION

As arid area, the Tunisian south has an acute scarcity in water resources. Jerba Island, faces an increase in water demand due to the huge touristy and demographic development. Indeed, three wastewater plants are currently working in the island in which one was recently constructed and a fourth one will be operated during 2007. At present, the operating plants produce 10,000 m^3/day of treated wastewater mostly disposed to the sea. This treated wastewater could be reused mainly for watering, irrigation, and aquifer recharge to attenuate water penury. Besides increasing water availability, wastewater reuse can reduce sea pollution hazards.

Meanwhile, the aquifer, almost only source of irrigation in the region, is suffering since many years ago from an overexploitation for fulfilling human and agricultural needs. This overrated groundwater consumption caused a water quality degradation, which is intensified by seawater intrusion. SAT has been practiced in various dry land and semiarid areas with high performance by reducing environmental risk and replenishing the aquifer to enable wastewater use for nonrestrictive crop irrigation.

M.K. Zaidi (ed.), Wastewater Reuse–Risk Assessment, Decision-Making and Environmental Security, 65–72.

In Jerba, as practised at oases regions, irrigation concerns three level of crop farming (vegetables, fruit trees, and palms). These oases risk to be damaged by unsuitable groundwater quality. In these areas, the secondary treated wastewater use in irrigation is prohibited and SAT becomes an interesting practice as tertiary treatment to improve its quality. In addition, applying SAT in coastal areas creates water barrier against seawater intrusion, and therefore reduces groundwater salinity. Nevertheless, SAT success is conditioned by an appropriate site selection, correct design, and continuous maintenance.

1.1. Study Area

Jerba Island is located at the southeast part of Tunisia and belongs to Medenine district. It covers 514 km^2 of surface area with 130 km of coasts, around 27 km length and 19 km width. It counts about 120,000 inhabitants. The climate is arid with 220 mm as mean annual precipitation, and 20°C as mean temperature. The island is very flat, lacking hydrographical network. The altitude varies between 0 m at coastlines and 50 m near Oued Ezzebib village. The geology is mostly from Quaternary with some mio-pliocene formations. The depth of the aquifer varies between 5 and 31 m and tourism is the dominant economic activity. Three wastewater treatment plants (WWTP) are operating in the area: Houmet Essouk, Sidi Mehrez, and Aghir and one is under construction at Ajim (Table 1).

Table 1. Characteristics of WWTPs located in the study area (from ONAS)

WWTP location	Year of operation	Nominal flow rate (m^3/d)	Pollutant load (kg BOD/d)	Treatment technique
Ajim				Under construction
Aghir	2001	15,750	3,325	Extended aeration
Houmt Essouk	1991	3,500	1,500	Aerated lagoons
Sidi Mahrez	1981	4,000	900	Aerated lagoons

All the treated wastewater of Houmt Essouk plant is disposed to the sea whereas Sidi Mehrez wastewater is mostly reused to irrigate the golf field of Jerba located at some kilometers from it. The excess wastewater is bypassed to Aghir plant. The major part of the treated wastewater produced by Aghir plant is disposed to the sea and only a little volume is reused for irrigation.

2. METHODOLOGY

The methodology followed to identify potential sites for SAT is an integration of AHP method in a GIS model. AHP, developed by Thomas Lorie Saaty since the

1970s [1], is a method used to determine the priority of different decision alternatives via pairwise comparisons with respect to a common criterion.

2.1. Criteria Selection

Several criteria were established taking into consideration simultaneously different aspects. They were grouped in three main criteria: technical, socio-environmental, and cost.

2.2. Technical Criteria

2.2.1. Aquifer depth

It is one of the most important technical criteria to consider because it represents the thickness of the vadose zone in which the treated wastewater will have a complementary purification before reaching the groundwater. Hence, suitable sites have to continuously assure 5 m of unsaturated zone even after mounding.

2.2.2. Geology

It is as important as aquifer depth for SAT because it depicts the content of a large part of the vadose zone. Geology formations of suitable sites should have an optimal clay fraction in order to fulfil contradictory conditions as having high hydraulic conductivity and performing good complementary wastewater purification. No sandy, sandy loam nor loamy sand formations will be discarded.

2.2.3. Soil texture

Suitable soils should have a very low clay fraction. No sandy, sandy loam nor loamy sand grain size distribution will be discarded.

2.2.4. Groundwater salinity

Obviously, wastewater salinity has to be lower than groundwater's one to prevent groundwater quality deterioration. Groundwater with salinity under 3 g/l will be discarded.

2.2.5. Soil salinity

Suitable soils must have low salinity in order to avoid aquifer salinization by leaching. Halomorph soils will be discarded.

2.2.6. Slope

Infiltration percolation plant site should be feasibly constructed in low slope to prevent basin safety risking and expensive refilling costs. Terrains with more 15% slope will be discarded.

2.3. Environmental Criteria

The environmental criteria is crucial to respect a safeguard distance in order to avoid direct human contact. Areas located at under 200 m of distance from settlements are discarded.

2.4. Cost Criteria

The cost criteria considered correspond to water transport costs (adduction and pumping) from wastewater treatment plant to SAT basins. The first criterion is potential site farness from WWTPs, which affects pipes length and afferent costs. The second criterion is potential site elevation regarding to WWTPs levels. This criterion reflects necessary wastewater pumping devices and related costs.

2.5. Criteria Layering by GIS

Spatial analysis for suitable SAT sites identification begins by representing each selected criterion by a thematic layer in which each point takes a value or a qualification according to that criterion. In order to layer all the criteria, data are gathered from related satellite images and official sources at different forms available (digital and hard copy maps, tables and charts) and then analyzed and treated using GIS and geostatistical tools.

We will take into account from the existing plants only Houmt Essouk, Aghir plants, and the under-construction plant of Ajim. Sidi Mehrez plant is eliminated from the study because all its treated wastewater is used for a golf field sprinkling.

2.6. Constraints Map Establishment

Subcriteria of technical and environmental criteria are used first as constraints in order to discard the unsuitable areas for SAT. An acceptable range of each criterion regarding feasibility of SAT was set by a critical threshold values, corresponding to standards. These normative statements are established after consulting national experts and reviewing international guidelines and technical documents [2–9].

2.7. Criteria Prioritizing

Then each matrix consistency is checked out through the calculation of consistency ratio (*cr*) which is defined as the quotient between the consistency index (*ci*) and the random index (*ri*):

$$cr = \frac{ci}{ri}$$

The consistency index (*ci*) is given by:

$$ci = \frac{(\lambda_{\max} - n)}{(n-1)}$$

where $\lambda_{\max}$ is the maximum value of eigenvector and *n* is the criteria number.

And the random index (*ri*) is obtained from a table established by Oak Ridge National Laboratory for matrix with rows going from 1 to 15 [1]. For *cr* lesser than 0.1, the priorities assigned are considered satisfying, otherwise they are judged not consistent to generate weight and have to be revised and improved.

2.8. Subcriteria Standardization

The process of potential SAT sites choice deals with subcriteria of heterogeneous types (qualitative and/or quantitative), different forms (continuous or discrete), and concerning different domains of measurement. In order to combine these heterogeneous data, it is crucial to standardize all the subcriteria bringing them into a common domain of measurement.

2.9. The Composite Decision Value Calculation

A final composite SAT map for the study area was obtained by aggregating the weights (*w*) of all the levels in the hierarchy. The process constitutes in multiplying the subcriteria weights by one-third (1/3) which is the common weight given to second hierarchy criteria (Table 2). The composite decision value is represented by the score (*Ri*) of the pixel *i*, determined by summing its weighted standardized criteria (r_{ik}):

$$R_i = \sum\nolimits_k w_k r_{ik}$$

Hence, we obtain a map with areas ranked according to their *R* value which is ranging from 0 for the least suitable for SAT to 255 for the most suitable.

2.10. Land Requirement Estimation

The required land for SAT basins is established only for Houmt Essouk and Aghir WWTPs, which are currently in operation. Ajim plant, which is still under construction, is ignored. The methodology used to estimate the required land is introduced [10]. For the calculations, we used the nominal daily flow rate for Houmt Essouk plant (3,500 m^3/d) and the remaining flow rate after satisfying irrigation needs for Aghir Plant, estimated to 11,596 m^3/d. The soil permeability, as 11.1 cm/h, has been adopted for the whole areas [11]. The total area needed to infiltrate the available treated wastewater in SAT basins, is estimated to 22 ha (17 ha for Aghir plant and 5 ha for Houmt Essouk plant).

Table 2. Hierarchy weighing for potential SAT sites

Main criteria	Subcriteria	Total weight
Technical $w = 0.33$	Aquifer depth $w = 0.27$	0.0891
	Geology $w = 0.27$	0.0891
	Soil texture $w = 0.27$	0.0891
	Soil salinity $w = 0.08$	0.0264
	Groundwater Salinity $w = 0.08$	0.0264
	Slope $w = 0.03$	0.0099
Environmental $w = 0.33$	Distance from residential area $w = 1$	0.3300
Economical $w = 0.33$	Distance from WWPT $w = 0.67$	0.2211
	Difference in elevation $w = 0.33$	0.1089

3. RESULTS AND DISCUSSION

Using technical and environmental criteria as constraints, the total area suitable for infiltration percolation is 1,489 ha, which represents 3% of the total island area. This constitutes a quite large area that can absorb large amounts of treated wastewater and contributes to Jerba aquifer recharge. The total area is generated by enclaves ranging from 0.06 ha to 101 ha, following mainly the suitable geological formations.

Indeed, geology has the most restrictive criteria to wastewater SAT. The total area meeting this criterion represents only 9% of the island. Residential area farness which is the next restrictive criterion limits the suitable area to 40% of the island. Slope is the least restrictive criterion satisfied by all the study area. The rest of criteria have suitable areas between 88 and 96% (Table 3).

Suitable sites involve different areas with diverse landscapes: around urban areas, agricultural domains, and coastal dunes. Nevertheless, these areas have a large differential suitability and some of them are very far from the WWTPs, spreading out to more than 20 km.

3.1. Best Required Land Location

As previously estimated, the total area needed to infiltrate the available treated wastewater by Aghir and Houmt Essouk WWTPs are 17 ha and 5 ha, respectively.

However the best-suited site for Houmet Essouk plant has groundwater salinity around 7 g/l, which is quite high for irrigation. Hence, important volume of treated wastewater is needed to decrease the salinity to acceptable level. Whereas lesser amounts are needed for Aghir plant best-suited sites since groundwater salinity is less than 5 g/l. These later suitable sites are located in coastal area, so SAT could contribute in mitigating sea intrusion, in addition to providing water for irrigation.

Table 3. Suitable areas obtained for each criterion

Criterion	Area (ha)	% from the total area
Slope	51,120	100%
Soil texture	48,342	95%
Soil Salinity	45,975	90%
Geology	4,434	9%
Aquifer depth	45,189	88%
Groundwater salinity	48,932	96%
Urban, Tourist and hill reservoirs area	20,589	40%
Total	1,489	3%

4. CONCLUSION

This work represents a helpful technical support for decision makers for a better integrated water management in Jerba Island. Further developments could be achieved by multiobjective analysis, including other treated wastewater reuse as irrigation, urban gardening, and industrial recycling.

Acknowledgment: We are thankful to the Tunisian Water Resources Authority (DGRE) staff, especially Mr Habib Chaieb, the Tunisian National Wastewater Office (ONAS-Jerba District) staff, and Jerba Island Safeguarding Association (ASSIDJE) for their willingness to provide us hydrogeological data, reports, and technical help. We also thank our colleague Dr. Ali Amari for providing information about Jerba geology. This work is one of the outcomes of the project contracted by CERTE and Tunisian Ministry of Research, Technology and

Competences Development (2002–2006). NATO provided financial support to the author to present this paper in the Advanced Research Workshop in Istanbul, Turkey, during October 12–16, 2006.

5. REFERENCES

1. Saaty TL. (1980). The Analytic Hierarchy Process. NY: McGraw-Hill.
2. ESRI (2003). Using Arc GIS Geostatistical Analyst. Printed in the USA.
3. FAO (1992). Wastewater Treatment and Use in Agriculture. M.B. Pescod. FAO Irrigation and Drainage Paper 47, FAO, Rome.
4. IOWA (1979). IOWA wastewater facilities design standards. Chapter 21: Land application of wastewater, 16 pp.
5. Laboratoire d'Hydrologie et de Modélisation de Montpellier (1991). Epuration par infiltration percolation: Aspects réglementaires liés au rejet dans le milieu souterrain. Chargé d'étude: BURGEAP.
6. Laboratoire d'Hydrologie et de Modélisation de Montpellier (1993). Epuration des eaux usées urbaines par infiltration percolation: Etat de l'art et études de cas.
7. Ministère Français de l'Environnement et du Cadre de Vie (1978). Inventaire des sites favorables à l'infiltration d'effluents épurés le long du littoral ouest.
8. US Environmental Protection Agency, Process design (1981). Manual for Land Treatment of Municipal Wastewater. Published by US-EPA.
9. US Environmental Protection Agency, Process design (1984). Manual for Land Treatment of Municipal Wastewater. Supplement on rapid infiltration and overland flow.
10. Kallali H, Yoshida M, Jellali S, Hassen A, Jedidi N. (2005). Use of Rapid Infiltration Technique for aquifer recharge with Treated Wastewater: Design of Souhil Wadi (Nabeul-Tunisia) pilot plant. 1st ZER0-M Conference on Sustainable water management. Istanbul, 15–16 March, 2005.
11. EGS Engineering and General Services (2006). Projet de recharge artificielle de la nappe de Jerba à partir des eaux usées traitées de la STEP d'Aghir - Etude d'exécution.

WASTEWATER USE IN JORDAN: AN INTRODUCTION

Zein B. Nsheiwat
Albert Katz International School of Desert Studies
Um-Alsumaq, Amaan, Jordan

Corresponding author: zein_nsheiwat@gmail.com

Abstract:

Being one of the most water scarce countries in the world and in a region where water shortage is a major concern; Jordan faces a great challenge to meet water demands and manage its limited hydrological resources. Wastewater reuse can be an option for alternative water resource especially for the agricultural sector. Moreover, wastewater recycling and reuse when properly implemented is consistent, cost effective as well as environmentally sound. Therefore, Jordan has come to recognize the future key role that treated wastewater can play in the development of the country.

Keywords: Wastewater, water demand, treated wastewater, management, agriculture.

1. INTRODUCTION

Jordan is classified as one of the most water scarce countries in the world and is located in an arid to semiarid zone; with a land area of approximately 88,778.802 km^2. The surface water bodies to which it has access are about 482.540 km^2 in area, this includes the hyper-saline Dead Sea and the Gulf of Aqaba [1]. The arid areas constitute more than 80% of the country's land which receives less than 200 mm annual rainfall.

The Statistical Year Book of 2004 shows that the Jordanian population in 2004 was 5.35 million with a 2.5% growth rate [1]. The Jordanian population is extremely young; 70% of the population is below the age of 30 and about half the population is less than 19 years. Such percentages impose an economic burden on many Jordanian families.

The country already suffers from water shortage where the nonrenewable water resources are already depleted as a result of over-pumping. This contributes to, the degradation of water quality and increasing the water salinity of these resources. Jordan is classified as one of the most limited water resources country

M.K. Zaidi (ed.), Wastewater Reuse–Risk Assessment, Decision-Making and Environmental Security, 73–79.

with one of the lowest per capita water availability ratios (160 m^3/cap/year) and this amount is projected to drop as a result of population growth to about 90 m^3/cap/year by 2025. Therefore, according to the water stress index (the value of annual rainfall divided by the total population (m^3/capital/year)) Jordan is classified as having "absolute water scarcity" with less than 500 m^3/cap/year [2].

With such population growth and fast development, an increase in water demand is expected. For that reason, Jordan will need to depend on alternative water resources, one of which is wastewater which constitutes a critical option for agriculture.

1.1. Water Resources in Jordan

The stream of surface water resources fluctuates between seasons and years with an average annual flow of 690 million cubic meter (MCM) of which 505 million cubic meter (MCM) are useable. Currently, the main surface water source is the Yarmouk River that is shared with Syria with one-third of the surface runoff within Jordan. This is followed by the Zarqa River Basin where the main wastewater treatment plant (WTP), As-Samra, is located. Now, and as a result of over extraction of groundwater from the Zarqa basin, the flow has become reduced and the stream flow is almost exclusively effluent discharged from three sewage treatment plants that serve Amman. Moreover, the Jordan River that discharge water to the Dead Sea was once a main water resource before the diversion of flow from Lake Tiberias was undertaken by Israel in 1964.

Groundwater exhaustion has emerged as a serious water problem in Jordan. All told there are 12 groundwater basins, with a renewable annual supply of 280 MCM, in 2003. The total annual groundwater extraction from all basins was 520 MCM exceeding the sustainable yield. The water table in these aquifers is declining and consequently, the water quality is deteriorating. Disi and Jafr are the main fossil aquifers with annual yield of 70 and 18 MCM, respectively. Most aquifers are exposed to overexploitation resulting in water's quantity and quality deterioration.

Currently, treated wastewater originates from 19 treatment plant. Around 70–80 MCM per year of treated wastewater is discharged into various water bodies or used directly for restricted irrigation mainly in the Jordan Valley. The new plan is the expansion in the use of treated water for irrigation. Desalinated water may be regarded as a future source. Using desalination process, about 40 MCM water is produced and used for domestic supply and very little is used for agriculture. A study evaluated the nonconventional water resources in Jordan: treated wastewater, water harvesting, importation of water and desalination based on economic, technical, availability, reliability, and environmental criteria [3]. The analysis showed that desalination has the highest ranking as a future source followed by water harvesting. The research suggests that desalination of brackish water is likely to become a potential nonconventional water source for Jordan.

1.2. Water Demand

In the year 2003, the estimated total demand was 1,388 MCM and only 890 MCM were delivered to all users in Jordan [2]. Agriculture constitutes the main consumer of water with 63.5% of the total water use, followed by the domestic sector (32.5%) and only 4% for the industrial sector (Figure 1).

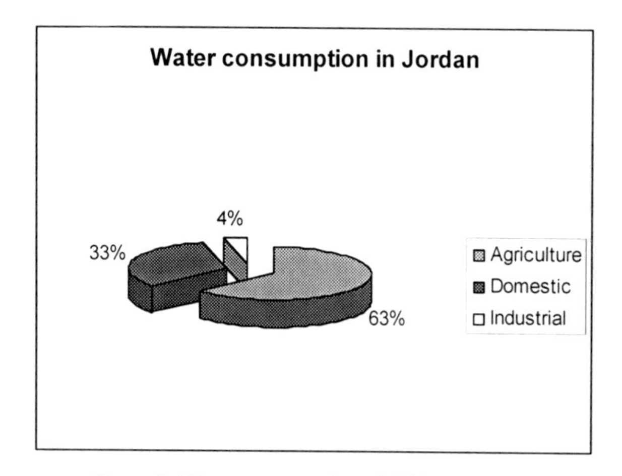

Figure 1. Water consumption of different sectors

It has always been hard to meet the actual domestic water demand. The Jordanian population uses an absolute minimum amount of water for domestic use. This creates a high awareness about the importance of water conservation. Water demand increases in the summer due to the dry hot weather conditions and growing tourism. Using inefficient traditional surface irrigation, such as open canals, makes the agriculture sector a larger consumer. In addition, urbanization has resulted in the utilization of the fertile highlands for residential purposes leaving relatively dry areas for cultivation. These drylands need more water as soil quality is poor and evaporation rates are higher [3].

Jordan's agricultural lands are either irrigated by private wells groundwater (privately managed farms) or by surface water from the Yarmouk River, side wadis, and recycled wastewater in publicly managed irrigation system. Drip irrigation with treated wastewater and flooding methods are used depending of the effluent quality, crops, and mixing water. For public health reasons, treated effluent is prohibited by law to be used in sprinkler systems [4].

Industrial water is allocated for two main industries: phosphates and potash; these industries and others are located outside Jordan's cities and therefore they need to secure their own water supplies. In practice, they either use surface water or drill private wells. Some industries receive their water through the domestic network system, while the government sells surface water and imposes resource tax on groundwater. The industrial sector is well aware of the challenges resulted from water shortage; therefore, it generally seeks to recycle its wastewater wherever feasible.

Water shortage constrains the expansion of certain industries. At present Jordan's energy industries and facilities as well as several major industries and future projects are located in Aqaba where seawater can be used for the cooling. A National Water Strategy was launched by the Ministry of Water and Irrigation in 1997 that recognizes the population growth pressure on the available water resources and the importance of managing these resources. The strategy includes 47 recommendations regarding resource development and management, public awareness, heath standards, research development, financing, private sector participation, performance, legislation, and shared water resources. It also includes policies on groundwater management, irrigation water, wastewater management, and water utility.

2. RECLAIMED WASTEWATER

Wastewater reuse is considered to be a promising solution in countries that suffer from unremitting water scarcity that scramble to meet escalating demands. Water reuse has been applied globally in many countries with some success as it has the potential to sustain the available resources [5].

The Jordanian Ministry of Water and Irrigation (MWI) struggles to achieve a sustainable balance between the growing demand and supply. The ministry is devoting considerable efforts to water reuse since it offers a sustainable, cost-effective, and environmentally sound solution [6]. Moreover, the Water Reuse and Environment Unit within the Water Authority of Jordan (WAJ) was established to maximize the reuse of treated wastewater to protect the environment and save freshwater for drinking purposes.

Limited wastewater collection has been practiced since 1930 and physical processes were applied to treat the water and gray water that was discharged to gardens. These practices resulted in substantial environmental problems, especially groundwater pollution. In the late 1960s, technology was introduced to collect and treat wastewater using the conventional activated sludge process.

Due to the overload of this system, operational and environmental problems appeared which negatively affected the quality of surface, ground, and irrigation water in the area. Consequently, in 1980 the government launched new plans regarding wastewater management.

Presently, 19 wastewater treatment plants are operating in Jordan, treating 88.5 MCM/year 2002, and the amount of treated wastewater is 73 MCM/year. About 60% of the Jordanian population is connected to sewage system, where As-Samra WTP being the main one (Table 1).

Table 1. Existing wastewater treatment plants, 2002 [5]

NO	WWTP	Operation	Gov.	Type of Trea.	Influent MCM	Effluent MCM
1	As Samra	1985	Zarqa	WSP	65.245	53.301
2	Abu Nuseir	1988	Amman	AS	0.722	0.712
3	Wadi Essir	1996	Amman	WSP+ Aeration	0.698	0.290
4	Wadi Arab	1999	Irbid	EA	2.579	2.516
5	Irbid	1987	Irbid	TF+AS	2.6	2.551
6	Ramtha	1988	Irbid	WSP	0.839	0.691
7	Salt	1981	Balqa	EA	1.425	1.299
8	Baqa'	1988	Balqa	TF	4.296	3.992
9	Fuhais	1996	Balqa	EA	0.556	0.381
10	Ma'an	1989	Ma'an	WSP	0.790	0.557
11	Wadi Mousa	2001	Ma'an	EA	0.316	0.08
12	Mafraq	1988	Mafraq	WSP	0.659	0.506
13	Jarash	1983	Jarash	EA	1.062	0.983
14	Kufranja	1989	Ajloun	TF	0.811	0.585
15	Madaba	1989	Madaba	AS	1.525	1.352
16	Karak	1988	Karak	TF	0.550	0.449
17	Tafila	1988	Tafila	TF	0.270	0.271
18	Aqaba	1987	Aqaba	WSP	3.406	2.655
19	Wadi Hassan	2000	Irbid	EA	0.155	0.075
	Total				**88.502**	**73.1**

The As-Samra plant began operating in 1985, and is located about 30 km northeast of Amman receiving wastewater from Amman, Zarqa, and Russaifa. The plant is based on a wastewater stabilization pond system. The collected influent is then treated to secondary treatment levels. The treated effluent is then discharged into Wadi Zarqa to join the runoff of this wadi. These effluents are captured at the King Tala Dam and its associated reservoir. Water from this dam is used for irrigation mainly in the Jordan Valley which is otherwise prohibited. As of September 2006, the average total flow into the As-Samra plant was about 150,000 m^3/day. Future plans at As-Samra focus on generating their own electrical power through establishing digesters that produce biogas using the sludge produced from the treatment process. The treated wastewater is intended to be used mainly for irrigation. Only 10% of the biological load of wastewater comes from industrial waste since the industrial discharge to the sewage treatment plant is low. Consequently, Jordan's wastewater is low in heavy metals and toxic organic compounds.

A survey was conducted to examine the farmers' willingness to pay for reclaimed wastewater in Jordan and Tunisia [7]. The results showed that farmers are willing to use reclaimed water but showed a clear preference for effluents that could be used for unrestricted irrigation. Yet, farmers were not willing to pay more for reclaimed water, mainly because they already have access to freshwater at low prices.

However, having standards and regulations for the use of treated wastewater is essential to guarantee the public health and acceptance and environmental safety. Therefore, in 1995 a set of comprehensive national standards was established by WAJ relying on the WHO standards, JS893/1995 [7]. The areas where wastewater reuse is permissible: irrigation of vegetables eaten cooked, tree crops, forestry, animal fodder, and industrial process, in addition to artificial recharge of aquifers, discharge to wadis and use in aquaculture, as well as discharge to recreational areas and parks [8, 9].

According to the JS893/1995 standards, treated wastewater is prohibited for use on irrigation crops eaten raw and during the last two weeks before harvesting. Likewise, treated water is prohibited for sprinkler irrigation and recharge aquifers that are used for drinking purposes.

Believing that the water reuse can be a solution for the water crisis in Jordan, and part of the National Water Strategy, the Wastewater Management Policy of 1998 requires that treated wastewater be integrated as a valuable water resource. The policy states that "priority shall be given to agricultural reuse of treated effluent for unrestricted irrigation" and "treated effluent shall be priced and sold to end users at a price covering at least the operation and maintenance costs of deliver." Furthermore, it stresses improving the quality of treated effluent by mixing it with higher quality water, crop selection according to irrigation water, and soil properties [5].

3. CONCLUSION

"Wastewater shall not be managed as waste." It shall be collected and treated to standards that allow its reuse in unrestricted agriculture and other nondomestic purposes, including groundwater recharge. Besides, supplying an additional water source and reducing the pressure on the potable fresh water, water reuse presents an alternative source for agriculture and industry, protects the environment and public health, reduces the amount of artificial fertilizers that is used, makes greening the areas possible, as well as allows for a variety of economic developments.

The Jordanians are used to the idea of water conservation, since it has to minimize water usage as well as take advantage of rainwater harvesting.

Jordan program is to raise people's confidence and awareness about the importance of this resource.

Acknowledgment: I am thankful to the NATO ARW committee to award financial support to attend the meeting and present this paper.

4. REFERENCES

1. DOS. (2004). The Statistic Year Book of 2004. Department of Statistics. Amman, Jordan. www.dos.gov.jo
2. Abdel Khaleq RA and Dziegielewski B. (2004). A National Water Demand Management Policy in Jordan. Ministry of Water and Irrigation. Amman, Jordan.
3. Jaber JO and Mohsen MS. (2001). Evaluation of Non-conventional Water Resources Supply in Jordan. Desalination 136 (2001), 83–92.
4. Malkawi SH. (2003). Wastewater Management & Reuse in Jordan. Ministry of Water and Irrigation. Amman, Jordan.
5. MWI. (1997). Water Strategy of Jordan. Ministry of Water and Irrigation. Amman, Jordan.
6. Ministry of Water and Irrigation. (2006). www.mwi.gov.jo
7. Abu Madi M, Braadbaart O, Al'Sa'ed R and Alaerts G. (2003). Willingness of Farmers to Pay for Reclaimed Wastewater in Jordan and Tunisia. Water Science and Technology: Water Supply 3(4), 115–122.
8. Nazzal YK, Mansour M, Al-Najjar M and McCornick P. (2000). Wastewater Reuse Law and Standards in The Kingdom of Jordan. Ministry of Water and Irrigation. Amman, Jordan.
9. McCornick PG, Hijazi A, Sheikh B. (2004). From Wastewater Reuse to Water Reclamation: Progression of Water Reuse Standards in Jordan. CBA International: Wastewater Use in Irrigated Agriculture. 153–162.

MUNICIPAL WATER REUSE IN TUCSON, ARIZONA, USA

Sharon B. Megdal
The University of Arizona, Water Resources Research Center
350 N. Campbell Ave., Tucson, AZ 85719, USA

Corresponding author: smegdal@cals.arizona.edu

Abstract:

Arizona's Groundwater Management Act limits the use of groundwater to meet growing demand for water by the municipal sector. The state's recharge and recovery program allows for water reuse through aquifer recharge and later recovery inside or out of the area of hydrologic impact. The paper discusses water reuse within the municipal sector in Arizona, with a special focus on the Tucson metropolitan area. Increased effluent utilization is playing a more prominent role in long-range planning efforts. The paper focuses on the role of reclaimed water in water management planning. It explains how the use of effluent is influenced by water quality considerations and institutional/legal arrangements.

Keywords: Effluent, recharge, management, planning, decision making.

1. INTRODUCTION

Communities in Arizona, a rapidly growing state in the semiarid southwestern United States, face significant challenges in meeting their water demands. In 1980, the Arizona Legislature enacted the Groundwater Management Act (GMA), which regulates the use of groundwater in areas of the state identified as active management areas (AMAs) [1]. AMA boundaries, shown in Figure 1, are based largely on hydrologic considerations. In the AMAs, large water users or water providers must report water use to the Arizona Department of Water Resources, the state agency charged with implementing and enforcing the GMA. A complex system of groundwater use regulations has resulted in increased reliance upon surface water to meet the growing water demands of municipalities [2]. With surface water sources nearly fully utilized, treated wastewater, also called effluent, is becoming an increasingly important component of municipal water supply portfolios [3].

It is estimated that water pumping and surface water diversions in the AMAs, where the vast majority of Arizona's population lives, totaled 3.8 million acre feet in 2003. That compares to an estimated 4 million acre feet of pumping and diversions for the rest of the state, resulting in total statewide water use of almost

M.K. Zaidi (ed.), Wastewater Reuse–Risk Assessment, Decision-Making and Environmental Security, 81–90.

8 million acre feet of water annually [4]. An acre foot is the amount of water required to cover one acre of land by one foot of water; it equals 325,851 gallons or 1.23 million liters. Agriculture is estimated to account for approximately 70% of this demand, although precise numbers are not available outside the AMAs due to lack of measurement and/or reporting requirements. Effluent use is estimated to be about 2% statewide, with the vast majority of reuse occurring in the AMAs [5]. In 2003, effluent was the source water for supplying approximately 5% of AMA water demand, or 189,800 acre feet [6], while in the same year, estimated use of effluent outside the AMAs was estimated to be less than 1% of total water demand in these areas.

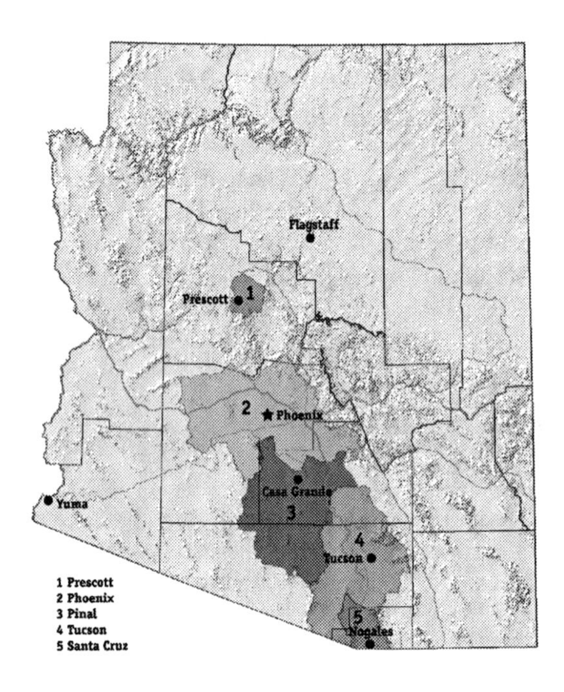

Figure 1. Arizona's Active Management Areas (Courtesy of the Arizona Department of Water Resources.)

Municipal wastewater in Arizona must be treated to certain water quality standards before it is discharged from treatment facilities. Currently, after treatment, effluent is used for golf course and turf irrigation, agricultural irrigation, and industrial uses, including power plant cooling. Some of the remainder is recharged through basin or stream bed infiltration. The Central Arizona Project (CAP) is a 336-mile canal designed to deliver 1.5 million acre feet of Colorado River water annually to Central Arizona, where the largest cities in Arizona are located. CAP water is projected to be fully utilized in the next 30 or so years. Effluent, once considered a nuisance water supply—that is, something to be disposed of—is of growing interest to municipalities searching for ways to meet future water demands. The Tucson AMA, with a population nearing one million, is no exception. However, water quality considerations, infrastructure limitations, public acceptability, and institutional constraints make it difficult to project the extent to which effluent may be used in the future to meet potable water demands.

1.1. The Framework for Water Management in the Tucson Region

The water management goal for the Tucson AMA is safe-yield. As defined by Arizona Revised Statute, safe-yield is "a groundwater management goal which attempts to achieve and thereafter maintain a long-term balance between the annual amount of groundwater withdrawn in an active management area and the annual amount of natural and artificial recharge in the active management area" [7]. Meeting the critically important goal of eliminating overdraft of the aquifers drives the water management activities of the region [8].

The main water users in the Tucson AMA are categorized as municipal, agricultural, and industrial. Of the approximately 350,000 acre feet of water demanded in 2003, 30% was used by agriculture, 15% by mining and other large industrial users, and 55% by municipalities [9]. Numerous large and small water companies, termed "municipal water providers," provide water to communities in the Tucson AMA. Tucson Water, operated by the City of Tucson, is by far the largest, serving approximately 80% of the metropolitan area's population. Although known as "municipal" water providers, these water companies may be owned and operated by cities and towns, water districts, cooperatives, or private individuals/companies. They supply water not only to residences, but also to business and commercial establishments and to turf facilities in their respective service areas. In addition, many individuals and some businesses supply their own water through individually owned and operated wells. Particularly small wells are considered legally exempt from state regulation.

2. THE ASSURED WATER SUPPLY RULES AND USE OF RENEWABLE WATER SUPPLIES

As a result of the GMA, new residential growth in AMAs with the safe-yield water management goal can no longer depend upon mined groundwater to meet community water demand. To strengthen this requirement, the set of rules known as the Assured Water Supply (AWS) Rules, was adopted in 1995. This requirement was crucial for moving the region to less reliance on groundwater in favor of alternative supplies. The rules provide significant flexibility. They do allow new growth to be served with groundwater, should a demonstration be made that 100 years' supply of groundwater is legally and physically available, but groundwater use deemed "excess" by the formulas in the rules must be replenished with water supplies that are considered renewable, such as Central Arizona Project (CAP) water or effluent [10].

A "promise" to use renewable supplies will not suffice when an Assured Water Supply (AWS) application is undergoing review. Instead, the applicant must demonstrate to the Arizona Department of Water Resources how renewable water supplies will be used. There are several ways that renewable supplies can be utilized individually or in combination:

1. A renewable water supply under long-term contract, such as CAP water, may be treated and then delivered directly to customers. This mechanism is not currently being utilized in the Tucson AMA.
2. A renewable water supply, such as CAP water or effluent, may be recharged at a permitted location within the AMA and a credit accrued for that storage. The credit may be "redeemed" when the water is recovered through a permitted recovery well. This is often considered indirect use of the renewable supply because its use is through storage and recovery rather than through a treatment plant, as in the first option. The location of the recovery of the stored water may be distant from or adjacent to the location of storage. The storage and recovery option for use of CAP water is being utilized by several water providers in the Tucson AMA. Storage and recovery activities must follow highly complex laws, regulations, and rules [11]. Reviewing staff examine the storage and recovery plans carefully to determine if the facilities to be used are in fact permitted and available for use.
3. The Central Arizona Groundwater Replenishment District (CAGRD) is an organization that assumes the responsibility to replenish groundwater use that is in excess of allowable use per the groundwater allocations established by the AWS Rules. A development or water company may sign up for replenishment services by becoming a member of the Replenishment District. The replenishment district then has an obligation to replenish the groundwater used within the AMA by its members withdrawn within three years of use. This delayed replenishment approach contrasts with the storage and recovery option above. Both options make use of Arizona recharge and recovery statutes, but in different ways. State law requires a plan to be submitted to and approved by the Department of Water Resources every ten years, but it does not require the replenishment district to demonstrate it has firm contracts to water for 100 years.
4. Extinguishment of grandfathered agricultural water rights can enable the water supplier to utilize a percentage of this water right as "mined" groundwater—that is, groundwater exempt from replenishment obligation.

The percentage becomes less generous with time, making this an increasingly limited option going forward [12].

5. The use of special sources of groundwater, such as remediated groundwater, is not considered mined groundwater.

2.1. Water Planning by Tucson Water

The most detailed long-range water plan available in the Tucson AMA is the "Water Plan: 2000–2050", released by Tucson Water in November 2004 [13]. Tucson Water is the biggest municipal water provider in Tucson and is operated by the City of Tucson. The Tucson Water Plan (TWP) identifies the important and complex issues facing the region as it attempts to accommodate the water needs of a growing population, and includes a recommended water supply strategy. Tucson has long delivered effluent water subject to tertiary treatment through its reclaimed system. Tucson's reclaimed water is sold to golf courses, schools and parks at a rate of $610 per acre foot. In 2003 Tucson delivered 13,121 acre feet directly as reclaimed water, about 40% of the total effluent produced in Tucson that year. This amount represents about 8% of Tucson's overall water demand. One of the goals of the Tucson Water Plan is to maintain this level of direct use as the area continues to grow.

Effluent not directly utilized through the reclaimed water system (the other 60% of the effluent supply) is currently discharged into the Santa Cruz River. The second goal outlined in 2030 the TWP aims to increase utilization of effluent through recharge and recovery. Effluent will be stored underground through basin infiltration at specially constructed recharge basins or through stream bed infiltration and then recovered. Tucson Water will have difficulty meeting water demands beyond 2030 unless it uses effluent in this manner or finds other water supplies to meet projected demands.

The increased use of effluent indirectly through recharge has received much attention both locally and nationally. This has brought widespread recognition of effluent as a water supply of growing importance [14]. Effluent utilization plays an important part in the preferred long-range planning option. Nevertheless, whether Tucson Water will be able to use effluent to meet potable water demands to the extent projected remains to be seen. The determinants of future effluent use, as discussed below, are often regional in nature.

2.2. Future Use of Effluent in the Tucson Region

Several issues affect the extent to which the Tucson region incorporates effluent utilization into its water supply portfolio. They are discussed in turn.

3. INSTITUTIONAL COMPLEXITIES

Control of effluent is divided among multiple jurisdictions, making unified planning for its use difficult. The institutional context for wastewater treatment is unusual in Tucson. In the late 1970s, through an intergovernmental agreement, the City of Tucson took over responsibility for the bulk of community water provision, while Pima County, the county in which Tucson is located, became the regional wastewater provider. As a result, Tucson claimed ownership of 90% of the effluent produced, with the remaining 10% going to Pima County. This is unlike most other large cities in Arizona, where water and wastewater responsibilities reside with the city. In addition, there is a significant commitment of the effluent (28,200 acre feet per year) from the two large metropolitan treatment plans to the United States Secretary of the Interior for use in settlement of water rights claims of the Tohono O'odham Indian Nation.

At one time, Tucson Water was expected to serve as the regional water provider, but those plans did not materialize due to inter-jurisdictional conflicts and difficulties Tucson Water experienced with delivery of treated CAP water in the early 1990s. Subsequent to the 1979 agreement, Tucson Water has transferred ownership of a portion of its 90% share to other jurisdictions. These jurisdictions sometimes have conflicting goals regarding its utilization. For example, for several years, Pima County criticized the city for underutilization of the effluent. Currently, unutilized effluent is discharged into the otherwise dry Santa Cruz River (see Figures 2 and 3), and the Town of Marana, through which the Santa Cruz River runs, is concerned about future withdrawal of a significant portion of the effluent by the City of Tucson. In addition, stakeholders concerned about degradation of the riparian areas in the effluent-dominated portion of the Santa Cruz River have concerns about future removal of the effluent in the future for municipal use. A multijurisdictional, large-scale planning effort for the Santa Cruz River is underway but is not yet complete.

Figures 2 and 3 show the significance of the wastewater flows to riparian growth. Figure 2 shows the dryness of the Santa Cruz River bed upstream of the discharge point of the Roger Road Wastewater Treatment Plant. Figure 3 shows riparian growth in Marana, downstream of the Roger Road plant and a second, larger treatment plant. Future deliberations are likely to determine how much effluent water is removed from the Santa Cruz River for use.

3.1. Indirect Use of Effluent

Unknowns related to effluent water quality make it difficult to plan for its utilization. Effluent or reclaimed water has long been accepted as a source of water for municipal turf irrigation. However, the current supply of effluent easily exceeds projections of future demands for municipal turf irrigation [15]. Agriculture is unlikely to be a future user of effluent in the Tucson area. In the Tucson AMA, agriculture is giving way to residential and commercial development, a trend that is expected to continue. Although there had been some use of effluent by agriculture in the past, there is currently no agricultural use of effluent in the Tucson AMA.

Figure 2. Outfall of regional wastewater treatment facility (Courtesy of Pima County Regional Flood Control District.)

Figure 3. Effluent-dependent riparian segment of the Santa Cruz River, flowing through Marana, AZ (Courtesy of the Town of Marana.)

The potential for increased utilization rests with the potable system. Yet, water quality considerations associated with reuse of effluent for potable use are significant. Any discussion of the use of effluent in this manner is almost immediately connected by the media and others with the moniker "toilet to tap", a proposition that is unappealing to the general public [16].

The public has become aware that hormones and other pharmaceuticals have been found in water discharged from wastewater treatment plants, including those in the Tucson AMA [17]. In the near to intermediate term, no Tucson area water provider is considering treating effluent through membrane facilities for direct delivery through the potable water system. Arizona's recharge and recovery program provides an attractive alternative for future use [18].

3.2. Use of Effluent through Recharge

The water storage and recovery program, discussed in Section 2.1, provides the statutory framework for a multifaceted system of recharge of aquifers. This program has proven to be invaluable in assisting water providers and the state in meeting water policy objectives. Due to cost considerations, indirect utilization of effluent through recharge and recovery is likely to be the mechanism of choice for water providers.

The recharge program allows accrual of storage credits, with the credit holder entitled to withdraw water at a future time. Recovery requires the permitting of a well or wells, and must be in the AMA where the storage occurred, but not necessarily in the area of hydrologic impact of the recharge. Recovery within the area of hydrologic impact is beneficial from a water management standpoint in that the aquifer that has benefited from storage is the location of the recovery activity. However, that also means that the water recovered is likely to resemble effluent water, after the tertiary treatment associated with soil aquifer treatment. Nitrates and total dissolved solids are higher in effluent water than in situ groundwater. Water quality considerations are likely to be less of an issue if the recovery occurs at a site that is remote from the storage. Yet, that means water is withdrawn from a location that has not benefited from the increase in aquifer storage. There are tradeoffs associated with each approach. Clearly, water quality considerations suggest that the water suppliers should take advantage of the flexibility in the storage and the recovery program and recover outside the area of hydrologic impact. Aquifer storage considerations, on the other hand, suggest that recovery ought to occur within the area of hydrologic impact.

To date, the storage of effluent for storage credits in the Tucson AMA has been somewhat limited. That is because it took considerable effort for the jurisdictions with ownership of effluent to come together and apply to have the Santa Cruz river, where most of the effluent is discharged, permitted as a storage facility. Moreover, storage of effluent passively through stream bed recharge, as opposed to constructed basins (Figure 4), results credit accrual for only 50% of the water stored rather than 100%, as is allowed at constructed basins (Figure 4). Some contest this provision, stating that it provides owners of the effluent with an incentive to remove the effluent from the river, thereby reducing water available for the environment.

Figure 4. Effluent recharge basins at the Sweetwater Wetlands reclamation facility (Courtesy of Deirdre Brosnihan.)

3.3. Use of Effluent by the CAGRD

As discussed in Section 2.1, one of the means of complying with the AWS Rules requirement that renewable water supplies be utilized is through enrollment in the CAGRD. CAGRD is required to replenish groundwater that is used to meet municipal demands, pursuant to calculations established by the AWS Rules and at this time does not have contracts for firm or long-term renewable water supplies. Since its inception in the mid-1990s, it has been using CAP water that would otherwise be unused to meet its replenishment obligations. At the same time that its replenishment obligations grow, availability of excess CAP water will diminish, and this agency will be required to find other supplies of water to meet its replenishment obligations. In the Tucson AMA, the effluent has been considered a possible source of water for replenishment. The US Department of the Interior is one of the participants in the recharge of effluent in the Santa Cruz River. Over time, the US Department of the Interior may enter into a long-term financial arrangement with the Tucson AMA. There could well be competition for this effluent in the future.

3.4. Environmental Considerations

In an effort to increase the amount of water available for environmental interests, the City of Tucson and Pima County modified their intergovernmental agreement governing effluent in 2001. A key provision was the establishment of an environmental pool of effluent water. Ten thousand acre feet could potentially be used for environmental purposes, subject to certain conditions. Although none of this pool has been allocated to the environment, this represents an important source of water to satisfy some environmental demands in the future.

4. CONCLUSION

The GMA provides the regulatory framework for groundwater management in the Tucson AMA. However, effluent is not a regulated source of water. Its use is related to groundwater regulations because it is a growing, alternative water supply. Therefore, its importance as part of the water supply portfolio for municipalities throughout Arizona is expected to grow. Water management is a very decentralized activity in Arizona; although there is regulation at the state level, decisions regarding how to meet the regulations are made at the local level. Key decisions will be made by local elected officials, only after considerable gathering of information and public input.

Acknowledgment: I thank Taylor Shipman for research assistance. This work was supported by the University of Arizona Technology and Research Initiative Fund (TRIF) Water Sustainability Program.

5. REFERENCES

1. Colby B, Jacobs K. (2006). Arizona Water Policy: Management Innovations in an Urbanizing, Arid Region. Washington. RFF Press, 26–44.
2. Colby B, Jacobs K. (2006). Arizona Water Policy: Management Innovations in an Urbanizing, Arid Region. Washington. RFF Press, 26–44.
3. Colby B, Jacobs K. (2006). Arizona Water Policy: Management Innovations in an Urbanizing, Arid Region. Washington. RFF Press, 45–60.
4. Arizona Water Atlas, Draft (2006). Arizona Department of Water Resources, http://www.azwater.gov
5. Eden S, Megdal SB. (2006). Arizona's Rapid Growth and Development: Natural Resources and Infrastructure. Phoenix. Arizona Town Hall, 81–112.
6. Arizona Water Atlas, Draft (2006). Arizona Department of Water Resources, http://www.azwater.gov
7. Arizona Revised Statute §45–561
8. Megdal SB. (2006). Water Resource Availability for the Tucson Metropolitan Area. Tucson, Water Resources Research Center.
9. Seasholes K, (2006). Personal communication, Sept. 15, 2006.
10. Colby B, Jacobs K. (2006). Arizona Water Policy: Management Innovations in an Urbanizing, Arid Region. Washington. RFF Press, 26–44.
11. Colby B, Jacobs K. (2006). Arizona Water Policy: Management Innovations in an Urbanizing, Arid Region. Washington. RFF Press, 188–203.
12. Arizona Administrative Code, R12-15-703.
13. Tucson Water, Tucson Water Plan: 2000–2050, www.ci.tucson.az.us/water/waterplan.htm
14. Lee M. (2006). "Divisive water proposal advances", The San Diego Union-Tribune, July 27.
15. Kittle RA. (2006). "Yuck! San Diego should flush 'toilet to tap' plan." The San Diego Union-Tribune, July 24.
16. Colby B, Jacobs K. (2006). Arizona Water Policy: Management Innovations in an Urbanizing, Arid Region. Washington. RFF Press, 188–203.
17. Swedlund E. (2006). "Tucson Water tests the waters at testings." Arizona Daily Star, October 2.
18. Brown, A. (2006). "Decision H_2O is up to you: samples help you decide if a reverse osmosis plant is needed." Arizona Daily Star, October 1.

SESSION 3. CURRENT RISK MANAGEMENT PRACTICES–DEVELOPED COUNTRIES

DAIRY WASTEWATER TREATMENT WITH EFFECTIVE MICROORGANISMS AND DUCKWEED FOR POLLUTANTS AND PATHOGEN CONTROL

M.T. Rashid and J. West
Department of Land Resource Science
University of Guelph, ON, N1G 2W1, Canada

Corresponding author: trashid@uoguelph.ca

Abstract:

Wastewater originated from dairy operations may harbor human pathogens including *Escherichia coli* (*EC*). Excess nutrients present in dairy wastewater can also pollute surface and ground waters. Effective microbes (EM) and duckweed have shown a great promise in wastewater treatment. The duckweed growth and EM applications were tested. Combined application of EM and duckweed growth significantly reduced the ammonium nitrogen, total phosphorus, total suspended solids and biological oxygen demand after three months and is a very efficient way of dairy wastewater treatment.

Keywords: Effective microbes, duckweed, total suspended solids, biological oxygen demand, *Escherichi coli.*

1. INTRODUCTION

In the Walkerton, Ontario, Canada, tragedy of May 2000, contamination of the municipal water system with *EC* O157:H7 and *Campylobacter jejuni* resulted in 2,300 people (out of 5,000) requiring medical attention, 7 of them died [1]. Investigation into the causes of the microbial contamination of the municipal well water indicated that the most likely cause was transport of manure bacteria to the aquifer by infiltrating water, although direct entry of surface runoff into the well could not be ruled out [2]. Wastewater originated from dairy operation may harbor different bacterial species including human pathogens such as enterohaemorrhagic *EC* [3, 4]. Cattle are considered a major reservoir of *EC* [5], this pathogen may potentially infect the drinking water supply from cattle wastewater [6]. Outbreaks are usually associated with consumption of contaminated food or drinking water exposed to pathogen laden animal manure or contaminated irrigation water [7].

M.K. Zaidi (ed.), Wastewater Reuse–Risk Assessment, Decision-Making and Environmental Security, 93–102.

Discharging wastewater with high levels of phosphorus (P) and nitrogen (N) can result in eutrophication of receiving waters, particularly lakes and slow moving rivers. To prevent these conditions, regulatory agencies in many countries have imposed nutrient discharge limits for wastewater effluents. Recently restrictions on P discharge have become more stringent due to environmental problems. Dairy operations have been identified as a potentially significant source of nitrate [8] and phosphorus [9] contamination in groundwater.

Current mainstream technologies for wastewater treatment, such as the activated sludge process with N and P removal, are too costly to provide a satisfactory solution for dairy wastewater treatment specially in developing regions. Biological treatment processes are inexpensive and are known for their ability to achieve good removal of pathogens nutrients and organic pollutants. Duckweed-based pond system could be an attractive technology for wastewater treatment aiming at nutrient recovery and reuse [10].

Effective microorganism's technology (EM) was developed by Dr. Teuro Higa in 1970's at the University of Ryukyus, Okinawa, Japan. First solution contained over 80 microbial species from 10 genera, isolated from environments in Japan, however with time, the technology was refined to include only lactic acid bacteria, phototrophic bacteria, and yeast [11]. The innoculum includes high populations of lactic acid bacteria (*Lactobacillus* and *Pedicoccus*) at 1×10^8 cfu ml^{-1} suspension, yeast (*Saccharomyces*) at 2×10^6 cfu ml^{-1} suspension and phototrophic bacteria, 1×10^3 cfu ml^{-1} [12]. Application of EM in septic systems, lagoons, activated sludge systems, and other remediation projects has reduced water quality indicators such as biological oxygen demand and chemical oxygen demand [13]. Pig manure odor and coliform bacteria were drastically reduced when treated with EM [14].

In recent years, research has focused on duckweed and its role in wastewater treatment and potential for nutrient recovery [15]. The treatment of wastewater by duckweed reduces the wastewater contaminants either directly through the nutrient recovery or indirectly by release of oxygen in the water column [16]. Treatment systems with protein production using duckweed represent a comprehensive solution for wastewater treatment [17]. Duckweed wastewater systems have also been studied for dairy wastewater [18].

2. MATERIALS AND METHODS

A wastewater treatment pilot pond was constructed by dividing the existing dairy wastewater pond (16 × 25 m) into four equal portions at Dwany Dairy Farm,

Ontario, Canada" (43° 7′ 60N, 80° 45′ 0W; 298 m above sea level). Farmer stores dairy wastewater in this pond and it is used to dilute the liquid dairy manure during hauling by pumping it back to the dairy barn. The experiment was conducted in 2004–2005 for two years during June to August; temperature varies between 25 and 32°C). The treatment system was in an open field exposed to weather conditions. Each of the 4 blocks of the main wastewater holding pond was assigned to following treatments. Block-1 was kept as control (untreated), Activated EM was applied to block-2 after every 2 weeks for three months at 1:100 ratio (1 part EM and 100 parts water; 6 applications). Duckweed (*Lemna gibba*) plants were grown in block-3 and block-4 received both treatments (EM application and duckweed growth). Duckweed plants were transferred into block 3 and 4 manually and acclimatized for a month (May) with wastewater. Wastewater samples were taken after every 15 days before the application of EM.

All samples were sent to a testing laboratory for total suspended solids (TSS), biological oxygen demand (BOD), NH_4-N, NO_3-N, total P, *EC*, and total coliform bacteria. Duckweed was also harvested after every 15 days and was incorporated in nearby manure piles. Analysis of variance for the data on TSS, BOD, NH_4-N, NO_3-N, total P, *EC*, and total coliform counts were performed by CoStat 6.3 statistical analysis program [19]. Data was collected for two years to: (1) investigate the performance of effective microbes and duckweed for dairy wastewater treatment and (2) to determine the effect of effective microbes and duckweed on reduction of nutrient and pathogenic bacteria from dairy wastewater.

3. RESULTS AND DISCUSSION

Chemical and biological properties of pretreated dairy farm wastewater are presented in Table 1. Chemical and biological analysis have not shown much differences between dairy wastewater samples taken in 2004–2005 in pH, TSS, BOD, NH_4-N, NO_3-N, Total P, *EC*, and total coliform counts. Obvious reason for this consistency could have been the same source, same time of the years and same dairy farming practices performed during the years.

3.1. Wastewater pH, TSS, and BOD

The pH in dairy wastewater ranged from 7.00 to 7.70 (initial values were 7.5 and 7.45 for 2004–2005) and was not drastically changed in all treatments during the course of the experiment (Table 2). Dairy farm wastewater pH values of all samples from all treatment ponds generally were near neutrality or basic with a very small variation (+1.33 to –6.67% decrease or increase). The application of EM (block-2) reduced the pH from 7.5 to 7.0 but it was still in the pH range where microbes exhibit optimal growth. Most microorganisms exhibit optimal growth at pH values between 6.0 and 8.0 and most can not tolerate pH levels above 9.5 or below 4.0 [20]. The declining pattern in TSS concentrations due to different dairy wastewater treatments is illustrated in Figure 1. Maximum temporal decrease in TSS contents was observed in block-4 where EM application to dairy wastewater was combined with duckweed growth followed by block-3 (duckweed alone), block-2 (EM alone) and block-1 (control), respectively. Total suspended solids in block-4 decreased from 380 to 65 mg l^{-1} at the end of experiment and an average TSS concentration reduction of 83% was observed in (Table 2) which was significantly higher (p >0.05) compared to EM application, duckweed growth alone and control treatments.

Table 1. Initial chemical and biological analysis before the start of experiment

Parameter	2004	2005
pH	7.50	7.45
TSS (mg l^{-1})	390	375
BOD (mg O_2 l^{-1})	680	670
NH_4-N (mg l^{-1})	72.00	68.00
NO_3-N (mg l^{-1})	6.00	6.70
Total P (mg l^{-1})	19.62	20.72
E. coli counts (cfu 100 ml^{-1})	5,000	5,200
Total coliforms (cfu 100 ml^{-1})	10,000	10,500

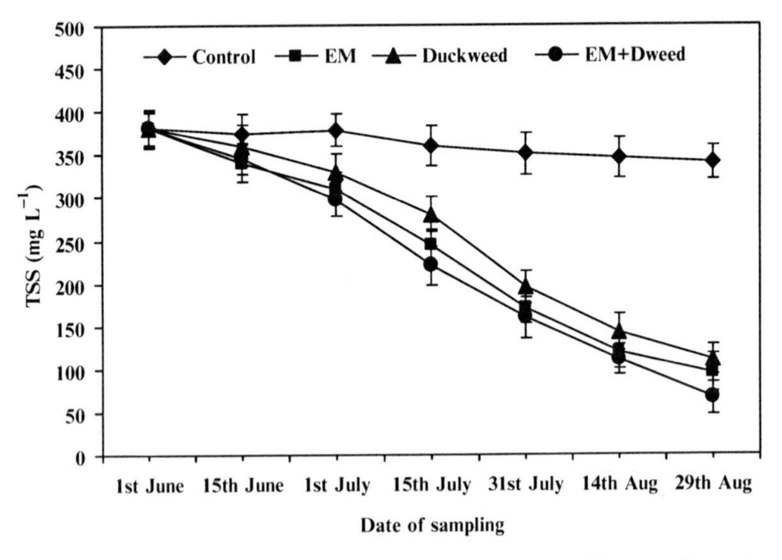

Figure 1. Reduction in TSS due to EM application and duckweed growth

The reduction in TSS in block-2 and block-3 was 75 and 71% respectively with no significant difference between these treatments. Least reduction in TSS contents (11%) was observed in control block. Duckweed removed 77% of TSS from a domestic wastewater pond [21]. Total suspended solids in duckweed ponds are mainly reduced by sedimentation process and biodegradation of organic matters, assimilation by duckweed roots and inhibition of algal growth [22].

Table 2. Effect on total TSS, BOD, NH_4-N, NO_3-N, total P contents, *EC*, and total coliform counts in water samples after the completion of the experiment

Chemical parameter	Control	EM application	Duck-weed growth	EM + Duck-weed	LSD
	(Percent reduction)				
pH	+1.33[a]	–6.67[b]	+1.33	–2.66	
TSS (mg l^{-1})	11 c	75 b	71 b	83 a	5
BOD (mg O_2 l^{-1})	20 c	94 a	77 b	95 a	7
NH_4-N (mg l^{-1})	16 d	44 c	72 b	86 a	7
NO_3-N (mg l^{-1})	37 b	43 b	57 a	67 a	11
Total P (mg l^{-1})	13 b	23 b	98 a	99 a	10
E. coli counts-cfu 100 ml^{-1}	29 b	71 a	75 a	75 a	6
Total coliforms-cfu 100 ml^{-1}	22 b	63 a	67 a	70 a	9

[a]Increase in pH of wastewater, [b]Decline in pH of wastewater

Duckweed plants in principle could also contribute to treatment process by direct assimilation of simple organic compounds, such as simple carbohydrates and various amino acids [23]. Largest reduction of TSS in this pond was probably due to transformation of complex organic molecules into simple organic compounds by yeast (facultative microorganisms) present in EM consortium [24] and then these compounds could have been assimilated by duckweed roots [16]. Biological oxygen demand (BOD) is defined as the amount of oxygen required to oxidize the organic content of wastewater and it is also the oxygen available to micro-organisms within the system. The patterns of a decline in BOD due to different dairy wastewater treatments are illustrated in Figure 2.

Maximum temporal reduction in BOD was observed in block-2 (94%) and block-4 (95%), where EM was applied alone or in combination with duckweed growth (Table 2). The reduction in BOD in both treatment blocks was statistically non significant, however, the reduction due to duckweed growth alone (77%) was significantly lower compared to EM application alone and EM application plus duckweed growth. However, reduction in BOD due to duckweed growth was significantly higher compared to control. The data regarding temporal reduction in BOD clearly show that application of EM was the most effective treatment for dairy wastewater treatment either alone or in combination of duckweed growth. Effective microbes are successfully being used in wastewater treatment in Japan and are becoming popular in wastewater treatment in many countries. The chemical oxygen demand (COD) and BOD of domestic wastewater were significantly reduced when treated with EM [25].

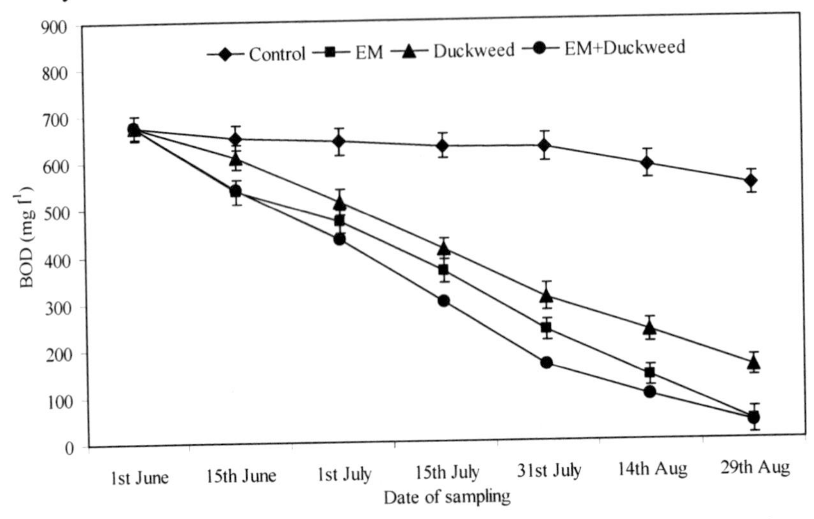

Figure 2. Reduction in BOD due to EM application and duckweed growth

Duckweed plants have been shown to release oxygen to wastewater at the rate of 3–4 g m^{-2} d^{-1} and this release of oxygen improves the oxygen supply. The wastewater with lower BOD always remains aerobic and duckweed can remove 70–96% of wastewater BOD [26]. Degradation of organic matter is enhanced by duckweed through addition of oxygen supply and additional surface for microorganisms responsible for organic matter decomposition [27]. Effective microbes might have been more active in block-4 due to the presence of duckweed plants as they provide more surfaces for the survival of microbes and this synergistic association might have accelerated the organic matter degradation.

3.2. *Escherichia coli* and Total Coliform Counts

The declining trends of *E. coli* and total coliforms in airy wastewater treatment blocks were illustrated in Figures 3 and 4, respectively. Maximum reduction in *EC* and total coliform counts in dairy wastewater initial counts were observed in block-4 (75 and 70%) where EM was applied in combination with duckweed growth. In this block the *EC* bacterial counts were reduced from 5,100 cfu 100 ml^{-1} to 1,300 cfu 100 ml^{-1}.

Coliform bacteria were reduced from 10,000 cfu 100 ml^{-1} to 3,000 cfu 100 ml^{-1}. Reduction in pathogenic bacteria in block-4 was not significantly different from blocks where EM was applied or duckweed was grown alone. These results show that duckweed and EM application were equally affective in the removal of pathogenic bacteria. The total coliform bacteria were reduced by 95% by growing duckweed for domestic wastewater treatment [28]. Pathogen removal is of utmost importance in case of dairy farm wastewater reuse for different purposes at the farm. Die-off of pathogenic bacteria is considered to be a complex phenomenon in waste stabilization ponds. Removal of *EC* and coliforms in duckweed ponds was probably through two main processes. First, the recovery of nutrients from the pond may have caused a deficiency in these nutrients required for microbial growth. Second, the adsorption of these bacteria to the duckweed followed by harvesting might have played a role in their removal [15].

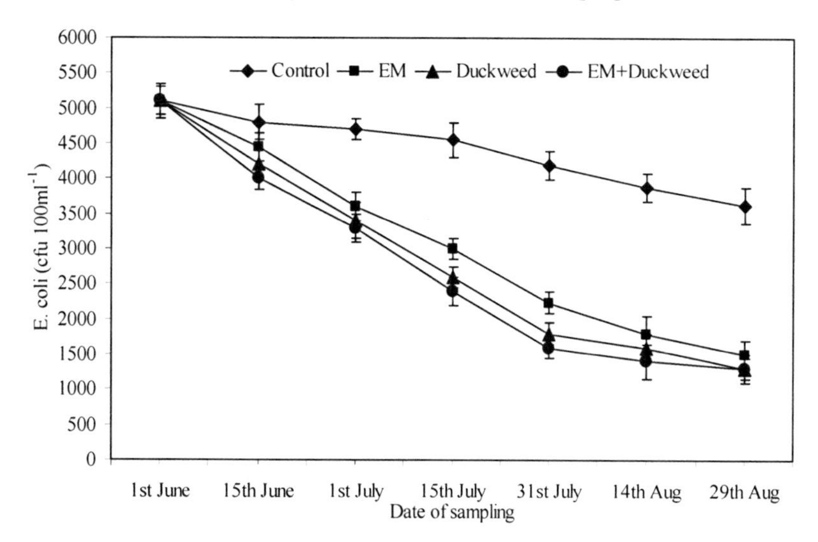

Figure 3. Reduction in *EC* counts due to EM application and duckweed growth

Effective microbe's innoculum have high populations of lactic acid bacteria (*Lactobacillus* and *Pedicoccus*), yeast (*Saccharomyces*) and phototrophic bacteria. Lactic acid bacteria present in EM produce lactic acid and other antimicrobial products as a result of carbohydrate metabolism [29]. During the biodegradation of organic particles in wastewater lactic acid bacteria might have produced antimicrobial products having antibacterial properties [30] and might have inhibited the growth of *EC* and total coliforms. Total and fecal coliforms in fish ponds receiving manure from EM-treated pigs were significantly lower than those from the control [14].

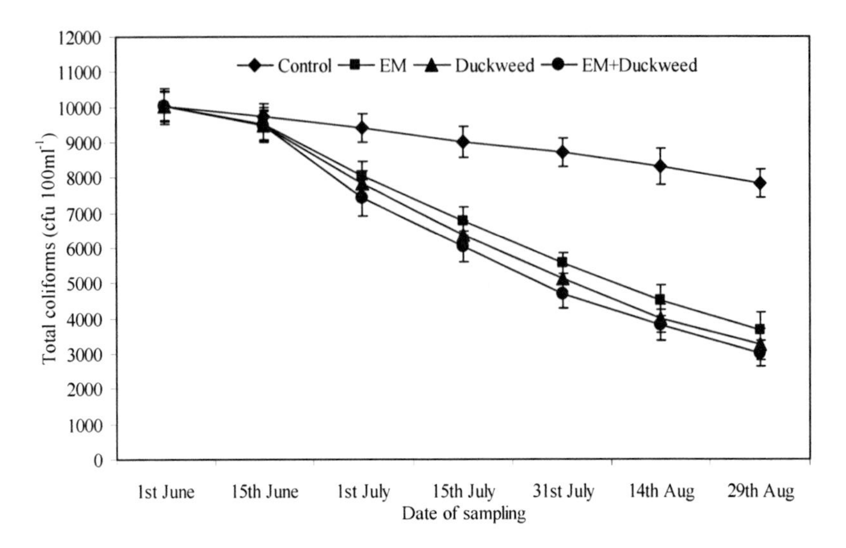

Figure 4. Total coliform counts due to EM application and duckweed growth

4. CONCLUSIONS

Combined application of EM and duckweed growth in dairy wastewater stabilization pond significantly reduced the NH_4-N, total P, and BOD compared to control treatment after three months. A threefold to fourfold reduction in total counts of *EC* and total coliforms also recorded after three months. As differences in the reduction of most of wastewater quality parameters due to EM application and duckweed growth were negligible, we suggest that any of the treatment option alone or in can be adapted for dairy wastewater treatment for its disposal to natural water streams or its reuse at the farm.

Acknowledgment: The authors are grateful to North Atlantic Treaty Organization (NATO) for financial assistance provided to attend the conference on Wastewater Reuse – Risk Assessment Decision Making, Environmental Security, held in Istanbul, Turkey 12–16, 2006 and present this paper.

5. REFERENCES

1. Livernois J. (2002). The economic costs of the Walkerton water crisis (Paper 14). In: Ontario Ministry of the Attorney General. The Walkerton Inquiry.
2. O'Connor DR. (2002). Part 1. Report on the Walkerton Inquiry. The Events of May 2000 and Related Issues. Queen's Printer for Ontario, Toronto, Ontario.

3. Chapman PA. (2000) Sources of *Escherichia coli* O157 and experiences over the past 15 years in Sheffield, UK. J. Appl. Microbiol. 88: 51S–60S.
4. Jones DL. (1999). Potential health risks associated with the persistence of *Escherichia coli* O157 in agricultural environments. Soil. Use. Mang. 15, 76–83.
5. Meng J, Doyle MP, Zhao T, Zhao S. (2001) Enterohemorrhagic *Escherichia coli*. In Food Microbiology: Fundamentals and Frontiers, 2nd ed. Doyle, M.P. Beuchat LR. and Montville, TJ. Washington, DC: ASM Press. pp. 193–213.
6. Ibekwe AM, Watt P, Grieve CM, Sharma VK, Lyons SR. (2002) Multiplex fluorogenic real-time PCR. Appl. Environ. Microbiol. 68, 4853–4862.
7. Park S, Worobo, RW, Durst RA. (1999). *Escherichia coli* O157:H7 as an emerging food born pathogen: Crit. Rev. Food Sci. Nutr. 39, 481–502.
8. Mackay DM, Smith LA. (1990). Agricultural chemicals in groundwater: monitoring and management in California. J. Soil. Water. Cons. 45, 253–255.
9. Hooda PS, Moynagh M, Svoboda IF, Thurlow M, Stewart M, Thomson M, et al. (1997). Soil and land use in Scotland. Soil Use Manage 13, 196–204.
10. Zimmo OR, van der Steen NP, Gijzen HJ. (2004). Nitrogen mass balance across duckweed-based WW stabilization ponds. Water. Res. 38, 913–920.
11. Sangakkara UR. (2000). Kyusi Nature Farming and EM for enhanced smallholder production. In Proceedings of the 13th International Scientific Conference of IFOAM. Alfoeldi, T. et al. (Ed). FiBL, Basal, Switzerland, 268.
12. Sajjad Z, Ahmad MS, Abbasi NA. (2003). Effect of phosphorus levels and effective microorganisms. Sarhad J. Ag. 19(2), 193–197.
13. Okuda A, Higa T. (1999). Purification of wastewater with effective microorganisms. Proc. 5th International Conf. Kyusi Nature Farming, Senanayake YDA and Sangakkara UR. (Ed) APNAN, Thailand, 246–253.
14. Hanekom D, Prinsloo JF, Schoonbee HJ. (2001). A comparison of the effect of anolyte and EM on the fecal bacterial loads in the water and on fish produced in pig cum fish integrated production units. Proc. 6th Int. Conf. Kyusi Nature Farming, South Africa, Senanayake YDA and Sangakkara UR. (Ed).
15. El-Shafai SA, El-Gohary FA, Nasr FA, van der Steen NP, Gijzen HJ. (2007). Nutrient recovery using a UASB system. Biores. Technol. 98, 798–807.
16. Alaerts GJ, Mahbubar MDR, Kelderman P. (1996). Performance analysis of a full-scale duckweed-covered sewage lagoon. Wat. Res. 30 (4), 843–852.
17. Hammouda O, Gaber A, Abdel-Hameed MS. (1995). Assessment of the effectiveness of treatment of wastewater - Enz. Microb. Tech. 17, 317–323.
18. Whitehead AJ, Lo KV, Bulley NR. (1987). The effect of hydraulic retention time and duckweed cropping rate on nutrient removal from dairy barn waste water. In KR Reddy and WH Smith (ed.) Aquatic plants for water treatment and resource recovery. Magnolia, Orlando, FL, pp. 697–703.
19. CoStat. (2001), 7 CoHort Software, 98 Lighthouse Ave. PMB 320, Monterey, CA 93940, USA.
20. Danalewich JR, Papagiannis TG, Belyea RL, Tumbleson ME, Raskin L. (1998). Characterization of dairy wastewater streams, current practices and potential for biological nutrient removal. Wat. Res. 32(12), 3555–3568.

21. Ran N, Agamib M, Orona G. (2004). A pilot study of constructed wetlands using duckweed (*Lemna gibba* L.) for treatment of domestic primary effluent in Israel. Water Research 38, 2240–2247.
22. Smith MD, Moelyowati I. (2001). Duckweed based wastewater treatment (DWWT): design guidelines for hot climates. Wat. Sci. Tech. 43(11), 291–299.
23. Hillman WS. (1976). Calibrating duckweeds light, clocks, metabolism, flowering. Science, 193, 453–458.
24. Higa T, Parr JF. (1994). Beneficial and effective microorganisms for a sustainable agric. and environ. Int. Nature Farm. Res. Center, Atami, Japan. p. 9.
25. Higa T, Kanal A. (1998). An Earth Saving Revolution-II: EM-amazing applications to agricultural, environmental and medical problems. 216–259.
26. Korner S, Lyatuu GB, Vermaat JE. (1998). The influence of *Lemna gibba* L. on the degradation of organic material. Water Res. 32, 3092–3098.
27. Korner S, Vermaat JE, Veenstra S. (2003). The capacity of duckweed to treat wastewater: Ecological considerations. J. Env. Qual. 32, 1583–1590.
28. Ran N, Agamib M, Orona G. (2004). A pilot study of constructed wetlands using duckweed (*Lemna gibba* L.) for treatment of domestic primary effluent in Israel. Water Research 38, 2240–2247.
29. Tannock GW (2004). A Special Fondness for Lactobacilli. Applied and Environ. Microbiol. 70(6), 3189–3194.
30. Barrett E, Hayes M, Fitzgerald GF, Hill C, Stanton C, Ross RP. (2005). Fermentation, cell factories and bioactive peptides: food grade bacteria for production of biogenic compounds. Aust. J. Dairy Technol. 60(2), 157–162.

OPTIMIZATION OF TREATMENT TRAIN FOR WATER REUSE SCHEMES IN THE CZECH REPUBLIC

Petr Hlavinek and Jiri Kubik
Brno University of Technology, Faculty of Civil Engineering, Institute of Municipal Water Management, Zizkova 17, Brno, Czech Republic

Corresponding author: hlavinek.p@fce.vutbr.cz

Abstract:

The design of water reuse schemes generates a big number of possible treatment trains, which has to match the quality and quantity requirements. The aim of this paper is to describe optimization techniques used by developed decision support tool. This will be followed by the case study to show the possible results of the optimization.

Keywords: Optimization, treatment train, water reuse, genetic algorithms.

1. INTRODUCTION

The decision support system used for optimization has been developed within the frame of AQUREC project supported by the European Commission under the Fifth Framework Program. This software is an output of Work Package 8 "Development and validation of system design principles for water reuse systems" of this project and it's called WTRNet (Water Treatment for Reuse and Network Distribution) [1–3].

1.1. Description of WTRNet Software

WTRNet is a simulation and optimization software which is used to evaluation of treatment trains and for sizing of distribution system in the selected reuse scheme.

1.2. Software Structure

Structure of the software WTRNet consists of four main components. These components are: control module, which is used as a coordinator between knowledge base model and both computational modules—treatment performance module and distribution system performance module. The control module is used as well as a graphical user interface (GUI) for control of input data and display of results.

M.K. Zaidi (ed.), Wastewater Reuse–Risk Assessment, Decision-Making and Environmental Security, 103–108.

2. KNOWLEDGE BASE MODEL

The knowledge base contains the following data: unit processes detailed information—suggested precursors or postcursors of unit process, pollutant removals for the basic pollutants, costing data (capital cost, EM cost, O&M cost), resource data (land requirements, labour requirements, sludge production, concentrate production, and energy consumption), and evaluation criteria scores. Next there are information for sizing and costing of distribution system and quality information of possible end uses. Unit processes included in the database are shown in Table 1.

Table 1. Unit processes included in WTRNet

Primary	Tertiary
Fine screen	Filtration over fine porous media
Sedimentation w/o coagulant	Surface filtration
Sedimentation w/o coagulant	Micro filtration
DAF w/o coagulant	Ultra filtration
Membrane filtration	Nano filtration
Actiflo	Reverse osmosis
Secondary	GAC
High loaded activated sludge + secondary sedimentation	PAC
Low loaded activated sludge w/o de-N + secondary sedimentation	Ion exchange
Low loaded activated sludge w/o de-N + secondary sedimentation	Advanced oxidation
Trickling filter + secondary sedimentation	SAT
RBC	Maturation pond
Submerged aerated filter	Constructed wetland—polishing
Stabilization pond	Flocculation
Constructed wetland	Disinfection
Membrane bioreactor	Ozone
EBPR	Paracetic acid
P-Precipitation	Chlorine dioxide
	Chlorine gas
	UV radiation

2.1. Treatment Performance Module

The treatment performance module is used for evaluation of treatment trains (TTs) by calculation of all possible costs and resources and evaluation criteria scores. Evaluation criteria are shown in Table 2 and these criteria are used for multiobjective optimization.

Table 2. Treatment train evaluation criteria

Type of criteria		Name of criteria
Quantitative		Effluent quality
		Cost
		Land required
		Sludge and concentrate produced
		Energy consumed and produced
Qualitative	Positive	Reliability
		Adaptability to upgrade
		Adaptability to varying flow rate
		Adaptability to change in water quality
		Ease of O&M
		Ease of construction
		Ease of demonstration
	Negative	Power requirements
		Chemical requirements
		Odour generation
		Impact on groundwater

2.2. Distribution Performance Module

The distribution performance module is used for optimal sizing of distribution system including pumping and storage facilities. More information about WTRNet and distribution performance module [1].

Two optimization techniques are used in the software of WTRNet. The first one is exhaustive enumeration and the second one is genetic algorithms (GA) optimization. Using exhaustive enumeration, all possible design alternatives are explored, with respect to the potential endusers and alternative treatment trains that satisfy their requirements [2–3]. Optimization by GA could be single objective or multiobjective. For single objective optimization, the net present value (NPV) was selected as the objective [4].

GAs are a set of guided search procedures based on Darwin's theory of natural selection. The basic idea of GA is maintain a set of solutions, which evolve over time through process of survival of the fittest similar to the population genetics in nature. The first step involved in running GA is the random generation of a population. A random number is generated by built in random generator of programming language. The next step is evaluation part, which involved calculation of fitness score of each solution. The fitness function is a means of rating how good or bad an individual is, as compared to others in the population. Better individuals have higher chances of survival and reproduction (Figure 1).

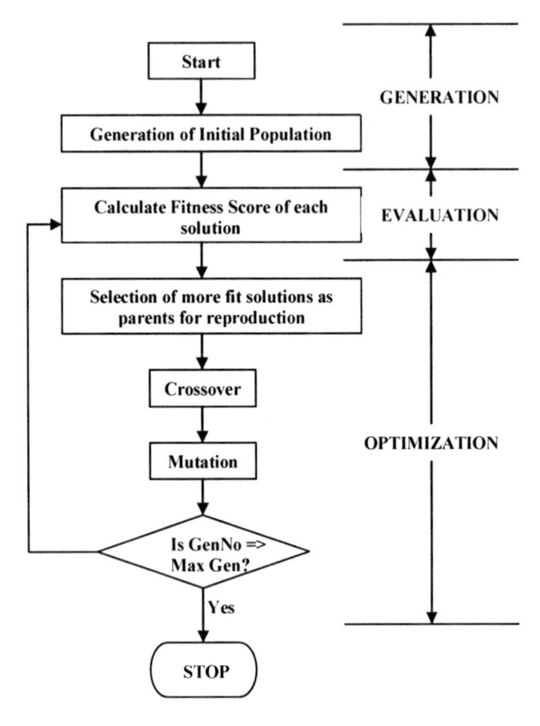

Figure 1. Graphical interface of WTRNet

2.3. Generation and Optimization of TTs Using GA

The problem of generation and optimization of TTs is to find the combination of TTs for given local conditions such as reuse, criteria which results in a maximum fitness score subject to following constraints:

- The TTs must meet the rules set out in the knowledge base
- The TTs must meet the reuse criteria specified by user
- The TTs must have minimum cost
- Influent quality to unit processes in TTs must not exceed the maximum allowable influent quality desribed in the database

- Land area required for TTs must not exceed the land available in the community
- TTs must have effectiveness measures of qualitative evaluation criteria

A binary coded string (0, 1) is used to represent a TT. The "0" represents absence of a particular unit process (UP). The "1" represents presence of the unit process. WTRNet is using OptiGA ActiveX control for the implementation of genetic algorithms.

3. CASE STUDY IN THE CZECH REPUBLIC

The Kyjov WWTP, which currently includes secondary treatment of wastewater, is being considered as a source for reclaimed water for several industries in the area. Location of the WWTP, potential endusers, and a preliminary distribution system that would deliver the reclaimed water to the potential endusers are shown in Figure 2.

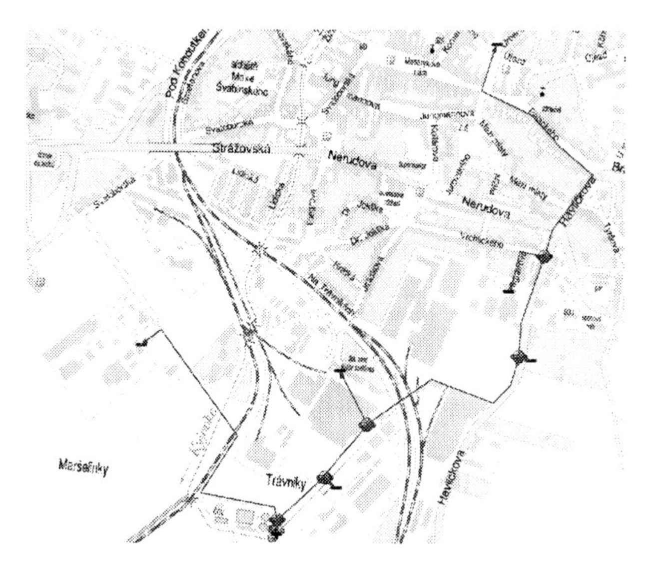

Figure 2. Kyjov case study situation plan

The main purpose of using this case study was to test the exhaustive enumeration technique to determine the optimal selection of endusers. The six potential endusers produced 63 different combinations. For each combination was followed to produce a least-cost design alternative, comprising of the treatment, distribution, and total costs. Potential endusers of Kyjov case study [5–6] are:

VETROPACK MORAVIA GLASS a.s.—producer of packaging glasses for food industry, ŠROUBÁRNA Kyjov spol. s r.o.—producer of connecting materials, ŠROUBÁRNA Kyjov spol. s r.o.—producer of connecting materials, MLĖKÁRNA Kyjov, a.s.—dairy, ŠEBESTA spol. s r.o.—producer and distributor technology for WWTPs and car wash, KM BETA a.s.—producer of roofing, EKOR s.r.o.—treatment of waste, waste disposal, technical services.

4. CONCLUSION

Wastewater reuse plays an important role in this time. The software tools, which are able to help to designers or to help decision-makers, are much on request now. WTRNet software brings an advantage in this decision. The case study shows an example of using of this decision support tool. Exhaustive enumeration could be used for smaller communities. Genetic algorithms are better choice for lager communities. The big advantage of this software is in calculation of integrated system for wastewater reuse (treatment train + distribution system). Only disadvantage is that it is not able to calculate the treatment train separately.

Acknowledgment: The authors acknowledge the European Commission for funding this work within the AQUAREC project on "Integrated Concepts for Reuse of Upgraded Wastewater" (EVK1-CT-2002-00130) under the Fifth Framework Program contributing to the implementation of the Key Action "Sustainable Management and Quality of Water" within Energy, Environment and Sustainable Development Thematic Program. Also supported by NATO ARW.

5. REFERENCES

1. Joksimovic D, Kubik J, Hlavinek P, Savic D, Walters G. Development of an Integrated Simulation Model of Reclaimed Water (2006), pp. 9–20.
2. Joksimovic D. Design Support Software. Aquarec report D14, October 2005.
3. Joksimovic D, Kubik J, Hlavinek P, Savic D, Walters G. Development of a Simulation Model for Water Reuse Systems (2004), Marocco.
4. Loughlin DH, Doby TA, Ducoste JJ, de los Reyes FL. System-Wide Optimization of WTPs Using Genetic Algorithms (2001), ASCE, USA.
5. Dinesh N. Development of a Decision Support System for Optimum Selection of Tech. for Wastewater Reclamation. (2002), U. Adelaide, Aust. p. 479.
6. Gaiker. Feasibility Study Methodology. Aquarec report M4.2, February 2005.

SESSION 4. RISK ASSESSMENT/INTERNATIONAL CONFLICTS

RANKING OF ECOLOGICAL RISKS RELATED TO WASTEWATER MANAGEMENT

David Alcaide Lopez de Pablo
Department of Statistics and Operations Research,
University of La Laguna, La Laguna, Tenerife, Spain

Corresponding author: dalcaide@ull.es

Abstract:

This paper is concerned with a fuzzy extension of the Borda ranking method. Main attention is devoted to the application for the ecological risk ranking under uncertain data. A way to integrate different opinions and knowledge of different experts is proposed. The fuzzy extension also considers the cases of incomparability of expert assessments and missing opinions, as well as indifferent opinions and calibration (semantic or linguistic variables). An application of the considered ranking methodology to study potential ecological risks in the coastal areas of the Island of Tenerife is considered.

Keywords: Ecological risk, risk ranking, wastewater, management.

1. INTRODUCTION

Reuse of wastewater and its disposal generate dangers and hazards leading to ecological risks. For each kind of wastewater, we consider five types of risks related to wastewater: risk for human population (human health), risk for animals, risk for plants, risk for the environment and natural resources, and risk for human-made technological infrastructures (buildings, bridges, streets, road and railway networks, …). Human-health risks depend on disease classes (cancer, skin illnesses, stomach illnesses, cardiorespiratory problems, etc.), groups of population (age, profession, men/women/children, place of living, human race, etc.), and sources of pollution (tourism, marine transport, marine farms, refineries, electrical power production plants, wastewater discharges, etc.).

Then, let us ask ourselves: Why is risk ranking important? The answer is clear: every risk management procedure needs to invest money, human power, and efforts to control and mitigate the risks. It is not possible to mitigate completely all the risks simultaneously and, consequently, we need to prioritize and rank the risks, that is, risk ranking is necessary [1–6]. These works give the evidence that it is necessary that each organization involved know exactly its tasks and duties in all risky situations. Everyone's tasks should be clearly identified, ranked, and

M.K. Zaidi (ed.), Wastewater Reuse–Risk Assessment, Decision-Making and Environmental Security, 111–120.

listed in the procedures form. When doing any ecological risk prevention planning, the decision makers start with an identification and ranking of the risks—they may not do it consciously but do it nevertheless.

A major obstacle to the wide application of risk analytical and ecological risk prevention techniques is the lack of management motivation. The expected value of this kind of techniques and the credibility of the results are oftentimes questioned. An important obstacle to the use of such techniques seems, also, to be the cost awareness. To meet this problem, emphasis should be given to a rapid and cheap risk ranking method. Risk ranking, or risk prioritization, is the ordinal or cardinal rank positioning of the risks in various alternatives, projects, or units. In this paper, we consider ecological hazards only, and, accordingly, the study focuses on the ecological risk ranking.

2. BORDA'S METHOD FOR RISK RANKING

The quantitative risk assessment literature consistently emphasizes the need to rank the magnitude of disaster consequences. An approach pursued in this paper is the Borda ranking method in combination with fuzzy expert if–then rules serving for processing fuzzy (interval-valued) weights. Let us remind in brief the basics of the Borda method [7]. Suppose that a finite number of criteria (experts, attributes, objectives, scenarios) are used to evaluate a finite number of alternatives (in our case, hazards, risks). Our goal is to aggregate information for each criterion and to obtain an overall ranking of the alternatives. It is well known that ranking methods can be placed into two basic categories: cardinal methods and ordinal methods. Cardinal methods find risk weights which permit to express the degree of preference of one alternative (risk) over another for each criterion; typical representative are multiple attribute utility theory and the analytic hierarchy process (AHP) methodology [8]. On the other hand, ordinal methods find only that the overall order (rank) of the alternatives basing on finite number of different orders supplied by different experts (criteria). Many ordinal ranking methods have been devised during the past two centuries, and they fall into several categories including positional voting, mathematical programming, and outranking techniques. This study will focus on the positional voting method devised by Borda and generalize it for uncertain and fuzzy data [9–10].

The method invented by Borda is a "positional" *method,* in that it assigns a score corresponding to the positions in which a candidate (alternative) appears within each voter's ranked list of preferences, and the alternatives are sorted by their total score. A primary advantage of positional methods is that they are computationally very easy: they can be implemented in linear time. They also enjoy the properties called anonymity, neutrality, and consistency in the social choice literature; they minimize the number of voting paradoxes. However, they cannot satisfy the Condorcet criterion, explained below. In fact, no method that assigns a weight to each position and then sorts the results by applying a function to the weights associated with each candidate satisfies the Condorcet criterion [11].

Suppose that there are N alternatives (different risk categories) and K criteria (experts), and that the kth expert has an associated weight W_k. First we assume that $W_k = 1$, however later we consider that each W_k is a positive, not necessarily integer, number. A preference order (sometimes called a permutation or total order) supplied by each expert ranks the alternatives from the most preferred to the least preferred without ties. A preference order in which x_1 is ranked first, x_2 is ranked second, and so forth, is written here as $x_1, x_2, ..., x_N$. We shall discuss an algorithm for aggregating criteria information and obtaining a consensus preference order.

Borda's voting method works as follows. Given N alternatives (risk categories), if points (grades) of $N - 1, N - 2, ...$, and 0 are assigned to the first-ranked, second-ranked, ..., and last-ranked risk category in each expert's preference order, then the winning alternative is the one with the greatest total number of points. In other words, if r_{ik} is the rank of alternative i by expert k, the Borda count for risk i is $b_i = \sum_k (N - r_{ik})$. The alternatives are then ordered according to these counts. The ties are handled by evaluating the rank for a tied alternative as the average of the associated rankings. That is, if an expert k have already ranked u alternatives and wishes to assign the same rank to v next alternatives then each of them is assigned grade $\sum_{q=1}^{v} \frac{r_{uk} + q}{v}$.

Although being simple and efficient, this method, however, does have some limitations, as the claim below shows. Remind that an alternative i is said to win a pairwise majority vote against an alternative j if more than one half of the voters (experts) have given preference of i over j (or, in other words, if more experts rank i above j than rank j above i). An alternative is called the Condorcet winner if it wins all pairwise majority vote elections against all other alternatives. A preference order $c_1, c_2, ..., c_N$ is the Condorcet order if c_i wins over any other alternative c_j, for $i < j$, in a pairwise comparison. If such an order exists, then necessarily c_1 is the Condorcet winner. The Borda vs Condorcet encounter [12]. If the Condorcet winner exists, Borda's method might not put it in first place. If the Condorcet order exists, Borda's method might not rank the alternatives in that order.

Example 1. Let the number of risk categories be $N = 4$, the number of independent experts be $K = 5$, and rank order of alternatives by experts be presented in Table 1. Table 1 shows that three expert estimates (EE) have ties: Risks A and B have the same scores for EE2, Risks B and C have the same score for EE4, and Risks C and D have the same score for EE5. For example, because Risks A and B are tied for first and second places for EE2, their average ranking is 1.5 for that column.

Table 1. Rank order of alternatives by each expert for Example 1

	EE1	EE2	EE3	EE4	EE5
Risk A	1	1.5	3	1	2
Risk B	2	1.5	1	2.5	1
Risk C	3	4	4	2.5	3.5
Risk D	4	3	2	4	3.5

Table 2 below gives the resulting number of Borda points for each combination of alternative and expert estimation, which is $N = 4$ minus the corresponding entry in Table 1, and these numbers are summed across the columns in each row to yield each Borda count b_i. Risk B is the "winning alternative," because it has the highest Borda count, namely, 12; this risk category is most dangerous. These Borda counts yield the preference order BADC as the unique consensus solution.

Table 2. Borda points and count for each alternative in Example 1

	EE1	EE2	EE3	EE4	EE5	b_i
Risk A	3	2.5	1	3	2	11.5
Risk B	2	2.5	3	1.5	3	12
Risk C	1	0	0	1.5	0.5	3
Risk D	0	1	2	0	0.5	3.5

As we see, the Borda method has several advantages as simplicity, computational efficiency, and minimum voting paradoxes, but Borda method has also some inherent limitations: (1) It is defined for uniform EE weights, $w_i = 1$, only. (2) It does not permit to handle inexact, uncertain, and fuzzy data, and, besides, it does not sensitive to how confident experts are in their estimates. (3) It does not satisfy the Condorcet principle [4]. In the next section, we present a generalization of the Borda method capable to overcome drawbacks (1) and (2), and, partly, (3).

3. RISK RANKING UNDER UNCERTAINTY: A FUZZY BORDA METHOD

García-Lapresta and Martínez-Panero [13] considered a fuzzy version of the Borda method. Their fuzzy way is based on a matrix of pairwise comparison. The matrix entries are numbers between 0 and 1 which represent their "degrees of preference." They consider incomparability of some expert estimation, but do not consider ties (indifference). In fact, they consider a finite set of alternatives, $X = \{x_1, \ldots, x_n\}$, and over each one of them, m experts show their preferences by means of fuzzy binary relations $R1, \ldots, Rm$. They use the membership function $\mu_{R^k} : X \times X \rightarrow [0,1]$ of Rk, $k = 1, \ldots, m$ and the number $r_{ij}^k = \mu_{R^k}(x_i, x_j) \in [0,1]$ indicates the level of preference intensity of expert k for x_i over x_j, as higher as closer to 1. In addition, they restrict themselves to fuzzy binary relation that they call reciprocal in the sense of $r_{ij} + r_{ji} = 1$ for all $i, j \in \{1, \ldots, n\}$ and for each expert $k \in \{1, \ldots, m\}$. For each expert k, and from his/her reciprocal fuzzy binary relation, García-Lapresta and Martínez-Panero establish an induced preference relation Rk over X that they defined as $x_i \succ_k x_j \Leftrightarrow r_{ij}^k > 0.5$. In this way, they construct, for each expert k, the corresponding $n \times n$ matrix $\left(r_{ij}^k\right)_{i,j=1,\ldots,n}$. Then, they define for each expert k a count to evaluate each alternative in according his/her preferences, i.e., in their approach, the expert k gives to the alternative x_i the value $r_k(x_i) = \sum_{x_i \succ_k x_j} r_{ij}^k$, which coincide with the sum of the entries greater than 0.5 in the row i. Then, they collectivize these individual counts to obtain a collective one with the formula $r(x_i) = \sum_{k=1}^{m} r_k(x_i)$. Finally, the collective Fuzzy Borda preference relation that Garcia-Lapresta and Martínez-Panero propose is given by $x_i P^{FB} x_j \Leftrightarrow r(x_i) > r(x_j)$, which is always negatively transitive, thus, they choose the highest second alternative.

Differing from the previous fuzzy Borda (PFB) method, the fuzzy Borda method we are proposing in this paper represents expert knowledge and estimations in the form of an acyclic graph with clusters (cactus graph). This scheme permits to take into account both: incomparability, indifference (ties) and missing data. And also permit us to make calibration, that is, we also provide a way to calibrate the levels in the case of indifferent alternatives (we allow to add semantic or linguistic variables, as example ties at "good" level are different from ties at "bad" level). In addition, this generalization release to both: fuzzy and no fuzzy cases. In the fuzzy case the generalization is done taking fuzzy in two directions: Fuzzy ranking with fuzzy numbers, i.e., instead of saying this alternative is ranking the first, the second, …, etc., it is allowed to the expert to say this alternative is fuzzy number 1, fuzzy number 2, …, etc. Then, in this case we use fuzzy arithmetic. And also, we introduce weights of experts and weights of estimations. These weights are fuzzy numbers or linguistic variables (very low, low, medium, high, very high, very bad, bad, normal, good, very good, etc.). Then, we use fuzzy weights and, consequently, in this case we use fuzzy Logic. The quantitative risk assessment literature consistently emphasizes the need to rank the magnitude of disaster consequences and the probability of those consequences within a given time frame. Therefore we can formally view the *risk* as the triplet (S_i, p_i, X_i), where S_i is the risk scenario, with each S_i having a probability (p_i) of occurring and a consequence (X_i) if it occurs. The first step in the strategic ecological risk prevention planning is then the process of identifying the risk scenarios, their probabilities, and consequences, and then investigating the effect of uncertainty on the probability and consequence estimates. We ask the following questions: What can happen (S_i)?, How likely is it to happen (p_i)?, What are the consequences should the risk scenario (S_i) occur ("severity" x_i)?, How confident are the experts in their estimates of p_i and x_i?

Typically, decisions are made with incomplete information or intrinsic variability, which leads to uncertainty. This uncertainty needs to be incorporated into the analysis in order to assess its impact on a decision. There are a number of traditional ways of incorporating uncertainty about parameter values and assumptions into models to help characterize risk: (1) Bayesian analysis, (2) stochastic cost-effectiveness and cost-benefit models, (3) decision/probability trees, (4) brainstorming and Delphi technique, (5) computer-aided Monte-Carlo analysis and simulation, (6) expert judgment and expert systems, and (7) multicriterion analysis.

3.1. Integrating the Borda Method with Fuzzy Logic

An approach pursued in this paper is different from the described above, we will integrate the Borda ranking method with fuzzy expert if–then rules for processing fuzzy (interval-valued) counts. Let us remind in brief some concepts of the fuzzy set theory and fuzzy interference [14–16].

Fuzzy inference is the process of formulating the mapping from a given input to an output using fuzzy logic. The process of fuzzy inference involves five steps: (1) Fuzzify inputs and outputs, (2) Apply fuzzy operators and formulate fuzzy if–then expert rules, (3) Apply the implication concept, (4) Aggregate the outputs, and (5) Defuzzify. In the method suggested, the fuzzy inputs will be the following: (a) Levels of risks A, B, C,... (low, medium, high, very high), (b) Positions of the risks (approximately 1, approximately 2,..., N), (c) Level of expert's confudence (low, medium, high), (d) Expert weights Wk, (e) Weights of risk categories wj. The fuzzy outputs are:

Fuzzy Borda's count Count (J)= N–(Level_of_Risk_J), J = 1,..., N.
Fuzzy operators: AND, OR. A single fuzzy if–then rule assumes the form if x is A then y is B, where A and B are fuzzy sets or linguistic variables or propositions. Examples of the if–then rules are:

If Level_of_Risk_J = low and Expert_K_Confidence = medium then Count (J) = N – (Level_of_Risk_J).
If Position_of_Risk_J = μ and Risk_J_Weight = wj and Expert_K_Confidence = medium then Count (J) = wj ($N - \mu$).

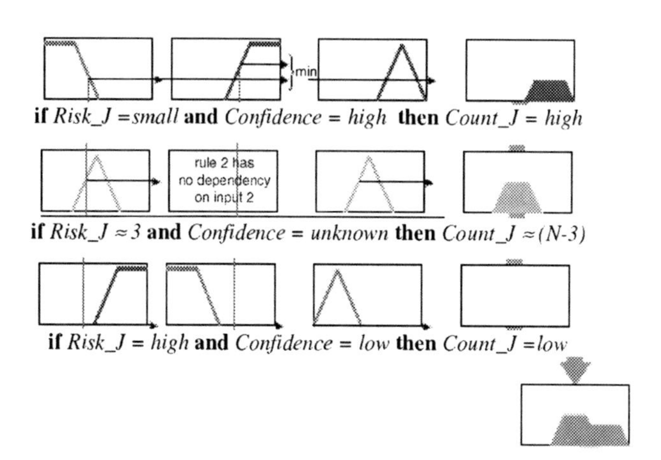

Figure 1. Examples of the expert rules

The rules may be also of the following forms:
If x is $A1$ and y is $B1$ then $z = k1$, where $k1$ is a fuzzy number or a linguistic variable. If x is $A1$ and y is $B1$ then $z = p1x + q1y + r1$, where $p1$ and $q1$ are constants (weights). Some of the rules are presented in Figure 1. Defuzzification is done in a standard [14].

4. CASE STUDY: RISK ASSESSMENT FOR THE POTENTIAL DANGERS ON THE COAST OF THE ISLAND OF TENERIFE

Tenerife Island in the Canary Islands (Spain is a tourist place with increasing density of population (412 people/km^2 in 2005), the large number of tourists and visitors and the corresponding increase of energy and resource consumption it is going to be a real potential risk for the human population.

The fuzzy Borda order is: CDBA.

We collect data from different experts (medical doctor, biologist, tourism economist, agriculture economist, farmer, building expert) to collect their estimation about how a potential wastewater or wastewater pollution could affect on several geographical areas of the island (north coast [NC], south coast [SC], west coast [WC], metropolitan area [MA], industrial area [IA], marine transport area [MTA], national park of Las Cañadas del Teide [NPT], sea mammals reserve [SMR], and natural forest park [FP]).

Table 3. Rank order of alternatives by each expert (e = extreme, s = strong, m = medium, a = acceptable)

	EE1	EE2	EE3	EE4	EE5
Risk in Region A	A	a	s	a	m
Risk in Region B	M	a	a	s	a
Risk in Region C	S	e	e	s	e
Risk in Region D	E	s	m	e	e

The expert opinions are expressed in semantic or linguistic variables: zero risk (z), negligible (n), very weak (vw), weak (w), medium (m), strong (s), very strong (vs), extreme (e), catastrophic (c). With this data, and doing a calibration in a similar way that Example 2, we construct the Table 3 and 4 that collects the opinions of the experts, that is, their answers to the question: "In according to your specialized knowledge: What is your opinion about the potential risk of the current status of the wastewater management in this area?" Note that the empty entries in this table denote that the corresponding expert has no or prefers to give no opinion about the corresponding area.

Table 4. Rank order of alternatives by each expert ($N = 9$ regions/alternatives, $K = 6$ experts) (zero risk (z), negligible (n), very weak (vw), weak (w), medium (m), strong (s), very strong (vs), extreme (e), catastrophic (c))

Risk in	EE1 Medical doctor	EE2 Biologist	EE3 Tourism economist	EE4 Agriculture economist	EE5 Farmer	EE6 Building expert
NC	m	m	m	m	s	n
SC	vs	m	m	s	vs	m
WC	s	w	m	s	vs	m
MA	vs	e	vs	–	–	vs
IA	vs	e	e	–	–	vs
MTA	s	vs	e	–	–	–
NPT	z	z	z	z	z	–
SMR	n	vw	n	–	–	–
FP	z	n	z	w	vw	vw

In Table 5, calibration is done according with a scale from 0 to 8 (zero risk = 0, catastrophic risk = 8), and all experts opinions are integrated in the composite total score.

Table 5. The calibration and the total score for each region/alternative

Risk in	EE1 Med. doctor	EE2 Biologist	EE3 Tourism economist	EE4 Agri-culture econo-mist	EE5 Farmer	EE6 Building expert	Total score
NC	m (4)	m (4)	m (4)	m (4)	s (5)	n (1)	22/6 ≈ 3.6
SC	vs (6)	m (4)	m (4)	s (5)	vs (6)	m (4)	29/6 ≈ 4.8
WC	s (5)	w (3)	m (4)	s (5)	vs (6)	m (4)	27/6 = 4.5
MA	vs (6)	e (7)	vs (6)	–	–	vs (6)	25/4 = 6.25
IA	vs (6)	e (7)	e (7)	–	–	vs (6)	26/4 = 6.5
MTA	s (5)	vs (6)	e (7)	–	–	–	18/3 = 6
NPT	z (0)	z (0)	z (0)	z (0)	z (0)	–	0
SMR	n (1)	vw (2)	n (1)	–	–	–	4/3 ≈ 1.3
FP	z (0)	n (1)	z (0)	w (3)	vw (2)	vw (2)	8/6 ≈ 1.3

5. CONCLUSION

From the practical point of view, in this paper we propose a way to integrate opinions and knowledge of different experts. An application of the considered ranking methodology to study ecological risks in the coast areas of the Island of Tenerife is reported. From the theoretical point of view, we extend the Borda method to a fuzzy case considering incomparability, missing opinions, indifference, and calibration (semantic or linguistic variables).

Acknowledgments: The author is very grateful to Professor Eugene Levner for his useful comments and suggestions. This research is partially supported by Spanish Government Research Projects DPI2001-2715-C02-02 and MTM2004-07550, which are helped by European Funds of Regional Development, and the NATO financial support to present this paper.

6. REFERENCES

1. Rao VR. (1992). Risk Prioritization: National Trends, Forecasts and Options for the Army, White Paper Series, Army Environmental Policy Institute, Champaign, Illinois.
2. NAS. (1983). Risk Assessment in the Federal Government: Managing the Process, National Academy of Sciences, National Academy Press.
3. Levner E and Proth JM. (2005). Strategic management of ecological systems: A supply chain perspective, in E. Levner, I. Linkov, J.M. Proth (eds), Strategic Management of Marine Ecosystems, Springer, Amsterdam.

4. Levner E, Alcaide D, Benayahu J. (2007). Environmental Risk Ranking: Theory and Applications for Emergency Planning, in Computational Models of Risk in Critical Infrastructure, D. Skanata, Ed., Elsevier NATO ASI/ARW Series. (In press).
5. Smith K. (2001). Environmental Hazards: Assessing Risk and Reducing Disaster, 3rd edition, Routledge, London, 389.
6. US EPA (Commission on Risk Assessment and Risk Management) (1996). Risk Assessment and Risk Management in Regulatory Decision-Making, Technical Report, Environmental Protection Agency, USA.
7. Lansdowne ZF. (1996). Ordinal ranking methods for multicriterion decision-making, Naval Research Logistics, 43, 613–627.
8. Saaty TL. (1980). The Analytic Hierarchy Process, NY. McGraw-Hill, USA.
9. Cook WD and Seiford LM. (1978). Priority ranking and consensus formation, Management Science, 24, 1721–1732.
10. Arrow KJ and Raynaud H. (1986). Social Choice and Multicriterion Decision-Making, Cambridge, MA, USA. The MIT Press.
11. Young HP. (1974). An Axiomatization of Borda's Rule. J. Economic Theory, 9: 43–52.
12. Fishburn PC and Gehrlein WV. (1976). Borda's Rule, Positional Voting, and Condorcet's Simple Principle, Public Choice, 28, 79–88.
13. García–Lapresta JL and Martínez–P M. (2002). A Fuzzy Borda count in Multi–person Decisión Making. Recent Developments, Springer.
14. Mamdani EH. (1977). Application of fuzzy logic to approximate reasoning using linguistic synthesis, IEEE Transactions on Computers, C-26, no. 12, 1182–1191.
15. Levner E, Ptuskin A and Friedman A. (1998). Fuzzy Sets and Systems: Theory and Applications, Russian Academy of Sciences CEMI Press, Moscow, 110. (Russian).
16. Ganoulis J. (1994). Engineering Risk Analysis of Water Pollution: Probabilities and Fuzzy Sets, Wiley-VCH, New York, 306.

RISK ASSESSMENT OF WASTEWATER CONTAMINATED WITH RADIONUCLIDES USING *ARABIDOPSIS THALIANA* L. PLANTS

Namik M. Rashydov, Valentina V. Berezhna, and Nataliya K. Kutsokon
Institute Cell biology and Genetic Engineering
NAS Ukraine, 148, Zabolotnogo street, Kyiv, 01148, Ukraine

Corresponding author: nrashydov@yahoo.com

Abstract:

Analysis of genotoxicity with plants assays is of a great importance due to pollution of the wastewater and soil by genotoxicants. The objective of these studies was to determine the approach to risk assessment contaminated with radionuclide different wastewater from 30-km zone of the Chernobyl and Kyiv region using the transgenic lines *Arabidopsis thaliana* L. We used three effective plant assays to evaluate the genotoxicity of Chernobyl and Kyiv region water reservoirs and soil sites. The high, total mutagenic effect was determined in wastewater taken from the lakes Glyboke and Telbin. Received data allow us to develop the approaches to risk assessment for subjective testing wastewater and contaminated with radionuclide water reservoirs and soil sites.

Keywords: Risk assessment, Arabidopsis-GUS-gene assay, Allium-assay, Tradescantia-SH-assay.

1. INTRODUCTION

Radiation hazard caused by the Chernobyl Atomic Power Station (ChAPS) accident in water and soil areas of Ukraine can be monitored by transgenic lines *Arabidopsis thaliana* L., Allium-assay and Tradescantia-stamen-hair (Tradescantia-SH) assay. At date many biological tests for determination of mutagenicity in our environment are proposed [1, 2]. Being easy for use, inexpensive, well developed and good correlated with others tests, plant test-systems have a lot of advantages in compared of others test-systems. A lot of plant assays were used to determine the mutagenicity of chemicals, physical mutagenic factors, polluted environment, waste etc. [3–7]. The experimentally accessible methods of an evaluation of numerical significance of probability of the main elementary events in the process of radiation-induced risk of formation are defined. These methods have their basis in utilization of special biotest system. The general view of the tree of risk depends on nature of targets on which influence of damaging agents are primarily directed. Deoxyribonucleic acid (DNA) in cell nucleus or mitochondria as well as membrane

M.K. Zaidi (ed.), Wastewater Reuse–Risk Assessment, Decision-Making and Environmental Security, 121–130.

systems serves as targets for ionizing radiation. Cell systems of another nature can play a role of targets in the cases of the influence of alternative factors too [8]. As a rule, the key difficulty in decision of this problem the calculation risk coefficient due to combined chemical (due to lead, cadmium, copper, zinc, mercury, DDT in food) and radiation (due to ^{137}Cs, ^{90}Sr in food) food contamination is necessity to estimate threshold and non-threshold functions the curves of the "dose-effect" in unified terms [9].

The Glyboke Lake is located in the 10 km area of ChAPS, is essentially a part of a former Prypyat riverbed separated by a dam. The lake accumulates 450 through 600 thousand cubic meters of water. Annual and seasonal variations of water mineralization in the range of 220 through 630 g/m^3 are caused by variations of atmospheric precipitates and snow-breaks.

After the ChAPS accident in 1986, a large quantity of radioactive debris fell upon the lake watershed and water level. At present, soil and bottom sediment contamination with ^{137}Cs and ^{90}Sr radionuclides is 17–28 MBq/m^2, i.e., it varies significantly over the area. Water contains about 2×10^4 Bq/m^2 of ^{137}Cs and 8×10^4 Bq/m^2 of ^{90}Sr. Some steady decline of the ^{90}Sr concentration has been observed over the last years, while variations of the ^{137}Cs are rather irregular. The lakes Telbin, Pushcha-Vodycja, Verbne, and Berizka are located in Kiev region content the concentration radioisotopes ^{137}Cs and ^{90}Sr less than the lake Glyboke [10]. The Prypyat River flows into the Dnipro River, where the level of contamination with radioactive isotopes and other pollutants is very low (Table 1). The aim of the present studies was to investigate discecting the risk assessment approach on base genotoxicity effects of contaminated with radionuclides ^{137}Cs and ^{90}Sr water and soil on transgenic lines Arabidopsis thaliana L. and to compare this data with Allium-assay and Tradescantia-stamen-hair-assay which methods have been widely adopted.

2. MATERIALS AND METHODS

In this experiments we used three lines of the transgenic plant *Arabidopsis thaliana* L. which content; RPD3-gene histonedeacetylase, SIR2-gene coded protein which deacetylate histones, SU(VAR)—gene suppressor variation and as reporter marker GUS-gene. Mentioned genes regulate structure of the chromatin in mitosis phases. In case of interchromatid recombination between sister homological chromatid reconstructed check interruption version GUS-gene become apparent after addition to etiolate plants substratum X-Gluc (5-Bromo-4-chloro-indolyl-β-D-glucoronide cyclohexylamine salt). The seeds was soaked and grown in water from lakes Glyboke, Telbin, from Prypyat River near Chernobyl and Dnipro River near Kyiv region. In ontogenesis phases rosette, budding and flowering used test-systems height of growth seedling and emergence blue spot as result interchromatid exchanges was measured. The yields of GUS-gene activity as blue spots in plants A. thaliana transgenic lines influenced by different concentrations of BrdU (5-Bromo-2′-deoxyuridine) also observed. In addition widely adopted in scientific research two plant assays for evaluation of genotoxicity were used. Waters genotoxicity in Allium-assay on induction of chromosome aberration was estimated and soils genotoxicity in Tradescantia-stamen-hair (Tradescantia-SH) assay on gene mutations was evaluated. Ten samples of both water and soil from different reservoirs in Kyiv were taken. Four samples were taken from lakes (Telbin, Verbne, Berizka, and Pushcha-Vodycja) and six samples were taken from sites on river Dnipro.

Study of water samples on total mutagenic activity was carried out on *Allium cepa* L. seeds. Seeds were growing in petri dishes on blotting paper moisten in studying water samples and cultivated in thermostat at 24°C. Distillate water as a control was taken. After 72 h moistening roots 4–9 mm long were fixed at Clarck's solution (ethanol:icy acetic acid at proportion 3:1). Energy of seeds germination on 72 h was determined as a percent of seeds which were grown. Temporary squashed acetorsein slides of meristems were prepared for microscopic analysis. The root tip cells of Allium for the cytogenetic effects study were used. Analysis was carried out by standard methods on aberrant anaphases and early telophases presence in first mitoses, mitotic index also was accounted [11, 12]. Frequency of aberrant ana–telophases and mitotic index were scored in percent.

Table 1. Contamination with radionuclides ^{137}Cs and ^{90}Sr different water sources in 30-km zone of the Chernobyl and Kyiv region, Bq/l

Source of water samples	Average value concentration radionuclide in water, Bq/l	
	^{137}Cs	^{90}Sr
The Glyboke Lake	6.27	135.00
The Telbin Lake	0.05	0.09
The Verbne Lake	0.02	0.05
The Berizka Lake	0.02	0.04
The Pushcha-Vodycja Lake	0.02	0.04
The Prypyat River	0.07	0.11
The Dnipro River	0.02	0.08

3. RESULTS AND DISCUSSION

Evaluation of total mutagenic activity of soils was performed with Tradescantia clone 02 that is heterozygous for blue/pink alleles in their floral parts. The blue floral parts including the stamen hairs cells is a dominant phenotype, and the pink floral parts is a recessive phenotype. Each flower includes six stamens, each stamen bears 30–40 hairs, and each hair contains the chain of the cells which normally are blue colored. Then mutated, one or more cells became pink colored, and they are considered as mutation event [13]. Tradescantia plants were growing in studied soil samples. Control plants of Tradescantia on unpolluted soil were grown. We carried out our investigation in three stages, and were separate controls in each experiment. After the initial lag stage (7 days) stamen hairs were scored for induced mutations during four weeks. The rate of mutations was expressed as a number of mutation events per 100 cells. All results were statistically processed; comparison between the experimental variants and controls were conducted by χ^2-method. High level of sister chromatid exchange was determined for water from lakes Glyboke and Telbin which depends on the level of contaminated with radionuclides ^{137}Cs and ^{90}Sr (Table 2). Similar datum received for samples soil from Kopachi and Chernobyl contaminated sites placed on 30-km zone alienation. (Table 3).

Table 2. Level of sister chromatid exchanges detected by GUS-gene activity in transgenic line BAR/BinAR/RPD3-9/5 of the *Arabidopsis thaliana* L. grown on several contaminated water samples

Samples of the waters	The line BAR/BinAR/RPD3-915 of transgenic plants *Arabidopsis thaliana* L.		
	The number of plants	The number of mutation	Yields mutations for per plant
The Glyboke Lake	33	65	1.97±0.10*
The Telbin Lake	53	28	0.53±0.05*
The Prypyat River	96	24	0.25±0.05
The Dnipro River (control)	67	14	0.21±0.05

* level of reliability $P < 0.05$

We also determined that the increase of sister chromatid exchanges (SCE) yield depended on level contaminated soil. The soil contaminated with ^{241}Am become more genotoxicity than the soil contamined with ^{137}Cs and ^{90}Sr. The yields of SCE for transgenic lines increased depending on influence of different BdrU concentrations from 0.1 until 5 mM observed (Figure 1).

As shown in Figure 1 BdrU increased the yield of GUS-gene expression dependently of concentrations of applied water solutions. It is very well known that compounds which lead to the formation of DNA interstrand cross-links are very strong inducers of sister chromatid exchanges [14]. We found that BrdU treatment transgenic lines of the plants *Arabidopsis thaliana* L.indeed results in a reduction of the damage induced similar by γ-radiation. The type of DNA damage for transgenic lines of the plants *Arabidopsis thaliana* L. is responsible for this enchanced effect and we have proposed this to be the DNA interstrand cross-link which may arise in cells that are labelled with BrdU for one round of replication and exposed from contaminated sites radiation.

Table 3. Level of sister chromatid exchanges detected by GUS-gene activity in transgenic line BAR/BinAR/RPD3-9/5 of the *Arabidopsis thaliana* L. grown on several contaminated soil sites.

Samples of the soils	The line BAR/BinAR/RPD3-915 of transgenic plant *Arabidopsis thaliana* L.		
	Number of plants	Number of mutation	Yields mutations for per plant
Control (Kyiv region)	114	24	0.21±0.05
Chernobyl (4.1±0.8 Bq/kg content ^{241}Am	56	14	0.25±0.05
Kopachi (491±74 Bq/kg content ^{241}Am)	21	43	2.05±0.15*
Soil contains only 10^5 Bq/kg ^{241}Am	34	18	0.53±0.08*

* level of reliability $P < 0.05$.

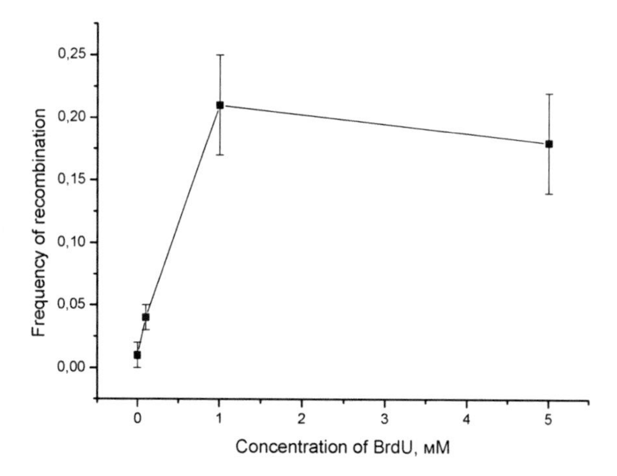

Figure 1. Frequency of recombination in Arabidopsis (line BAR/PTA/SIR2-1) under influence of 5-Bromo-2′-deoxyuridine (BrdU)

Growth reactions in different phases of the ontogenesis of Arabidopsis plants of dealt with γ-irradiation doses on the Figure 2 is shown. The tangency α angular of the curves "dose-effect" in Figure 2 becomes different depended of phases of the ontogenesis was observed which good agreed with other authors results [15]. The levels of sister chromatid exchanges dependent of the γ-irradiation doses in the Figure 3 are demonstrated. In case determinates manifestation GUS-gene the small dose given additional damages of the DNA. At the curve of relationship dose versus yield of genetic damages in seedling cells do not observed adaptive responses in any cases. The shoulder of the curves dose relations and tangency angles α in case GUS-gene tested decreased in spite of growth reactions of the seedlings.

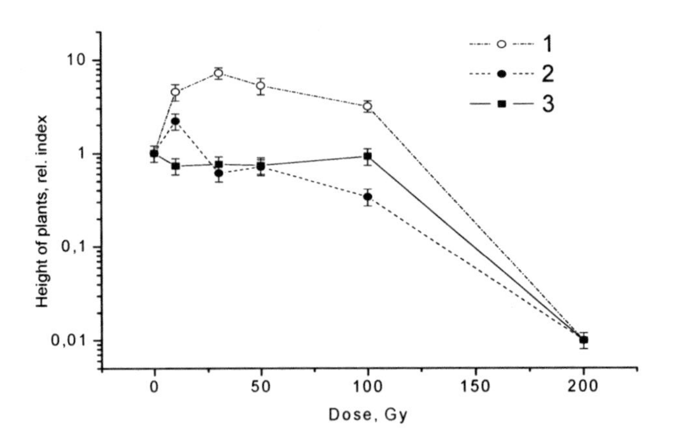

Figure 2. Influence of γ-irradiation on stem height of *Arabidopsis thaliana* L. plants (line BAR/BinAR/RPD3-9/5) in several phases of the ontogenesis: rosette (1), budding (2), and flowering (3)

For DNA damages threshold effect was not observed. But treatments by small dose which act at the physiological phase's ontogenesis *A. thaliana* L. often fail to be dose-modifying adaptive responses. Comparisons between the height of growth seedling and mutant cells with SCE show that absence of repair activity is rarely small dose 10 Gy modifying, because the "shoulder" of the curves is usually much reduced or absent (Table 4). It is apparently from the Figures 2 and 3 and Table 4 that adaptive response are shown in somatic parameter test—"plant height" only. However, in the same doses in the same somatic cells adaptation by genetic marker – "GUS-gene activity" – was not observed. In the case of GUS-gene determination even under low dose 10 Gy increase of damage was shown, and the curve shoulder became narrower. This fact indicates about decrease of repair processes, although increase of GUS-gene activity under low level doses is observed. So, recombination process in genome is one of the ways of repair process in plants. As the identity of fragments exchanged under high doses isn't possible, the increase of mutations level under high doses is referred with decreasing of SCE level.

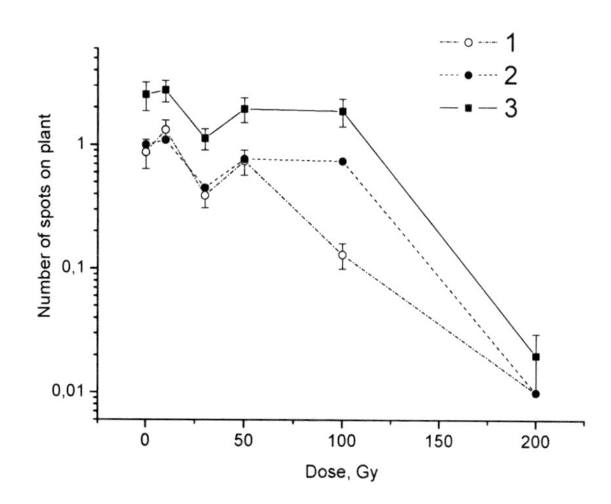

Figure 3. Influence of γ-irradiation on GUS-gene emergences in *Arabidopsis thaliana* L. plants (line BAR/BinAR/RPD3-9/5) in several phases of the ontogenesis: rosette (1), budding (2), and flowering (3)

Table 4. Parameters of "dose-effect" curves in Arabidopsis (line BAR/BinAR/RPD3-9/5) in different ontogenesis phases

The ontogenesis phases	Assays	tg, α	Shoulder, Gy
Rosette phase	Plant height	1	120
	GUS-gene activity	0.08	10
Budding phase	Plant height	0.1	26
	GUS-gene activity	0.4	10
Flowering phase	Plant height	0.6	100
	GUS-gene activity	0.5	10

This data indicate that adaptive responses observed by test-system height of growth seedling depended on phase ontogenesis of the plants, but in phases flowering *Arabidopsis thaliana* L is quite the opposite. By test-system with SCE as well as GUS-gene yields observed that threshold the curves "dose-effect" usually much reduced or absent. Our data evidence that under small doses recombinant processes dealt with repair of double strand breaks are possible, but they are unlikely under high doses. So, we supposed, that recombination repair as well as SCEs at the high doses γ-irradiation essentially decreased. In comparison with transgenic plant–two plant assays (Allium-assay and Tradescantia-stamen-hair assay) for evaluation of genotoxicity of the water samples also were used.

Table 5. Level of chromosome instability in *Allium cepa* L. cells induced by waters from reservoirs

Samples	Energy of germination, %	Frequency of aberrant anaphases ±Sp, %
Control	46.67	1.93 + 0.45
Telbin	45.24	3.52 + 0.57*
Verbne	41.90	5.85 + 0.75***
Berizka	47.62	2.35 + 0.56
Pushcha-Vodycja	39.05	3.65 + 0.65*
Dnipro 1	52.38	3.14 + 0.71
Dnipro 2	46.67	1.39 + 0.56
Dnipro 3	50.95	2.21 + 0.56
Dnipro 4	53.33	3.39 + 0.56
Dnipro 5	47.14	3.24 + 0.77
Dnipro 6	47.62	1.76 + 0.50

* level of reliability $P < 0.05$, *** level of reliability $P < 0.001$

3.1. Allium-assay

The spontaneous level of aberrations in the control was 1.93 ± 0.45%. We revealed a significant increase of chromosome damage rate in three samples of lake water (Table 5) (Telbin, Verbne, and Pushcha-Vodycja. The highest level of genotoxicity was induced by the sample taken from the lake Verbne where frequency of aberrant anaphases has threefold as many as a control –5.85 ± 0.75%, $P < 0.001$. We didn't reveal mutagenicity in others water samples, including all samples taken from Dnipro River and Lake Berizka. The energy of seeds germination was without significant changes in all variants.

3.2. Tradescantia-SH assay

The significant increase of mutation events level was demonstrated in five soil samples at that the highest rates were induced by the soils taken from the beaches on the lakes Telbin and Berizka, and Dnipro 1 site. In two variants where samples were taken from lake Pushcha-Vodycja and Dnipro 5 site, the rates of pink mutations event were increased also but some lower. There was no significant increasing in the soil samples taken from the beaches on lake Verbne and from Dnipro 2, 3, 4, 6 sites ($p > 0.05$) (Table 6).

Table 6. Mutations induced in Tradescantia-SH assay by soil samples

Variants	SH frequency ± Sp, %
Control 1	0.035 ± 0.024
Telbin	0.296 ± 0.082**
Verbne	0.018 ± 0.018
Berizka	0.249 ± 0.051**
Control 2	0.014 ± 0.010
Pushcha-Vodycja	0.074 ± 0.026*
Dnipro 1	0.109 ± 0.045**
Control 3	0.081 ± 0.047
Dnipro 2	0.147 ± 0.055
Dnipro 3	0.145 ± 0.051
Dnipro 4	0.204 ± 0.062
Dnipro 5	0.346 ± 0.122*
Dnipro 6	0.104 ± 0.035

* level of reliability $P < 0.05$, ** level of reliability $P < 0.01$.

4. CONCLUSIONS

We used three effective plant assays (Arabidopsis GUS-gene, Allium chromosome aberrations and Tradescantia-stamen-hair mutations) to evaluate the genotoxicity of water reservoirs and soil sites from Chernobyl and Kyiv regions. Our data demonstrate that all samples taken from lakes demonstrated mutagenicity in Allium or Tradescantia assays. At the same time only two of six probes of river soils were mutagenic, we supposed these effects were determined by local chemical pollution or relief features. Studying the Arabidopsis GUS-gene-assay we found that in water samples from lakes Glyboke and Telbin genotoxicity appeared. Five samples taken from the river Dnipro didn't demonstrate mutagenic effects in all assays. Our experiment demonstrates that the lakes are more polluted by mutagenes than river Dnipro sites. The data allowed us to develop an approach to risk assessment for objective testing for Glyboke Lake, where contamination with radionuclides was very high. Using the transgenic plant can be received at same time data which mentioned influence physical factors to somatic cells and genetic system too. The curves "dose-effect" dependently from biological test-system may estimate threshold and nonthreshold probability which gives appraisal of results to calculated risk coefficient due to combined chemical and radiation factors.

Acknowledgment: We express thanks to Dr. A. Orel for providing the transgenic lines of the *Arabidopsis thaliana* L. plant seeds for this investigation. We are also thankful to the NATO ARW committee to award financial support to attend the meeting and present this paper.

5. REFERENCES

1. Fiskesjo G. (1988). The Allium test - an alternative in environmental studies: the relative toxicity of metal ions. Mutation Research 197, 243–260.
2. Georgievsky V. (2003). Radioecological risk and remediation. "Quaternary changes Ukrainian environmental: Risk for health", Kiev, 14–41.
3. Gichner T, Veleminsky J. (1999). Monitoring the genotoxicity of soil extracts from two heavity polluted sites in Prague using the Tradescantia stamen hair and micronucleus (MCN) assays. Mutation Res. 426, 163–166.
4. Grant WF. (1994). The present status of higher plants for the detection of environmental mutagens // Mutation Research 310 (2), 175–185.
5. Grodzinsky DM. (2003). Ecological risks and generalized indexes of harm. "Quaternary changes Ukrainian environmental: Risk for health", Kiev, 5–13.
6. Guide to short-term tests for detecting mutagenic and cancerogenic chemicals. Geneva. WHO. 1985.
7. Kutsokon N, Rashidov N, et al. (2004). Biotesting of radiation pollutions. "Radiation safety problems in the Caspian region", Kluwer Acad. Pub. 51–56.
8. Kutsokon NK, Rashidov NM, Grodzinskii DM. (2002). Cytogenetic effects of ^{241}Am in the Allium-test. Radiation Biology. Radioecology 42 (6), 675–677.
9. Kuzmenko MI. Radionuclides in the water ecosystem, Kiev, 2001.
10. Ma TH, Cabrera GL, Cebulska-Wasilewska A. (1994). Tradescantia stamen hair mutation bioassay. Mutation Research 310 (2), 211–220.
11. Mohammed KB, Ma TH. (1999). Tradescantia-micronucleus and -stamen hair mutation assays on genotoxicity. Mutation Research 426, 193–199.
12. Rank J, Nielsen M. (1997). *Allium cepa* anaphase-telophase root chrom. aberration assay *N*-methyl-*N*-nitrosourea. Mut. Res. 390/1-2, 121–127.
13. Rank J, Nielsen M. (1993). A modified Allium test as a tool in the screening of the genotoxicity of complex mixtures. Hereditas 118, 49–53.
14. Zhuravskaja AN, Voronov IV, Prokopiev IA. (2006). The influence of the variability physiological-biochemical characteristics on a radiosensitiveness ecoforms *Atriplex patula* L. Radiation Biology. Radioecology 46 (1), 71–76.
15. Wojcik A, Bochenek A, et al. (2006). DNA interstrand crossslinks are induced in cells prelabelled with 5-bromo-2 deoxyuridine. Radiation Research, Kyiv, 149.

MULTICRITERIA DECISION-MAKING TOOL FOR OPTIMAL WATER RESOURCE MANAGEMENT

Konstantyn Atoyev
Glushkov Institute of Cybernetics, Kiev, Ukraine

Corresponding author: atoe@isofts.kiev.ua

Abstract:

On the basis of formalization of interrelations between global and hydrological changes the mathematical model for optimal water resource management was elaborated. Its distinctive feature is the possibility to investigate regimes with sudden, discontinuous changes or phase transitions as a result of small, continuous changes in variables that influence the water system. Model permits to determine the conditions for sustainable development and the possible mechanisms of chaotic behavior that causes the instability in the area of environmental security.

Keywords: Reducing water stress, decision-making tool, water resource management, risk assessment, theory of catastrophes.

1. INTRODUCTION

Water Stress is defined by the long-term average of the annual withdrawal-to-availability ratio (WTA) [1]. This ratio describes how much of the average annual renewable water resources of a river basin are withdrawn for human purposes (in the household, industrial, agricultural, and livestock sectors). In principle, the higher this ratio is, the more intensively the waters in a river basin are used; this reduces either water quantity or water quality, or both, for downstream users. According to this variable, water stress increases when either water withdrawals increase and/or water availability decreases. On the basis of experience and expert judgments, it is assumed that if the long-term average WTA-ratio in a river basin exceeds 0.4 (or 40%), the river basin experiences severe water stress. Currently, some 2,400 million people worldwide (more than 40% of the total population) are estimated to live in river basins under severe water stress. Under various scenarios, the total number of people in areas with severe water stress is expected to increase. This makes water stress one of the most pressing environmental problems of sustainable development and, according to any of the scenarios, will remain so for the foreseeable future.

M.K. Zaidi (ed.), Wastewater Reuse–Risk Assessment, Decision-Making and Environmental Security, 131–144.

Water stress, which is pressure on the quantity and quality of water resources, exists in many places throughout Europe, resulting in serious problems of water shortages, flooding, pollution, and ecosystem damage. The amplifications of hydrological changes due to longer-term links and a great extent of causal chains in global world leads to essential shifts in the area of environmental, economic, epidemiological, and social security of society. The deformation of security's area causes the crises of control systems and leads to sharp increasing natural and technogenic disasters and decreases the horizon of forecasting for extreme events, especially connected with power low that decreases more slowly than Gaussian probability low. That is why during the past decade, the rare events take place with increasing frequency (tropical cyclones and hurricanes in USA, typhoons in East Asia). The high susceptibility to initial conditions becomes a common feature of hydrological systems in "epoch of globalization" and causes impossibility to predict all long-term consequences of various technogenic impacts on environment. Uncertainty and principle unpredictability became the keywords that characterize various aspects of water resource management. The cost of mistakes in management becomes very high, as we see on the examples of Aral Sea and some other ecological catastrophes of last time.

1.1. Interrelations between Global and Hydrological Changes

Global change not only creates the world with principle new structural features (strengthening destroy power of rare events, increased possibilities that some minor disturbances can cascade into a disaster with unpredictable ecological, economic, and political consequences, increased degree of global mega threats and their damage, increasing role of systemic risks, strengthening longer-term links and a great extent of causal chains, high susceptibility to initial conditions), but also generates unprecedented anthropogenic impacts on nature. One of consequences of such impacts is climate change. Various scenarios of Intergovernmental Panel of Climate Changes (IPCC) indicate some possibility of irreversible catastrophic effects that could wipe a great part of the world population due to global warming.

Short list of different social and economic losses caused by global warming includes following items: (1) direct health effects (casualties during floods, storms, typhoons, hurricanes, and other natural disasters, the frequency of which will increase as a result of global warming; increased morbidity and mortality from ischemic heart disease, respiratory diseases, diseases of the nervous system, kidneys, etc. during hot summer days. The frequency and intensity of extreme hot summer temperatures will increase as a result of global warming); (2) indirect health effects (increased incidence of infectious and parasitogenic diseases, due to increased rainfalls, water logging, changes of areals of natural-loci infections; higher risk of intestinal infections because of breakdowns of water supply and sanitation networks; increased morbidity and mortality from suspended particulates in air (TSP) and other air pollutants during forest fires); (3) losses for agriculture (loss of fertility because of water erosion, increased soil density, desertification, lack of mineral nutrients, increased salinity, water logging, soil pollution; restructuring of soil biota, loss of land productivity; lack of water resources in drought-prone regions; growing number of floods in flood-prone regions and water logging in regions with high precipitation; introduction of new pests, spread of traditional pests and microbes); (4) losses for forestry (introduction of new pests, massive spread of traditional pests; loss of biodiversity); (5) other types of losses (unprecedented natural disasters will lead to human casualties and destruction of infrastructure and consequent economic losses; protection from extreme temperature variations will require additional stocks of fuel and energy resources; permafrost melting will lead to destruction of buildings, industrial facilities, communication networks; changes of vapor transfer patterns in Earth's atmosphere and changes in albedo of Earth's surface; loss of environmental equilibrium of biosphere as a whole) [2].

Future water supplies may also be adversely affected by climate change. Potential climate change resulting from anthropogenic emissions of carbon dioxide and other greenhouse gases is a major environmental issue. The latest assessment of the IPCC is that the balance of evidence suggests a discernible human influence on global climate. Climate change is expected to increase global mean sea levels, with some levels in Europe rising and some possibly falling, according to some studies; to change the run-off patterns of water courses; to increase the frequency of droughts and flooding, and to shift the patterns of agriculture, forests, precipitation, and biological diversity.

The rate of soil degradation is expected to rise as a result of warming, with a reduction in soil water storage capacity mainly in southern Europe. Studies of potential climate changes have led to predictions of a wetter climate in northern Europe and a drier climate in southern Europe. There is also the possibility of a recession of mountain glaciers and increased precipitation over parts of northern Europe. Changes in climate may also have a significant effect on the availability of freshwater and hydrological changes can have a potential impact on climate.

1.2. Water Stress in Europe

River basins identified to be experiencing severe water stress are among others— the Don, the Seine, the Meuse, the Thames, as well as most river basins in Southern Italy, Spain, Greece, Southern Ukraine, and Turkey. The levels of severe water stress (Percentage of national territory in which water consumption exceeds 40% of

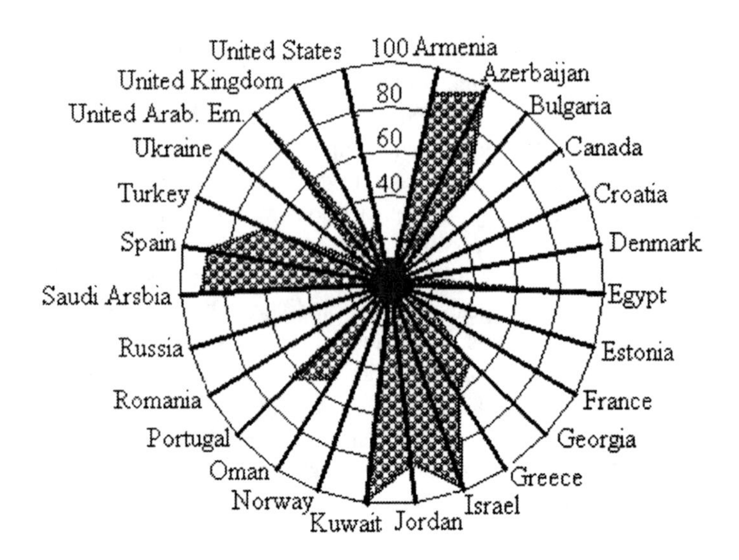

Figure 1. Percentage of country under severe water stress

available water) and some other parameters that determines quality and quantity of water resources and Environmental Sustainability Index (ESI) for several NATO and partners countries are presented at Figure 1 and Table 1 on the basis of Yale University ESI2005 Report's data [3].

The integrated analysis of global change impacts on European river basins under various scenarios indicates several important trend that will influence future changes in water stress in Europe: annual water availability generally increases in Northern and Northeastern Europe and decreases in Southern and Southeastern Europe. Growing demand for water in the domestic, industrial, and agricultural sectors has led to increased withdrawals, and may lead to even higher withdrawals in the future. At the same time, climate change may reduce water availability at some locations [4].

2. MATHEMATICAL MODELING OF WATER AVAILABILITY AND WATER USE: WHY OUR PROGNOSIS BECOME LESS RELIABLE

The water-global Assessment and Prognosis (WaterGAP) model, a global model estimating water availability and water use, was developed at the University of Kassel, Centre for Environmental Systems Research (CERS) in Germany with the

Table 1. Variables determined ESI for several countries

Parameters & Units	Israel	Turkey	Ukraine	USA
ESI (rank)	62	91	108	45
GDP/Capita ($)	17.30	5.86	4.75	32.48
Freshwater availability per capita (thousand cubic meters per person)	0.62	2.85	1.93	8.43
Internal groundwater availability per capita (thousand cubic meters per capita)	0.07	0.97	0.42	4.43
Industrial organic water pollutant emissions per available freshwater (AF) (metric tons of daily emissions per cubic km of AF)	15.63	0.97	5.03	0.88
Percentage of national territory in which water consumption exceeds 40% of available water (%)	97.62	64.36	16.88	30.66
Death rate from intestinal infectious diseases (deaths per 100,000 population)	1.4	0.69	0.80	0.03
Percentage of population with access to improved drinking water source (%)	100	93	98	100

cooperation of the National Institute for Public Health and the Environment of the Netherlands (RIVM). WaterGAP comprises two main components, a global hydrology model and a global water use model. Block-scheme of WaterGAP model is presented at Figure 2, as it was proposed [3].

The aim of the model is to provide a basis: (1) to compare and assess current water resources and water use in different parts of the world; (2) to provide an integrated long-term perspective of the impacts of global change on the water sector. WaterGAP belongs to the group of environmental models that can be classified as "integrated" because they seek to couple and thus compile knowledge from different scientific disciplines within a single integrated framework [3].

The global hydrology model simulates the characteristic macroscale behavior of the terrestrial water cycle to estimate water resources, while the global water use model computes water use for the household, industrial, irrigation, and livestock sectors.

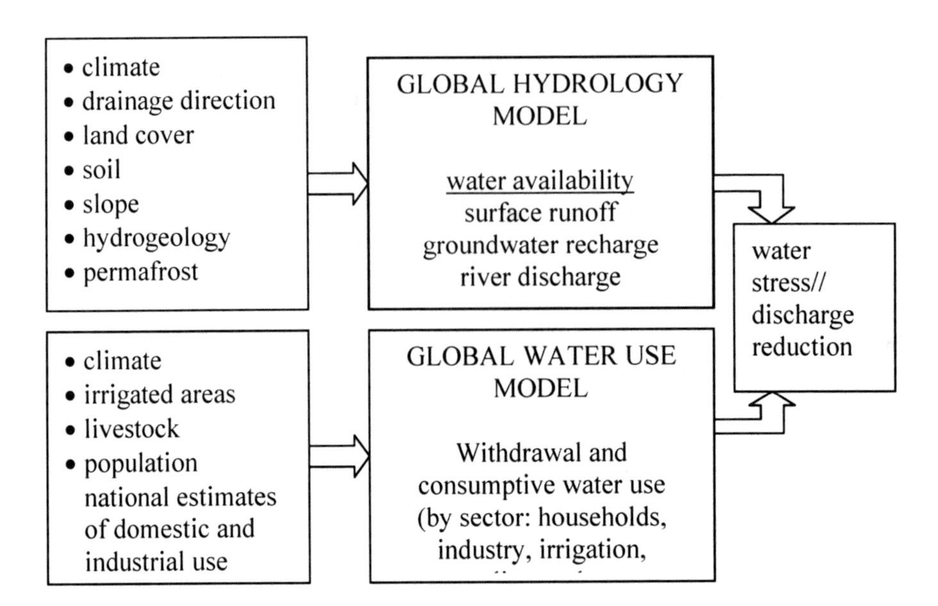

Figure 2. Schematic representation of WaterGAP 2, a global model for water availability and water use [3]

All calculations made apply to the entire land surface of the globe (except the Antarctic) and are performed for cells on a 0.5° by 0.5° spatial resolution. At the moment this is the highest feasible resolution for global hydrological models since climate input data is usually not available at higher levels of detail.

The WaterGAP model requires several scenario-specific driving forces on climate change and key socioeconomic developments to compute the future situation for water resources. The current water stress situation, along with the broad implications of developments under different scenarios up to the 2030s, have been outlined in [1] on the basis of results obtained from the WaterGAP model. The forecasting of freshwater availability (percentage of population living in areas with severe water stress) under four scenarios (markets first, policy first, security first, sustainability first) for Africa (AFR), Asia and the Pacific (A&P), Latin America and the Caribbean (LAC), West Asia (WA), Europe (EU), North America (NA) examined in [1] is presented at Table 2, where ☝—increase in problem and/or persistent large problem, ☞—some improvement or deterioration, ☟—decrease in problem.

However, it should be noted that all model-based assessments have unavoidable uncertainties linked to underlying model assumptions and available input data.

For example, water-use data by sector is scarce and often incomplete. In the absence of higher resolution data, it is assumed that regional trends in structural changes apply to all countries within the region, while, in fact, the trends may vary considerably from country to country. Another example of uncertainty of the model output is that the hydrology model is attuned to matching measured long-term runoff for some 700 drainage basins, which only cover about half the globe's terrestrial surface outside Antarctica. For the remaining river basins, model parameters are derived from attuned river basins, making estimates of water availability more uncertain. An additional uncertainty is related to the water stress variable.

Table 2. Synopsis of scenario impact estimate

Scenario	AFR	A&P	EU	LAC	NAM	WA	World
Markets first	☟	☟	☟	☟	☞	☟	☟
Policy first	☞	☟	☝	☟	☝	☟	☟
Security first	☟	☟	☞	☞	☞	☟	☟
Sustainability first	☞	☞	☝	☞	☝	☟	☞

Furthermore globalization leads to more complex and interconnected network of modernized society, including its ecological component. The behavior of such complex system sometimes becomes unstable and shows sudden, discontinuous changes or phase transitions as a result of small, continuous changes in variables that influence the system. Small causes can lead to large consequences and generate chaotic oscillations of vulnerability, since the amplifications of global change play important role in appearance of the supported chaotic regimes of environmental variables, including water resources. One of such chaotic oscillation is presented at Figure 3 [5].

How can we decrease the degree of uncertainty in forecasting of water resources dynamics? The "integrated" global models, estimating water availability and water use (for instance WaterGAP) are not efficient for systems with high susceptibility to initial conditions, for which some minor disturbances can cascade into a disaster with unpredictable ecological consequences.

To solve above-mentioned problem it is necessary to elaborate models that increase our capacity to investigate regimes with sudden, discontinuous changes or phase transitions as a result of small, continuous changes in variables that influence the water system. It permits to increase the horizon of forecasting and efficacy of water resource management.

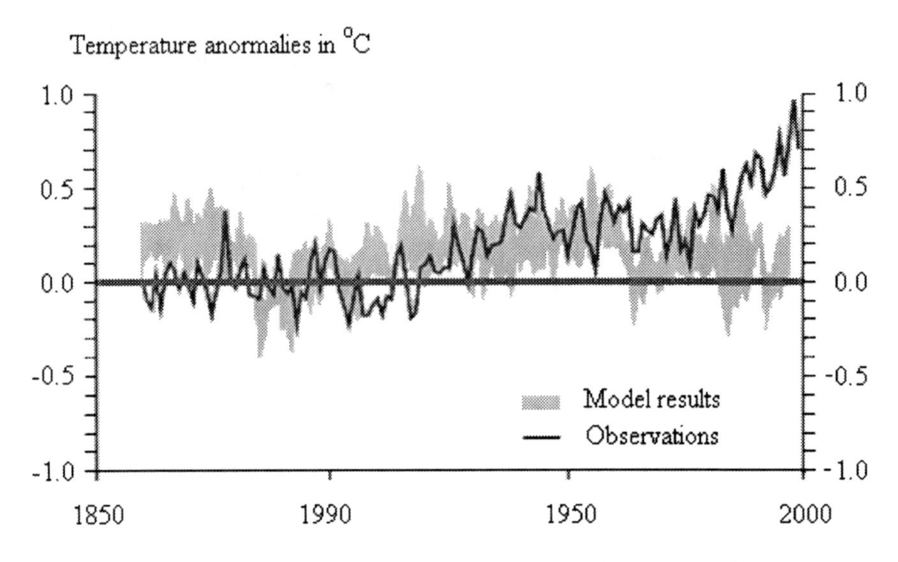

Figure 3. Comparison of modeled and observations of temperature rise

3. THE ADVANCED MODELING METHOD FOR WATER RESOURCE MANAGEMENT

The advanced modeling method for water stress assessment and water resource management has been elaborated. A state of hydrological system is considered as complex interrelationship in a set of variables, which characterize global (climatic, socioeconomic, bioecological) and hydrological changes (quality and quantity of water resources, deteriorations of water quality by pollution from agriculture, industry and households sectors, levels of basic contaminations: metals, phosphorus, pesticide, and nitrates). The formalism of catastrophe theory was utilized.

Let us propose that hydrological system satisfies all the requirements of potential system and can be described by some potential function U(X) of ecological vulnerability (water stress) - X. This function has some steady states; some of them are stable, others—unstable. The transformations of system from one steady state to another, or transformation of a character of a steady state (for instance, from stable to unstable) are functions of control parameters A. These parameters control both a movement of system's current position upon surface U, and transformations of the surface itself. To make it simple, it is possible to put forward following suppositions. Potential function U(X, A) has three stable steady states. The first one characterizes normal condition. The level of X in this state is minimal ($\mathrm{WTA} < 0.2$). The second one characterizes state with a medium level of X ($0.2 < \mathrm{WTA} < 0.4$). The third one characterizes crisis with high level of X ($\mathrm{WTA} > 0.4$). The transformations of security are defined by one of the universal deformations of catastrophes' theory - a "butterfly":

$$- \partial U(X, A)/\partial X = X5 + A1X3 + A2X2 + A3X + A4, \qquad (1)$$

where Ai (i = 1,4) are control parameters, characterizing various aspects of the environmental and technogenic impacts on climatic (i = 1), socioeconomic (i = 2), bioecological (i = 3), and hydrological (i = 4) variables.

The probability of sudden and unexpected transformation from one steady state to another may be performed using the theory of catastrophes allowing the calculation of bifurcation values, curves, and surfaces of control parameters. The probability of transformation is estimated as an extent of a control parameter approaching these bifurcation values characterizing system's transition from one steady state (norm) to another (catastrophe) [6]. The main advantage of this approach is a determination of water stress transformation as a function of dynamic variables of the investigated system. It also allows identifying the weakest link of the system under examination and the areas in need of an improvement.

The following algorithm of water stress assessment can be proposed: (1) collecting of information characterizing the above-mentioned indices input from contemporary databases; (2) determination of indices characterizing appropriate group of parameters calculation by means of developed mathematical models using inputted data; (3) calculation of the bifurcation parameters values at which the number of system states is changing; (4) estimation of restoration possibilities of each considered systems by the remoteness of the parameter characterizing appropriate index from its bifurcation value.

Environmental and technogenic impacts may cause transformation of hydrological system. Bifurcation values of climatic, socioeconomic, bioecological, and hydrological parameters can be calculated by means of traditional mathematical models (WaterGAP). Reaching such critical values abruptly increases the probability of a transfer from one functional state to another. Thus, for a given state of hydrological system it is possible to determine ranges of functional parameter values corresponding to the above-mentioned levels of water stress.

Several authors have developed statistical methods to fit catastrophe models. Guastello's polynomial regression technique uses reverse hierarchical entry (entering higher order terms first) in computing the regression equation and compares the model with two alternative linear regression models [7]. A multivariate methodology for estimating catastrophe models and developed a program called Gemcat II [8]. It is claimed that the program can fit any of the catastrophe models to data. A catastrophe fitting technique was developed that is based on a stochastic interpretation of catastrophe theory [9]. Taking into account all limitation of above methods that were analyzed we choose to modify Gemcat II to fit catastrophe model.

The decision-making tool has been elaborated for water resource management in Ukraine on the basis of proposed approach to minimize the level vulnerability of hydrological systems. The elaborated methodology includes following steps:

1. Determination of potential function U(*X*, *A*) on the bases of experimental data set with the help of technology proposed in [8].
2. Determination of location and scale parameters that transform the initial data set into the set of catastrophe control parameters.
3. Calculation of the bifurcation surfaces that separate parameters sets, corresponding to various levels of vulnerability and their projections on planes of chosen control parameters (Figure 4).
4. The mathematical model for investigation of global change impact on water resources and security.

Deficiency of freshwater can have a potential impact on sustainable development and security. Future water market may be similar to contemporary fuel market with its adverse competition and struggle in "hot points" of global world. It is necessary to have simple decision-making tool for investigation of the relationships between levels of globalization, water resources, and vulnerability of society.

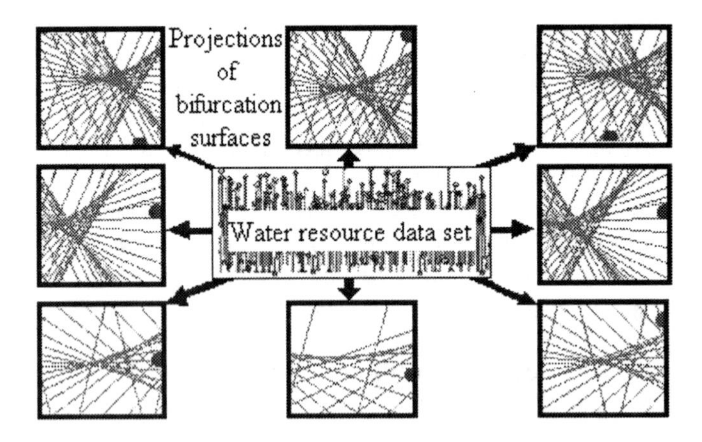

Figure 4. Transformation of the initial data set into the set of catastrophes controls

Let *X*—the level of globalization, determined by technology, economic integration, personal contacts, and political Enlargement, *Y*—the level of water resources and Z—the level of instability, connected with the struggle for water resources [10].

The following postulates may be taken as a basis of model development in terms of demand and supply. The level of

1. Water wealth determines demand for globalization.
2. Globalization determines the supply of globalization.
3. Economic integration determines the demand for water resources.
4. Water resources determines the level of its supply.
5. Threats also limits the development of water sources. Levels of threats and globalization determine this limitation. A negligible level of threat may have catastrophic consequences, if the levels of economic integration, technological advances, and political enlargement (globalization) are high.
6. Water resources and globalization determine the speed at which the threat arises.
7. Some mechanisms (political, economic, and military) which limit threat increases and secure world stability. The activity of these protective mechanisms is proportional to the level of threats.

The relationships between globalization, water resources, and threat of the instability of world order are determined by the mathematical model that may be transformed to Lorents' model of metastable chaos [11, 12].

$$\mathrm{d}X/\mathrm{d}t = \sigma(Y - X), \ \mathrm{d}Y/\mathrm{d}t = rX - Y - XZ, \quad \mathrm{d}Z/\mathrm{d}t = XY - bZ, \tag{2}$$

where σ, b, r characterize the rates of processes. Parameter r is a function of following variables: (1) normalized current demand ($b1$) and current supply ($b2$) on level of globalization; (2) normalized current demand ($c1$) and current supply ($c2$) on water resources. Moreover, the following limitations take place:

$$\partial r/\partial b1 > 0, \ \partial r/\partial c1 > 0, \partial r/\partial b2 < 0, \partial r/\partial c2 < 0.$$

The rise of demands leads to increasing of parameter r. The rise of supply leads to its decreasing. There is interval of demand–supply correlation corresponding to different regimes of metastable chaos (Figure 5).

The chaotic behavior of water resource deviation obtained with the help of model (2) is presented at Figure 5a (variable and time t in conventional units). From year 1000 to year 1860, variations in average surface temperature of the Northern Hemisphere analyzed in [5] are shown at Figure 5b. The line shows the 50-year average. Small fluctuations of parameter r case the essential transformation of system dynamics. At the upper boundary of above-mentioned interval of demand–supply correlation such fluctuations may lead to catastrophic consequences connected with transition from stability to instability, from a state with a low level of vulnerability to a state with high one. The gray region at the Figure 5b marked such bifurcation area, when system cardinally changes own dynamics.

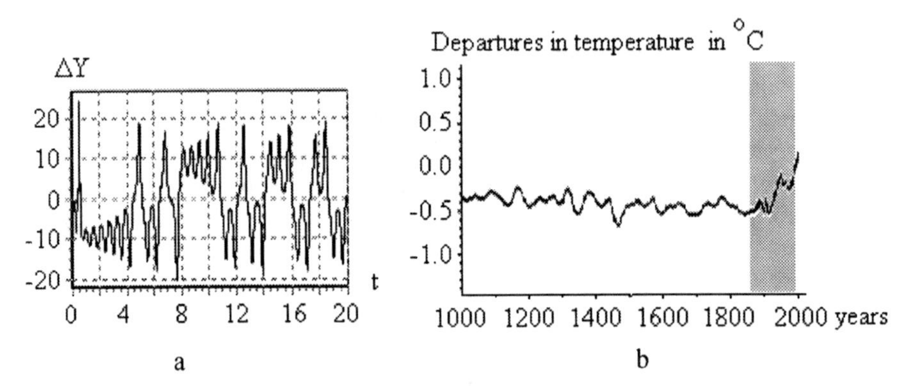

Figure 5. Chaotic behavior of water resource deviation (ΔY) from initial state (a) and variations of the Earth's surface temperature: years 1000 to 2100 (b)

4. CONCLUSION

There are factors which essentially decrease the efficiency of water resource management. One of them is generation by global change the principle new risks that leads to more unpredictable world with high susceptibility to small fluctuations. The effects of a disturbance could impact areas far from the initial source due to existing longer-term links and a great extent of cause and effect in global world. Environmental risks are those insidious risks emerging slowly but surely over time. Any environmental impact increases possibility of the case where some fluctuation can cascade into a disaster with unpredictable ecological consequences. Above-mentioned reasons decrease the horizon of forecasting and efficiency of management. The capacity of decision-making tool may be increased with the help of above-examined modeling methods. They permit to investigate impacts of small, continuous changes in variables on system dynamics and determine the demand–supply correlations corresponding to sustainable development and transformations to turbulent regimes.

Acknowledgment: I would like to express my deep appreciation for Mr. Mohammed K. Zaidi, who brought water reducing stress to my attention. I am very grateful to Prof. Eugene Levner for the most fruitful discussion and development of the water reuse assessment using mathematical modeling. I would like to thank all organizers of workshop and especially Prof. Nava Haruvy and Dr. Alper Baba for high level of scientific program and their cordial attention to all participants of workshop that became in rewarding professional event. Full financial support was provided to present this paper by the NATO ARW organizing committee.

5. REFERENCES

1. UNEP/RIVM (2004). José Potting and Jan Bakkes (eds.). The GEO-3 Scenarios 2002-2032: Quantification and analysis of environmental impacts. UNEP/DEWA/ RS.03-4 and RIVM 402001022.
2. Round Table of Non-Governmental and Social Organizations "A Social Forum on Climate Change" World Climate Change Conference, 2003, Moscow) A. Golub, V. Zakharov (eds). Group "Reform-Press", Moscow.
3. Esty DC, Levy M, Srebotnjak T, de Sherbinin A. (2005). 2005 Environmental Sustainability Index: Benchmarking National Environmental Stewardship. New Haven: Yale Center for Environmental Law & Policy.
4. Alcamo J, Döll P, Henrichs T, Kaspar F, Lehner B, Rösch T, Siebert S. (2003). Global estimates of water withdrawals and availability under current and future "business-as-usual" conditions, Hydrol. Sci. J. 48(3), 339–348.
5. Climate Change 2001 - Impacts, Adaptation, and Vulnerability. IPCC Third Assessment Report (2001). JJ. McCarthy, Osvaldo F. Canziani, NA Leary, DJ Dokken, KS White (eds.). Cambridge University Press, Cambridge.
6. Atoyev KL. (2005). The Challenges to Safety in East Mediterranean: Mathematical Modeling and Risk Management of Marine Ecosystem, NATO Science Series: IV: Earth and Environmental Sciences 50, 179–197.
7. Guastello SJ. (1988). Catastrophe Modeling of the Accident Process: Organizational Subunit Size. Psychol. Bull., 103, 246–255.
8. Lange R, Oliva TA, McDade SR. (2000). An Algorithm for Estimation Multivariate Catastrophe Models: GEMCAT II, Studies in Nonlinear Dynamics and Econometrics, 4:137–168.
9. Cobb L, Zacks S. (1985). Applications of Catastrophe Theory for Statistical Modeling in the Biosciences, J. Am. Stat. Assoc. 80, 793–802.
10. Let X - the level of globalization, determined by Technology, Economic Integration, Personal contacts and Political Enlargement, Y - the level of water resources and Z - the level of instability, connected with the struggle for water resources. an der Maas HLJ, Kolstein R, van der Pligt J. (2003). Sudden Transitions in Attitudes, Soc. Math. Res. 32 (2), 125–152.

11. Danilov-Danilyan V. (2005). Freshwater Deficiency and the World Market, Water Resources, 32(5), 625–633.
12. Weizsacker E. von, Lovins AB, Lovins LH. (1997). Factor Four. Doubling Wealth – Halving Resource Use. Club of Rome. Earthscan Publ. London.

RISK-BASED DECISION SUPPORT OF WATER RESOURCE MANAGEMENT ALTERNATIVES

Paul D. West and Timothy Trainor
United States Military Academy, West Point, NY, USA

Corresponding author: Paul.West@usma.edu.

Abstract:

Enhancing public welfare through the deliberate management of water resources is vital for every society. Pollution, overuse, and consumption challenge a society's ability to develop and sustain water supplies for municipal, agricultural, industrial, and recreational use while protecting fisheries and wetlands. Water resource management decisions are complex and involve risk. This paper identifies a risk taxonomy to help managers identify where those risks are and their severity. It is presented in the context of the Susquehanna River Basin that spans three states in the United States, with management interests at the state, regional, and national levels.

Keywords: Risk, taxonomy, decision support, management, utility.

1. INTRODUCTION

Demand for water grows as populations increase and new uses are found and prioritized. Water management infrastructure is costly to build in both time and money and must be sustainable for decades in the face of uncertain future requirements. Comprehensive water management planning must account for risks not only to physical elements of the system, but also to those elements that enable the system to meet changing needs and uncertain times.

Managing the Conowingo "pond" in the northeast United States highlights these challenges. The pond, a 9,000-acre (3,642-hectare) reservoir spanning 14 miles (22.5 km) in Pennsylvania and Maryland, was created in 1928 with the completion of the Conowingo dam [1]. The Conowingo system gradually outgrew its intended purpose of solely providing hydroelectric power, and by the dawn of the twenty-first century, a complex system of users was dependent on the pond for its survival. Key stakeholders faced this new reality in 2002 with the creation of the Conowingo Pond Workgroup of the Susquehanna River Basin Commission. Their goal was to develop a resource management plan that provides for current and future users

M.K. Zaidi (ed.), Wastewater Reuse–Risk Assessment, Decision-Making and Environmental Security, 145–156.

while meeting existing state and federal regulations. This complex decision scenario is used to illustrate how a new, comprehensive risk-based decision support system can help decision makers choose between competing alternatives in both short-term and long-term projects. The approach is to quantify exposure to sources of operational risk, identify measures for assessing their effects, and determine the utility of various alternatives based on the decision maker's sensitivity to each of the risk categories. The result is an analysis of alternatives that reflects the decision maker's assessment of risk and willingness to accept it.

1.1. The Conowingo Pond Problem

The Conowingo Pond region is at the southern terminus of the Susquehanna River Basin, shown in Figure 1, which spans much of Pennsylvania and portions of New York State to the north and Maryland to the south.

The Conowingo Dam is one of four hydroelectric projects on the lower Susquehanna River. All are regulated by the Federal Energy Regulatory Commission (FERC), whose oversight includes minimum flow requirements to maintain a reliable energy source. However, as populations and uses grew, so did competing requirements. By 2002, the Conowingo Pond was a source of water for

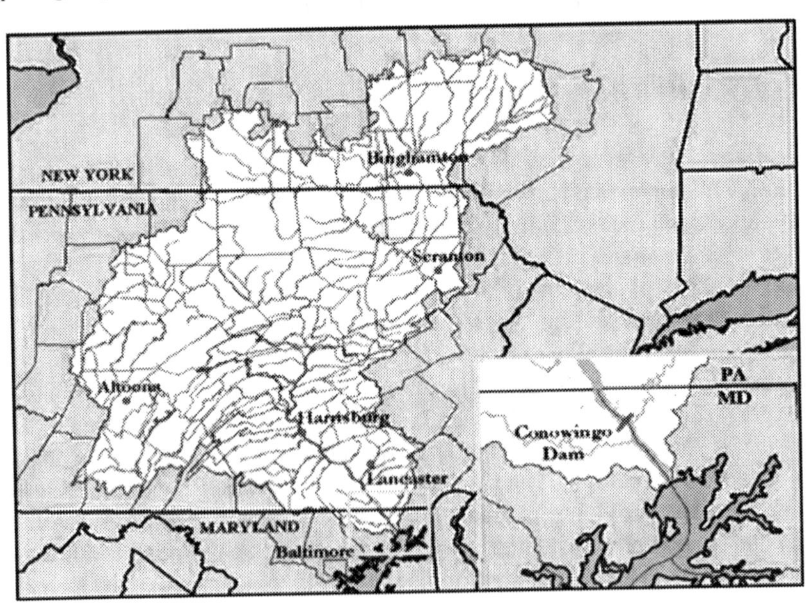

Figure 1. The Conowingo Pond region

- Conowingo Hydroelectric Station
- Muddy Run Pumped Storage Facility
- Peach Bottom Atomic Power Station
- Baltimore, Maryland, municipal water supply
- Harford County, Maryland, municipal water supply
- Chester Water Authority (southeast Pennsylvania and northern Delaware)
- Recreational use
- Sustained stream flows downstream of the dam.

The Muddy Run facility stores water pumped from the pond during low energy requirements to resupply the pond during high-use periods. The Peach Bottom facility requires a constant source of water for coolant, and a sustained stream flow is essential to supply downstream users, support fish and wildlife, and control salinity.

1.2. Risk and Risk Management

Risks exist at all stages of a system's life cycle—from establishing the need and developing the system concept, to designing and producing the system, to deploying and operating the system, to its retirement. Opportunities for failure are ever present. This paper focuses on operational risks that can be "designed out" early in the system development process.

Risk is often expressed in terms of expected value—the probability and severity of adverse effects [2]. It is measured as the combined effect of the probability of occurrence and the assessed consequences given that occurrence [3]. Identifying risks comes in the form of determining sources of risk events and situations under which they may occur [4]. From a system operations point of view, the management of those risks can be defined as, "the process of identifying, assessing, and controlling risks arising from operational factors and making decisions that balance risk costs with mission benefits" [5]. US Department of Defense guidance for risk management in the acquisition process specifies that "Program risk includes all risk events and their relationships to each other. It is a top-level assessment of impact to the program when all risk events at the lower levels of the program are considered." It continues, "One of the greatest strengths of a formal, continuous risk management process is the proactive quest to identify risk events for handling, and the reduction of uncertainty that results from handling actions"[6].

Several risk taxonomies have been proposed to capture risk events and their relationships to each other [7]. However, these tend to focus on specific applications or remain broad in scope. Taxonomy described the operational risk factors from a system-level view that forms the basis of a risk-based decision support tool. For taxonomy to be useful to the decision maker, it must be comprehensive, measurable, and relevant. These attributes, according to Keeney and Raiffa, [8] have the following qualities:

- It is comprehensive if, by knowing the level of an attribute in a particular situation, the decision maker has a clear understanding of the extent that the associated objective is achieved.
- It is measurable if it is reasonable to both
  - Obtain a probability distribution or to assign a point value, and
  - Assess decision-maker's preferences for different attribute levels.
- It is relevant to the particular courses of action under consideration.

2. A SYSTEM-LEVEL RISK TAXONOMY

A "systems" approach to risk management requires that the scope of risk assessment be extended to account for a comprehensive range of factors. Such a framework must be sufficiently robust to apply to all systems while being adequately specific to provide a quantifiable assessment. Taxonomy described below is based on decomposition of total system risk to a point at which relevant measures can be obtained. The complete structure is shown in Figure 2 and described in detail in the following sections.

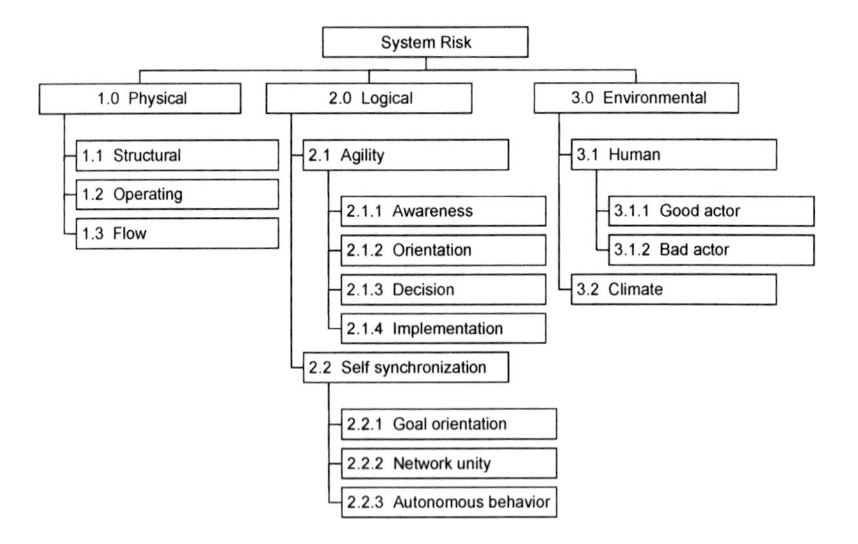

Figure 2. System risk taxonomy

2.1. Risk Domains

Top-level elements that contribute to total system risk are the physical, logical, and environmental domains in which a system operates.

- Physical factors are the tangible components of the system.
- Logical factors include all cognitive functions, whether by software or human intervention.
- Environmental factors are external factors that affect system operation.

2.2. Physical Factors

These consist of the structural components that normally do not change during the life of a system, operating components that process material to make the system function, and flow components, which are the materials processed through the operating components.

In the Conowingo Dam system, the dam itself is a structural component; turbines, floodgates are related equipment are operating components; and water and lubricants are flow components.

2.3. Logical Factors

Agility and self-synchronization are primary drivers of the logical domain.

Agility is the process by which a superior information position is turned into a competitive advantage. It is the quality that enables a system to efficiently adapt to changing conditions, and is essential for long-duration systems such as water management systems to avoid obsolescence. Agility is characterized by

- Awareness, the degree of comprehending the common operating picture.
- Orientation, the degree of comprehending the situation given a level of training, education, and experience.
- Decision, the degree to which cognitive comparisons can be made. It is the "irrevocable allocation of resources to affect some chosen change or the continuance of the status quo."
- Implementation, the degree to which an action can be taken as a result of a decision.

Self-synchronization is the ability of a well-informed system to organize and synchronize complex activities from the bottom up.

- Goal orientation is the degree of comprehending the desired end state—the result, or effect—of the process. It is the decision maker's intent and includes not only the mission, but also key tasks to be accomplished so that subelements understand the intermediate goals and can act autonomously when unexpected situations arise. Intent is a clear, concise statement of what the system must do to succeed. It does not include the why, the how, or the level of acceptable risk related to the process.
- Network unity is the degree to which nodes in the system can function collectively to achieve the goals of the system by maintaining the integrity of the network. This provides the unity of effort.
- Autonomous behavior is the degree to which nodes in the system can function independently to achieve the goals of the system given a clear understanding of the mission, a common operating picture, clear goal orientation, and a clear set of rules to bound the decision space.

2.4. Environmental Factors

These are the external factors, both human and nonhuman, that can affect the system.

Human factors include all interactions with people, regardless of motivation.

- Good actor considerations include the degree to which well-intentioned humans may adversely affect the functioning of the system. These include incorrect responses to events, carelessness, and accidents.
- Bad actor considerations include the degree to which mal-intentioned humans may adversely affect the functioning of the system. Bad actors include disgruntled or co-opted insiders, criminals, terrorists, or hostile nation states.

2.5. Mapping Stakeholder Needs to Risk Domains

Stakeholder involvement is critical to the success of the systems decision process [9]. Stakeholders ensure that decision makers have the appropriate frame for a decision, and provide reliable and credible information. Stakeholders comprise the set of individuals and organizations that have a vested interest in the problem and the solution [10]. Besides decision makers, stakeholders can include customers, system operators, system maintainers, bill payers, owners, regulatory agencies, sponsors, manufacturers, and marketers [11]. Stakeholder input is generally gained through interviews, focus group meetings, or surveys. The Conowingo Pond Work-group consisted of representatives of 27 stakeholder groups who met in 17 sessions over a 4-year period. They identified the 11 major concerns shown in Table 1.

Table 1. Conowingo stakeholder concerns

Hydroelectric power generation	Multipurpose use benefits
Public water supply	Anadromous fish restoration
Upstream consumptive use	Upstream reservoirs
Minimum flow requirements	Environmental resources
Minimum dissolved oxygen	Cooperative management
Summer minimum pond levels	

These concerns reflect the interests of a disparate group of stakeholders and may conflict. For example, concerns for upstream reservoirs would be less of an issue if hydroelectric power generation did not require a water flow. Reconciling these concerns is accomplished by value modeling, in which both qualitative and quantitative models are developed. The result is a coherent method for assessing solution alternatives. In the proposed risk-based approach, the risk taxonomy provides the core for the quantitative value model, while the qualitative model is developed directly from stakeholder input.

The modeling and assessment that follows is illustrative and was not conducted with the Conowingo Pond Workgroup. It is intended to show how the methodology can be used to support decision making for complex water management projects.

2.6. Qualitative Value Modeling

Solution design is a deliberate process for composing a set of feasible alternatives for consideration by a decision maker [12]. It follows, but overlaps with the problem definition phase of the system design process, and it is essential that stakeholder needs, wants, and desires are understood for feasible alternatives to be developed. Qualitative value model develop consists of the following five steps:

- Identify the fundamental objective. This is a clear, concise statement of the primary reason for addressing the problem. For the Conowingo project, it may be stated as to "develop a long-term management plan that ensures water availability for municipal, industrial, and recreational users and sustains the natural environment."
- Identify functions that provide value. These may include "provide hydroelectric energy" and "provide municipal water supply."
- Identify objectives that define value. Objectives provide a statement of preference, such as "minimize salinity encroachment" and "maximize summer pond levels."
- Identify value measures. Value measures indicate how well a candidate solution meets an objective. For example, salinity may be measured in the

concentration of salt in the water in parts-per-million (ppm) or in percentage. Seawater, with salinity of about 35,000 ppm, may also be considered as being about 3.5% salt. Another measure may be "practical salinity units," in which seawater is about 35 and freshwater (1,000 ppm) is 1. The choice of measure depends on how well it informs the decision maker.

- Discuss the value model with key stakeholders. Feedback and buy-in on the appropriateness of the model is critical.

2.7. Quantitative Value Modeling

Quantitative value models identify how well an overall candidate solution attains stakeholder values. They consist of two basic parts: a weighting mechanism to prioritize competing attributes, and a utility function that indicates how much utility an attribute's value has, given the decision maker's preferences. Utility functions are used to convert values with different units of measure to a single scale, which can then be summed across all measures to attain an overall alternative utility score. The concept is that the alternative that provides the greatest utility to the decision maker is the preferred choice.

A risk-based decision support system considers the decision maker's risk preference for each value. Risk preferences are generally categorized in one of the four shapes shown in Figure 3.

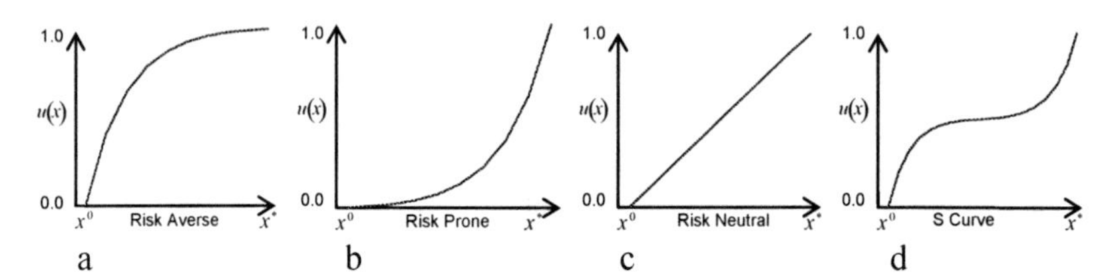

Figure 3. Utility curve shapes

The curve at Figure 3a shows a sharp increase in utility when risk is low, but less utility as it increases—indicating a risk-aversion. Figure 3b shows a risk-taking attitude, where more utility is gained when the risk is greater, while Figure 3c shows risk neutrality. Figure 3d indicates a change in risk tolerance—there is an aversion to risk at low levels, but a willingness to accept higher risks.

The idea of risk and utility is often discussed in terms of decisions involving risk and reward. Consider a choice between two lotteries. In the first, there is a 99%

chance of receiving $10 and a 1% chance of receiving nothing. In the second, there is a 60% chance of receiving $100 and a 40% chance of receiving nothing. A risk-averse person may choose the first option since there is a greater chance of getting something. A risk-taking person may choose the second, since the potential reward is greater. Someone with an S-curve preference may choose the first option initially, but if the reward were sufficiently great may choose the riskier option.

The weighting mechanism for the risk-based decision support system is derived from the risk taxonomy described earlier. Total system risk is aggregated in the top-level node. Each tier beneath it reflects the degree each element contributes to the higher level. When summed, the risks associated with the physical, logical, and environmental domains represent total system risk. Therefore, a local weight (LW) can be assigned to each of the three domains that when summed equals one, as shown in Figure 4. Individual weights must be elicited from key stakeholders. The same strategy applies to each sublevel progressively down the tree. A final, global weight (GW) value for each pathway is found by multiplying all local weights along a pathway. This process reveals the distribution of total risk from all 13 risk factors. The sum of all global weights will also be one.

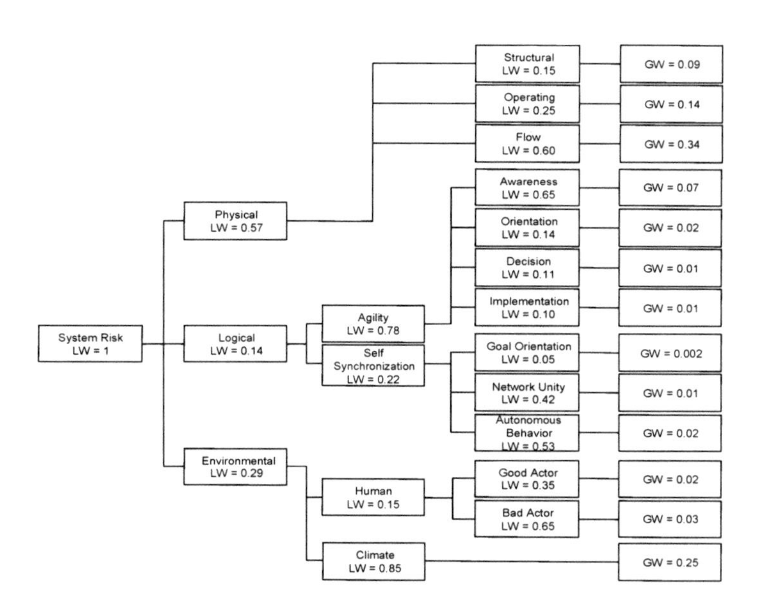

Figure 4. Risk taxonomy as a value hierarchy

3. ANALYSIS OF ALTERNATIVES

Quantifiable measures based on stakeholder values are developed for each of the lowest-level factors. Alternative solutions are then developed that reflect qualitative and quantitative stakeholder values. Alternatives are scored in each of the 13 risk areas based on empirical or simulation-based data.

Standard multiattribute utility (MAU) methods are then used to determine total utility scores for each alternative. Figure 5 illustrates the process for two possible alternatives: maintain level storage and automatically waiver levels outside of limits.

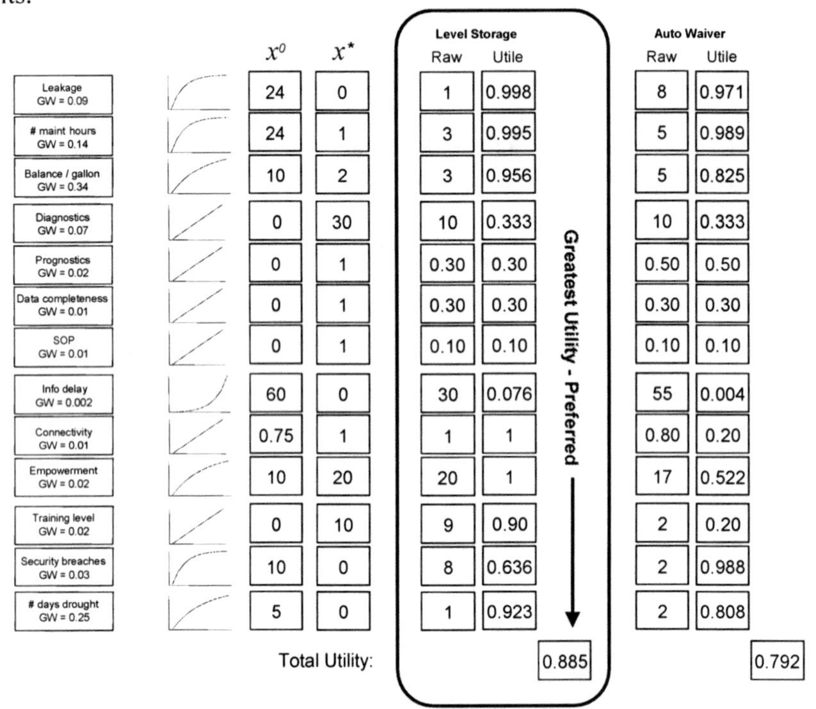

Measure	x^0	x^*	Level Storage Raw	Level Storage Utile	Auto Waiver Raw	Auto Waiver Utile
Leakage GW = 0.09	24	0	1	0.998	8	0.971
# maint hours GW = 0.14	24	1	3	0.995	5	0.989
Balance / gallon GW = 0.34	10	2	3	0.956	5	0.825
Diagnostics GW = 0.07	0	30	10	0.333	10	0.333
Prognostics GW = 0.02	0	1	0.30	0.30	0.50	0.50
Data completeness GW = 0.01	0	1	0.30	0.30	0.30	0.30
SOP GW = 0.01	0	1	0.10	0.10	0.10	0.10
Info delay GW = 0.002	60	0	30	0.076	55	0.004
Connectivity GW = 0.01	0.75	1	1	1	0.80	0.20
Empowerment GW = 0.02	10	20	20	1	17	0.522
Training level GW = 0.02	0	10	9	0.90	2	0.20
Security breaches GW = 0.03	10	0	8	0.636	2	0.988
# days drought GW = 0.25	5	0	1	0.923	2	0.808
Total Utility:				0.885		0.792

Figure 5. Multiattribute utility analysis of alternatives

Stakeholders determine the minimum acceptable threshold value (x^0) and the ideal value (x^*). The local utility of an individual score (called a utile) is determined by where the raw value falls along the utility curve for that measure. The utile value is then weighted by the risk factor weight to determine the weighted utility score for that measure and alternative. This is done for each measure, then the weighted utility scores are added to find the total utility of the alternative. The alternative that has the most utility for the decision maker should be the preferred choice.

Raw scores are found for each measure and alternative. A raw score of 1 is shown for the first measure. This is very near the desired value of 0 and therefore has a high degree of utility for the decision maker. Utiles are normally determined mathematically based on where they fall on the curve. In this case, the raw value is 99.8% of the way between the minimum acceptable threshold of 24 and the ideal value of 0, given the shape of the utility curve. Although this number is high, the weight assigned to that measure by stakeholders is low (0.09). The local utility of this measure, then, is also low (0.08982).

The total utility of the first alternative sums to 0.885, while that of the second is 0.792. This tells the decision maker that based on key stakeholders' risk assessment—derived from the risk taxonomy, their risk tolerance, and their minimally acceptable and ideal values—the level storage alternative provides the greatest overall utility and should be the preferred choice.

4. CONCLUSION

Water management decisions affect diverse and changing populations, and enhancing public welfare through the deliberate management of water resources is vital for every society. This paper presents a process for segmenting risk into a manageable set of factors that affect the operation of a system. This risk taxonomy provides a structure for assessing key stakeholder values to support management decisions.

This comprehensive risk assessment provides input for traditional multiattribute utility analysis whereby otherwise feasible alternatives are evaluated by the total utility they offer stakeholders. The product is a values-based decision support tool to design and management complex projects that can contribute to the success of a water management plan.

Acknowledgment: Partial financial support was provided by the NATO ARW to present this paper at the Istanbul meeting.

5. REFERENCES

1. Workgroup, Conowingo Pond (2006). Conowingo Pond Management Plan. Harrisburg, PA, Susquehanna River Basin Commission: 153.
2. Haimes, Yacov Y. (1998). Risk Modeling, Assessment and Management. New York, John Wiley & Sons, Inc.
3. Blanchard BS, Fabrycky WJ. (1998). Systems Engineering and Analysis. Upper Saddle River, NJ, Prentice Hall.
4. Sage AP. (1992). Systems Engineering. New York, John Wiley & Sons, Inc.
5. Army, United States (1998). FM 100-14: Risk Management. Washington DC.

6. Smith EA. (2001). "Network-Centric Warfare; What's the Point?" Naval War College Review Winter 2001.
7. West P. (2003). Dynamic Risk Management of Network-Centric Systems. Ann Arbor, MI, ProQuest.
8. Keeney RL, Raiffa H. (1976). Decisions with Multiple Objectives. New York, John Wiley & Sons.
9. Parnell G, West P. (2006). Systems Decision Process. Decision Making for Systems Engineering and Management. G. Parnell and P. Driscoll. New York, John Wiley and Sons.
10. Sage AP, Armstrong, JE(Jr). (2000). Introduction to Systems Engineering. New York, John Wiley & Sons, Inc.
11. Trainor T, Parnell G. (2006). Problem Definition. Decision Making for Systems Engineering and Management. G. Parnell and P. Driscoll. New York, John Wiley and Sons.
12. West P. (2006). Solution Design. Decision Making for Systems Engineering and Management. G. Parnell and P. Driscoll. New York, John Wiley and Sons.

SESSION 5. RISK ASSESSMENT METHODS: HAZARD IMPACT OF TREATMENT METHODS

ASSESSMENT OF THREE WASTEWATER TREATMENT PLANTS IN TURKEY

Orhan Gundaz and Celalettin Simsek
Department of Environmental Engineering and Department of Drilling, Dokuz Eylul University, Izmir, Turkey

Corresponding author: orhan.gunduz@deu.edu.tr

Abstract:

The techniques that involve such comparative assessments are especially convenient for water resources with different quality characteristics. The reuse of the treated effluent of wastewater treatment plants in irrigation is a commonly applied method in semiarid regions such as the Mediterranean where water is a scarce commodity. However, these effluents must be comparatively assessed with other sources of irrigation waters to evaluate the pros and cons of their application particularly in irrigating edible crops.

Keywords: Wastewater, irrigation, Irrigation water quality index, comparative assessment.

1. INTRODUCTION

The ever-increasing demand for water has directed the focus of water managers towards the potential reuse of domestic wastewater particularly in arid and semiarid areas such as the Mediterranean basin. Regardless of the level of treatment efficiencies achieved in advanced treatment plants, humans still have a psychological barrier towards reusing the treated wastewaters for domestic water supply. Nevertheless, these waters could be used effectively for agricultural and recreational irrigation purposes. Particularly, the treated effluents demonstrate ideal alternatives to freshwater resources for irrigating nonedible crops such as cotton and tobacco as well as watering green spaces in parks and recreational areas.

Regardless of the origin of the water used in irrigation, it is now widely accepted that it has to satisfy certain quality criteria such that it would not create problems with respect to the soil and the plant it is applied by Ayers and Westcot [1]. In this regard, the water used for irrigation is to be assessed with regards to a series of potential risks including salinity hazard, infiltration, and permeability hazard,

M.K. Zaidi (ed.), Wastewater Reuse–Risk Assessment, Decision-Making and Environmental Security, 159–167.

specific ion and trace element toxicity as well as some miscellaneous impacts on particular type of plants and crops [2]. These quality issues become particularly important when the source of irrigation water is the treated wastewater.

In the light of these fundamentals, this study focuses on the possibility of reusing the treated effluents of three different wastewater treatment plants of the City of Izmir, Turkey (i.e., Cigli, Guneybati, and Havza) that demonstrate distinct quality characteristics. The study implements the irrigation water quality (IWQ) index method and proposes a new approach for comparing the quality of the available alternatives from a quantitative point of view.

1.1. The City of Izmir and Wastewater Administration

The City of Izmir is one of the largest metropolitan areas of Turkey with a total population of about 3.5 million [3]. It is also a busy commercial and industrial center and a gateway between the Aegean and Central Anatolia. The city is located on the Aegean coastline and is situated at the tip of the bay that bears its name (Figure 1). It is under the influence of typical Mediterranean climate with hot, dry summers and warm, wet winters. Being situated in a semiarid region, the city of Izmir receives an average of 682 mm precipitation that falls mostly between December and February and 1393 mm evaporation that typically occurs between May and October according to the 26 year long records (1979–2005). The negative surface water budget is compensated by groundwater reserves to serve the domestic water demands of the city.

The Izmir metropolitan area is mainly situated along the coastline of the Bay of Izmir and demonstrates unique characteristics with respect to wastewater collection and sewerage. Currently, three wastewater treatment plant (WWTP) serve the metro area and its neighborhoods (Figure 1).

The Cigli WWTP is located to the northern parts of the city and is the largest of all three. It serves more than 80% of the city and is also amongst the largest plants in Turkey. The Guneybati WWTP is a small-to-medium sized unit that serves the southwestern suburbs of the city. The recently constructed Havza WWTP, on the other hand, is outside the city limits and is located to the south of the town. This plant primarily serves the District of Menderes and its immediate vicinity. It should be stressed that all plants implements biological treatment with nutrient removal. Currently, the treated effluents of Cigli and Guneybati plants are discharged to Izmir Bay, whereas that of Havza plant is disposed to a drainage channel, which eventually discharges to Kucuk Menderes River. All three plants are operated by the Izmir Water and Sewerage Administration (IZSU) of the Izmir Greater City Municipality.

2. CIGLI WASTEWATER TREATMENT PLANT

The Cigli WWTP is the largest treatment facility that serves the majority of the Izmir metropolitan area. The wastewaters originating from Izmir metropolitan area are collected and transferred via an interceptor canal that runs parallel to the shoreline and spans the entire inner bay. This canal carries the wastewaters originating from a large area including Balcova–Uckuyular, Konak, Gaziemir, Buca, Bornova, and Karsiyaka–Cigli neighborhoods as shown in Figure 1.

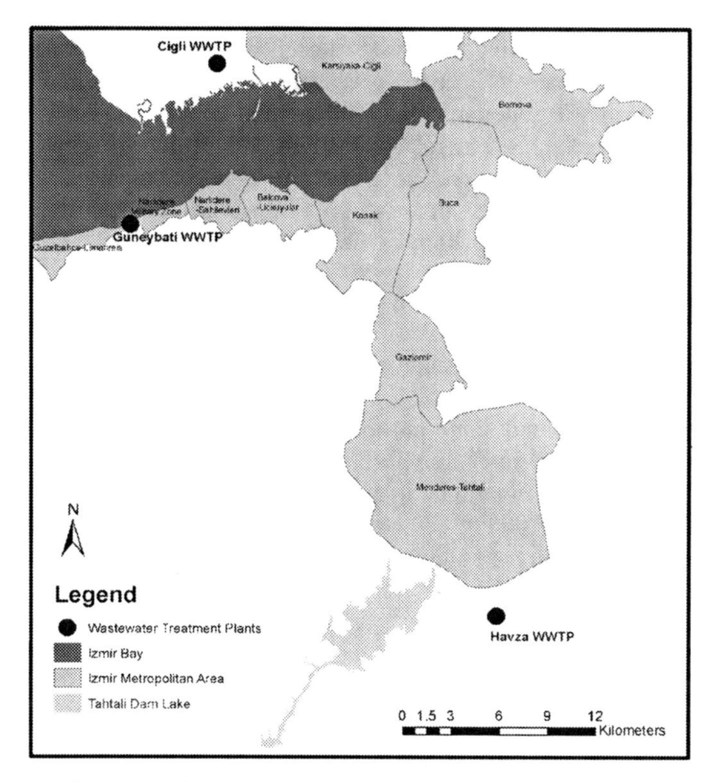

Figure 1. Izmir metropolitan area and wastewater treatment plants

The plant not only receives about 80% of all the domestic wastes of about 3.5 million population but also the pretreated wastewaters of numerous industrial establishments situated within the city [3]. The Cigli WWTP is constructed at the end of the main interceptor canal and is situated on the old Gediz River delta located to the north of the Bay of Izmir. It covers a total area of 300,000 m^2. The plant implements activated sludge biological treatment system with advanced phosphorus removal to treat an average flow rate of 7 m^3/s under dry weather conditions. The biological treatment unit is a combination of anaerobic biophosphorus tanks and aerobic oxidation tanks. The biophosphorus tanks decompose phosphorous compounds and convert them to biologically consumable forms. The oxidation tanks are then used to remove the phosphorus as biomass together with the organic matter. The oxidation tanks contain oxic and anoxic zones to allow nitrification–denitrification processes for further removal of nitrogen from wastewater. The treated effluents of the plant are discharged to the bay by a 2.4 km-long open channel.

The interceptor canal carrying the wastewater to the plant mostly operates under atmospheric pressure except for a short pressurized section. The canal follows the shoreline and primarily operates under submerged conditions. This situation is the main reason for high electrical conductivity values due to seawater seeping to the line from failing pipes and improper pipe connections along gravity flow sections of the canal. The high salt content of the wastewater negatively influences its reuse potential in irrigation within the nearby Menemen Plain [3].

2.1. Guneybati Wastewater Treatment Plant

The Guneybati WWTP is a small-to-medium sized facility that serves the southwestern suburbs of the city including the Narlidere military zone, Guzelbahce–Limanreis and some portions of Narlidere–Sahilevleri and Balcova neighborhoods as shown in Figure 1. The plant is the second treatment facility that serves the Izmir metropolitan area. It has a total capacity of 21,600 m^3/day and covers an area of 15,000 m^2. The plant implements biological treatment with nutrient removal by using a combination of anaerobic biophosphorus tanks and aerobic oxidation tanks. The Guneybati facility performs organic carbon and nitrogen removal via biological methods and phosphorus removal via chemical and biological methods. The treated effluents are discharged to the Bay of Izmir through a 600 m-long deep sea outfall.

Similar to the Cigli plant, the Guneybati plant is also fed by an interceptor channel that runs along the shore line. This canal also runs below the mean sea level along certain portions of the line and significant amounts of seawater seeps into the line from failing pipes and improper pipe connections. Thus, the plant receives an influent with high salt content. This characteristic of the influent sometimes reaches to such high levels that its corrosive nature influences the normal operation of the mechanical equipment in the plant.

2.2. Havza Wastewater Treatment Plant

The Havza WWTP is also a small-to-medium sized facility designed to serve the southern suburbs of the City of Izmir as well as the District of Menderes and its vicinity.

Menderes city is located within the watershed of Tahtali Dam which is currently the primary water resource of Izmir metropolitan area (Figure 1). The wastewaters originating from these areas are transferred out of the Tahtali watershed and are diverted to Havza plant. Thus, this treatment facility is extremely important to protect the water quality in Tahtali Lake. With respect to the treatment units selected and the general layout implemented, Havza plant is designed identical to the Guneybati facility and has the same total capacity of 21,600 m^3/day.

The Havza plant is located in the middle of fertile agricultural fields and its effluents could potentially be used in irrigating these fields. Currently, the biologically treated effluents of the plant are UV-disinfected and ultimately disposed to a drainage channel in the plain. A major portion of these effluents infiltrates into the ground and recharge the groundwater reservoir underneath the plain. Presently, no official reuse plan is effective in this plain regarding the effluents of Havza plant.

3. METHODOLOGY

This study implements the Irrigation Water Quality (IWQ) Index method that was originally developed by Simsek and Gunduz [2] based on the quality criteria laid out by Ayers and Westcot [1] and Crook [4]. It lies on the principle of linearly incorporating the influence of numerous water quality parameters that are responsible for several hazards of importance in irrigation waters. The technique simultaneously includes five hazard groups associated with irrigation water quality (i.e., salinity hazard, infiltration and permeability hazard, specific ion toxicity, trace element toxicity, and some miscellaneous effects). The index computation is performed by using the formulae:

$$\text{IWQ Index} = w_1 r_1 + w_2 r_2 + \frac{w_3}{3}\sum_{i=1}^{3} r_i + \frac{w_4}{N}\sum_{j=1}^{N} r_j + \frac{w_5}{3}\sum_{k=1}^{3} r_k \quad (1)$$

where w_i and r_j stand for the weighing coefficients of the particular hazard group and the suitability ratings of quality parameters and N is the total number of trace elements included in the analysis. The classification ranges of the quality parameters and the associated ratings are in Simsek and Gunduz [2]. Once the final index value is computed according to 1, it is used to assess the ultimate suitability of the water resource. This assessment is performed based on the values given in Table 1.

In the table, low suitability waters are regarded as a resource which should normally not be used in irrigation unless no other alternative is available and resistant crops are cultivated on the field. On the other hand, highly suitable waters are considered to be convenient for irrigation. Still, the final decision should be done based on detailed analysis with regards to the particular soil type and crop pattern that these waters would be applied to.

Table 1. Final assessment based on IWQ index method [2]

IWQ Index	Suitability of water for irrigation
< 22	Low
22 – 37	Medium
> 37	High

4. RESULTS AND DISCUSSIONS

The technique discussed in the previous section is implemented to assess the suitability of wastewater treatment plant effluents serving the City of Izmir for irrigation purposes. The results of these analyses are presented in Table 2.

The instantaneous samples were collected from Cigli, Guneybati, and Havza plant influents and effluents are analyzed for the quality parameters of the IWQ index technique. Of the standard physicochemical parameters, pH and EC are measured on the field by portable devices whereas bicarbonate, nitrate–nitrogen, and chloride are measured in Dokuz Eylul University labs using standard analytic and spectrometric methods. The analysis of sodium, calcium, magnesium, boron, and trace elements are measured in Canadian ACME laboratories by ICP-MS method.

The results clearly demonstrate the influence of seawater seepage into the collector lines of Cigli and Guneybati WWTPs. The electrical conductivity values of the influent reaching these units are 5 and 15 times greater than the electrical conductivity value of the influent reaching the Havza plant, respectively. The extremely high concentrations of sodium, magnesium, manganese, and chloride ions further verify this condition. Consequently, the computed specific absorption rate (SAR) of Cigli and Guneybati effluents are 18.4 and 38.0, respectively, that make these waters extremely problematic as an irrigation water resource. On the other hand, the effluents of Havza plant are low in all of the parameters mentioned above as this plant receives wastewater from the hinterland of the City of Izmir away from the bay.

Based on the values given in Table 2, the IWQ index values are computed and presented in Table 3. Accordingly, the Cigli and Guneybati WWTP effluents got an index value of 29 where as the Havza WWTP effluents received an index score of 37. Therefore, it could be concluded that Cigli and Guneybati effluents demonstrate medium suitability, which correspond to water that should be used with caution and upon further study.

Table 2. A comparison of influent and effluent water quality in three wastewater treatment plants of Izmir City

Quality parameter	Cigli WWTP		Guneybati WWTP		Havza WWTP	
	Influent	Effluent	Influent	Effluent	Influent	Effluent
pH	7.29	7.06	7.27	7.00	7.65	7.62
Temp. (°C)	25.6	26.4	23.4	24.4	23.3	25.6
EC (µS/cm)	8690	7920	23200	19780	1613	1491
Salinity (‰)	4.9	4.4	14.0	11.8	0.6	0.6
Cl (mg/l)	2579.9	2399.9	15699.6	15399.6	230.0	250.0
HCO_3 (mg/l)	1094	886	1278	922	1008	856
NO_3–N (mg/l)	6.5	17.0	20.0	13.5	10.5	14.5
Ca (mg/l)	123.0	118.2	239.9	208.9	90.1	91.4
K (mg/l)	65.7	62.7	168.0	142.6	72.4	72.2
Mg (mg/l)	159.4	152.1	443.4	376.0	29.0	26.5
Na (mg/l)	1368.1	1284.6	4632.1	3976.5	143.5	141.9
SAR	19.2	18.4	41.0	38.0	3.4	3.4
Al (µg/l)	26	24	<10	<10	22	11
As (µg/l)	21.6	22	19.6	18.2	15.2	7.7
B (µg/l)	890	1045	1315	1231	590	539
Be (µg/l)	<0.5	<0.5	<0.5	<0.5	<0.05	<0.05
Cd (µg/l)	<0.5	<0.5	<0.5	<0.5	0.44	0.11
Co (µg/l)	0.91	0.48	0.38	0.37	0.55	0.35
Cr (µg/l)	39.9	16.7	<5	<5	7	2.1
Cu (µg/l)	12.5	12.3	26.3	23.5	6.2	8.9
Fe (µg/l)	209	<100	107	<100	210	17
Li (µg/l)	37.1	42.5	58.4	51.3	21	18.4
Mn (µg/l)	166.3	32.0	351.0	323.6	74.6	46.4
Mo (µg/l)	3.6	2.2	2.7	2.8	0.6	1
Ni (µg/l)	41.9	26.2	<2	<2	8.9	12.3
Pb (µg/l)	2.9	1.4	<1	<1	1.1	1
Sb (µg/l)	0.54	1.39	<.5	0.52	0.36	1.4
Se (µg/l)	22.9	21.1	76.2	60	1.3	0.9
V (µg/l)	9.9	13.6	35.9	35.7	3.1	4.1
Zn (µg/l)	105.1	89.2	31	95.6	196.7	82.9

On the other hand, Havza plant effluent has obtained an index score of 37 that makes it highly suitable for use in irrigation purposes. Particularly, the UV disinfection process implemented in this facility as the final treatment unit makes its effluents especially convenient even for use in the irrigation of edible crops that are widely cultivated in Torbali plain where this plant is situated.

Table 3. Suitability of effluent waters of Izmir WWTPs for irrigation

Water resource	IWQ index	Suitability of water for irrigation
Cigli	29	Medium
Guneybati	29	Medium
Havza	37	High

5. CONCLUSION

This study provides the details of a comparative assessment of the effluent water quality in the three wastewater treatment plants currently serving the City of Izmir, Turkey. The assessment was conducted to achieve a qualitative evaluation of the effluent characteristics of the plants for their possible reuse in irrigation. An index technique which was originally developed for characterizing the ambient groundwater quality that is used for irrigation purposes is applied to the effluent water of the treatment plants. According to the results of the analysis, it has been found out that the Havza WWTP effluents are comparably more convenient for irrigating the agricultural lands in the nearby Torbali Plain. The effluents of Cigli and Guneybati plants, however, were found to have medium suitability for irrigation. These effluents have very high electrical conductivity levels due to seawater intrusion to the sewerage system from the Bay of Izmir. This phenomenon causes elevated levels of certain parameters, which make the water fairly unsuitable for irrigation purposes.

Acknowledgment: NATO provided financial support to present the paper at the ARW – Wastewater Reuse – Risk Assessment, Decision making and Environmental Security held in Istanbul, Turkey, during October 12–16, 2006. The authors would also like to express their gratitude to the Izmir Water and Sewerage Administration for its support throughout this study.

6. REFERENCES

1. Ayers RS and Westcot DW. 1985, Water Quality for Agriculture, FAO Irrigation and Drainage Paper No. 29, Rev. 1, U. N. FAO, Rome.
2. Simsek C and Gunduz O. 2006, IWQ Index: a GIS-integrated technique to assess irrigation water quality, Environmental Monitoring and Assessment.

3. Gunduz O, Turkman A and Doganlar DU. 2005, Alternative formulations for the reuse of treated wastewater in Menemen Plain Irrigation Scheme. Integrated Urban Water Resources Management, edited by P. Hlavinek, pp. 259–268.
4. Crook J. 1996, Chapter 21: Water Reclamation and Reuse, LW. Mays (ed.), Water Resources Handbook, McGraw Hill, USA. 21.1–21.36.

POLLUTION POTENTIAL OF THE SHALLOW AQUIFERS IN JORDAN

Nizar Al-Halasah and Bashaar Y. Ammary
Royal Scientific Society, Amman, 11941, Jordan
Balqa Applied University, Huson College, Huson, 21510, Jordan

Corresponding author: halasah@rss.gov.jo

Abstract:

Eight wells were monitored for a period of seven years. These wells were located upstream, under direct influence of AWSP, along the watercourse where AWSP effluent is discharged. Wells located under the direct influence have shown higher concentrations of pollutants in comparison with the other monitored wells. Comparison of the water quality of the wells with the AWSP influent and effluent quality has shown that AWSP is not the main source of pollution. Additionally, agricultural activities and overpumping practices are the major potential sources of pollution.

Keywords: Groundwater, wastewater treatment, waste stabilization ponds, wastewater reuse, quality.

1. INTRODUCTION

As-Samra waste stabilization ponds (AWSP) were constructed in 1985 to serve the cities of Amman, Zarka, and Ruseifaa. It is located about 40 km northeast of Amman and is considered as one of the largest wastewater stabilization ponds in the world. The system of treatment consists of three parallel trains, each train consists of a series of two anaerobic, four facultative, and four maturation ponds. The effluent from the maturation ponds of the three trains is collected in a channel, then discharged into Wadi Dhuleil, which joins Zarka River and flows to King Talal reservoir about 40 km and subsequently used for irrigation in the Jordan Valley area. The design of surface area of AWSP is such that the ponds occupy 181 ha of land with an effective volume of 2.8 million m^3. AWSP currently treats around 77% of the domestic wastewater in Jordan [1].

AWSP is located above one of the most important groundwater aquifers in Jordan, the Amman–Zarqa basin. This basin is one of the main sources of domestic water supply for Amman and Zarka cities, where more than half of the population of Jordan reside. The hydrogeology of the basin is controlled by the geological setup, which also controls the piezometry, occurrence and movement of groundwater, and

M.K. Zaidi (ed.), Wastewater Reuse–Risk Assessment, Decision-Making and Environmental Security, 169–174.

the distribution of productive areas in the aquifers region. Four major aquifer units were identified in the basin, Amman-Wadi Sir (B2/A7) is the upper one, Hummar (A4), Naur (A1/2), and Kurnub sandstone (K) is the lower one [2]

The Amman-Wadi Sir aquifer (B2/A7), which is known as the upper aquifer, is the most utilized and important aquifer in the Zarqa River Basin. It is a carbonate aquifer overlain with wadi fill deposits (sands and gravels) along the Zarqa River. All these formation are in hydraulic connection and thus constitute a single aquifer. This highly fractured, carbonate aquifer is well jointed and fissured with solution channels and karstic features. In general, the intensity of weathering and fracturing of the aquifer increases in the vicinity of faults. The average thickness of the aquifer is around 175 m in Amman and Ruseifa areas. The variation in thickness of the Amman-Wadi Sir aquifer is well known within the Zarqa River but not known elsewhere in the study area. Based on data used for groundwater modeling of the B2/A7 aquifer (conducted by BGR (1997): Groundwater resources of southern Jordan report, Technical Cooperation Projects, Federal Institute for Geosciences and Natural Resources. Hannover, Water Authority of Jordan), the hydraulic conductivity of the B2/A7 aquifer ranges between 1.3×10^{-4} and 7×10^{-3} m/s, which equivalent to that of fine sand. Vertical hydraulic conductivity ranges between 2×10^{-6} and 1×10^{-4} m/s. Specific storage was determined to be 10^{-5} /m and the specific yield is 0.05 or 5%. The specific yield can be used as an estimate for effective porosity of the aquifer. The A4 and A1/2 are located beneath this aquifer and are separated from each other by small beds of clay materials. The lower aquifer is the Kurnub (K) sandstone which crops at the lower parts of the basin at King Talal dam (KTD) [3].

The water table at AWSP area is close to the surface. As mentioned above, the upper aquifer is unconfined and consists in its upper portions of highly permeable wadi deposits. The surface water feeds the underground water. Consequently, the water table has risen since the construction of AWSP by some of 20 m in the immediate surroundings of the treatment plant, as a result of the formation of a recharge mouth. This rise dies off gradually in a radial manner. The groundwater quality along Zarka River has greatly deteriorated due to this recharge with polluted water from AWSP effluent and from the seepage of irrigation water used in the area; it is thus no longer suitable for most use [4].

The inflow of AWSP exceeded design flow of 68,000 m^3/day in 1987, with BOD_5 effluent concentration around (120 mg/l). By 1990 however, both hydraulic flow and organic loading were much higher than the design values, the hydraulic load reached around (96,000 m^3/day). The BOD_5 concentration in the effluent of the plant exceeded (105 mg/l) [5].

In 2000, the hydraulic load reached around (170,000 m^3/day) and BOD_5 concentration in the effluent of the plant around (106 mg/l). The average number of thermotolerant feacal coliform leaving the plant were around (20,000 MPN/100 ml) [1].

Jordan is about 90,000 km^2 in area, and lies among the dry and semidry climatic zones which are characterized by their minimal rainfall and high percentage of evaporation. Its climate is a mix of Mediterranean and dry desert climate. The temperature varies from a few degrees below zero in the winter to around 46°C in the summer season. Annual precipitation ranges from 50 mm in the desert to 600 mm in the northwest highlands. Only 9% of Jordan's area receives more than 200 mm of the rainfall annually. Approximately 92.2% of the rainfall evaporates, 5.4% recharges the groundwater and the rest 2.4% goes to the surface water.

Recently, the problem of water shortage in Jordan has been exacerbated as a result of high natural population growth, influxes of refugees and returnees to the country in response to political situation in the Middle East area, rural to urban migration and increased modernization and higher standards of living. Consequently, Jordan is facing a future of very limited water resources among the lowest in the world on a per capita basis. Available water resources per capita are decreasing as a result of population growth, they are projected to decline from more than 160 m^3 per capita per year for all uses at present to only 91 m^3 per capita per year by 2025, putting Jordan in the category of having an absolute water shortage [1].

2. MATERIALS AND METHODS

Monitoring was conducted by taking a sample from each well every 2 months for 7 years. All analyses were conducted according to standard methods [6]. Sulfate and nitrate concentrations were measured using ion chromatography method. Total and fecal coliform counts were conducted using the most probable method (multiple tubes). Boron was measured using inductivity coupled plasma method.

3. RESULTS AND DISCUSSION

The piezometric level at the region of AWSP for the upper aquifer, indicates that the water level is around (505 m.above.sea.level) while the altitude of the AWSP varies between (550–575 m.a.s.l), so the water depth in the study area varies between 45 and 70 m. The movement of the groundwater is from the east to the west which coincides with the surface drainage of Wadi Dhuleil [5].

To assess the impact of AWSP on the shallow aquifer and the water quality, a monitoring program study has started in 1994. Eight wells were monitored in this study; well 1 was chosen as a reference point for the water quality in Wadi Dhuleil basin, it is located around 40 km northeast of AWSP, well 2 is located 16 km upstream east of AWSP. Wells 3, 4, and 5 are located nearby Wadi Dhuleil and downstream AWSP. Moreover, wells 6, 7, and 8 are located around (8 km) southwest AWSP as shown in Figure 1.

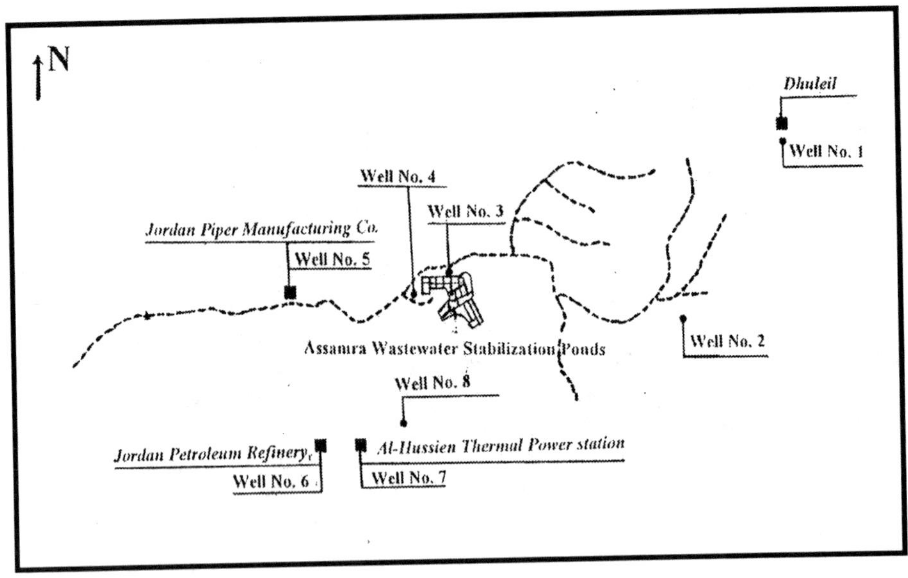

Figure 1. Locations of the monitored wells

Yearly average results of analyses of the monitoring program are discussed below: the data collected address the parameters: total dissolved solids (TDS), sulfate, nitrate, total hardness (TH), boron, and total coliform counts (TCC), and thermo-tolerant coliform Counts (TTCC).

Wells 3 and 4 that are under direct AWSP infiltration effect have much higher levels of TDS, SO_4, NO_3, TH, and B than all other monitored wells. The concentration of TDS, SO_4, and hardness in these wells, however, are much higher than the concentration of these parameters in AWSP influent or effluent. The infiltration of water from AWSP to these wells should have a beneficial effect on their water quality. In fact, the quality of well number 3 has improved (TDS has decreased from 9570 in 1994 to 5343 in 2001, for example) during the monitoring period. This may be due to infiltration from AWSP, as this well has not been in use for a number of years due to its unacceptable water quality. This gives a clear indication that AWSP has not and is not the main source of contamination for these wells.

The levels of TDS, SO_4, NO_3, and TH for well 1, which is chosen as a reference point for the water quality in Wadi Dhuleil basin, were less than their levels in all other monitored wells at the beginning of the monitoring program. However, their concentrations have increased dramatically during the monitoring period. For example, TDS, NO_3, and hardness concentrations have more than tripled during the monitoring period (1994–2001), and became higher than the other monitored wells, except wells 3 and 4. Most of the increases have occurred after 1997, the year where the concentrations of TDS, NO_3, and TH have shown a sudden increase. This is another proof that AWSP is not the main source of pollution as well number 1 is located 40 km upstream of AWSP, and it is unlikely that it is receiving seepage water from AWSP. Due to its good quality, the well has been used extensively for agricultural activities in the surrounding area. These two reasons (over pumping and agricultural activities) should be the sources of pollution for this well.

Wells 6, 7, and 8 located in an industrial zone with little agricultural activities have shown a gradual increase in TDS, SO_4, and hardness concentrations. This increase is mainly due to over pumping. Unlike well 1, 6, 7, and 8 have not shown an increase in NO_3 concentration due to the low agricultural activities in this region.

Well 5 located along Wadi Dhuleil has similar concentrations of TDS, SO_4, NO_3, TH, and B to wells 6, 7, and 8. It has much lower concentrations than Wells 3 and 4. At the end of the monitoring period it has even lower TDS, NO_3, and hardness concentrations than well 1. This suggests that well 5 is not greatly affected by the passage of AWSP effluent in Wadi Dhuleil.

Boron concentrations in the monitored wells have not shown any sign of change during the monitoring period. Boron concentrations in the wells under direct infiltration of AWSP have the highest boron concentrations among the monitored wells. Boron concentration in AWSP influent and effluent are around 0.6 mg/l during the study period, much lower than boron concentration in wells 3 and 4.

The high nitrate concentrations in wells 3 and 4 and the almost similar nitrate concentrations in the rest of the monitored wells suggest that wells 3 and 4 are the only wells affected by seepage from AWSP. Well 1 has shown a steady increase in NO_3 concentration due to the high agricultural activities in the region. The wells 3, 4, 5 which are located nearby Wadi Dhuleil, showed that they have microbiological pollution TCC and TTCC higher than the other wells, this is mainly due to the wastewater flow in that Wadi. The well 2 has not shown any significant change in all the monitored parameters during the study period. Except for boron, it has the lowest concentration of the tested parameters at the end of the study period. It has the second lowest (after well 1) boron concentration.

4. CONCLUSION

This paper shows that although AWSP has contributed to the pollution of the aquifer, it is not the main pollution source. It was noticed that the wells adjacent to Wadi Dhuleil contains higher concentrations from TDS, SO_4, NO_3, TH, and B, in addition to TCC and TTCC from the wells that located upstream and south-east the plant. The quality of the wells adjacent to AWSP is much worse than the quality of AWSP in a number of parameters. This is a clear indication that other sources of pollution are causing this deterioration in the wells' quality. Over-pumping, agricultural, and industrial activities may be responsible for such deterioration. In addition, the data shows that microbial contamination in the AWSP has been greatly attenuated through the soil before reaching the wells.

Acknowledgment: I am thankful to the NATO ARW committee to award financial support to attend the meeting and present this paper.

5. REFERENCES

1. Al-Halasah N and Mashakbeh O. 2004. "Water Quality Monitoring in Assamra WSP and Zarka River and Its Impact on the Groundwater", Environmental Research Centre, Royal Scientific Society, annual report (1994–2003).
2. Arabtech Consulting Engineers and CES Consulting Engineers (Salzgitter GmbH), "Review of Water Resources Development and Use in Jordan", 1993.
3. Ministry of Water and Irrigation, 1998. Water Quality Improvement and Conservation Project, Ruseifa Area Water Quality Study Training Model, 1998.
4. Rimawi O, Awwad M and Shatanawi M. 1993. "Effect of Unintended Artificial Recharge on the Groundwater Quality of the Shallow Aquifer in Vicinity of Khirbet As Samra Stabilization Ponds", Water and Environmental Research and Study Centre, University of Jordan.
5. Ministry of Water and Irrigation, Open Files. Ministry of Water and Irrigation, 2001. "Jordan's Water Budget for the years 1999 and 2000". MWI, 1998a, Groundwater Management Policy. Ministry of Water and Irrigation, Jordan.
6. ABHP 1992, www.abhp.org

EFFICIENCY OF OXIDATION PONDS FOR WASTEWATER TREATMENT IN EGYPT

Hussein I. Abdel-Shafy and Mohammed A. M. Salem
Water Research & Pollution Control Dept., National Research Centre, Cairo, Egypt
Serabium Wastewater Treatment Plant, Ismailia, Egypt

Corresponding author: husseinshafy@yahoo.com

Abstract:

Oxidation pond techniques have become very popular with small communities because of their low construction and operating cost and offer a significant financial advantage over other recognized treatment methods. However, in the arid and semiarid regions the salination due to the evaporation is one of the main problems. The present year long study evaluates the use of aerated lagoons system for the treatment of municipal wastewater in Ismailia, Egypt. The treated effluent is presently used for irrigating lumber trees forest around the wastewater treatment plant on the sandy desert soil.

Keywords: Wastewater, treatment, oxidation ponds, aerated lagoons, heavy metals.

1. INTRODUCTION

Employing oxidation pond technique is world wide used for wastewater treatment. Ponds have become very popular with small communities because their low construction and operating costs offer a significant financial advantage over other recognized treatment methods [1]. One of the most significant advantages is the simplicity in construction and operation [1, 2]. Ponds are also used extensively for the treatment of wastewater that is amenable to biological treatment. These ponds are usually classified according to the nature of biological activity that is taking place such as: aerobic, aerobic–anaerobic and anaerobic. However, the disadvantage is that the biomass concentration is relatively low (25–50 mg/l) compared with activated sludge process (3000–5000 mg/l) [3]. In addition, the specific volume reaction rate is low; therefore larger areas are necessary for sufficient effluent quality.

Aerated lagoons treatment system was developed from the traditional waste stabilization ponds (WSPs), where mechanical aeration was installed to increase the oxygen supply inside the ponds system. It was recommended that WSPs be placed

M.K. Zaidi (ed.), Wastewater Reuse–Risk Assessment, Decision-Making and Environmental Security, 175–184.

after the aerated lagoon to permit the microbial solids to settle and be stabilized. The inorganic elements remaining will stimulate the algae to grow up in the WSPs. The biochemical oxygen demand (BOD) and total suspended solids (TSS) in the effluent will be determined by the algae and not by the microbial solids discharged from the aerated lagoon. It was concluded that a combination of 24 h aerated lagoon and WSP has an area approximately 40% of the area of a traditional WSPs for treating the same volume of sewage. The aerated lagoons have more advantages than the traditional WSPs, in terms of eliminating odor nuisance and less land requirement [4]. In contrast, the aerated lagoons are little more costly than the traditional WSPs and lower in performance, with respect to pathogen removal [5, 6].

2. MATERIALS AND METHODS

2.1. Plant Location and Design Description

The plant located 15 km to the south of Ismailia city. It is in operation since 1996 to serve 450,000 population. The capacity of the plant is to treat 90,000 m^3/day wastewater. The maximum receiving capacity is 180,000 m^3/day. It is designed as a secondary treatment plant on an area of 860 feddan (about 361 ha) as an aerated oxidation ponds. The plant is equipped with flow meter, three mechanical screens to retain solids of more than 2.2 cm, then two sandgrit removal lines in parallel series. Each line is designed to receive 45,000 m^3/day. Each one consists of an aerated lagoon, an aerated facultative lagoon, and a polishing lagoon (Table 1). The bottom and walls of the lagoons are sealed with clay and a synthetic liner to prevent any leakage of sewage to the groundwater [7].

Table 1. Design description of the aerated lagoons in Ismaila

Lagoons	HRT (d)	Volume (m^3) of each	Dimensions (m) of each	No. of aerators
Aerated (A1, A2)	1	45,450	70 × 114 × 4.3	Dual speed
A. Facultative (F1, F2)	5	225,360	400 × 114 × 4.3	Dual speed
Polishing (P1, P2)	5	225,150	490 × 118 × 3.5	–

The aerated lagoon was designed as a complete mixing reactor, where all contents of the lagoon were held in suspension. Operating 10 aerators with homogeneous distribution through the lagoon was enough to reach the conditions of complete mixing and full aeration. Dissolved oxygen (DO) contents were kept at the range of 3–5 mg/l. The aerated facultative lagoon was designed as an incomplete mixing reactor. The aerators are used to transfer the oxygen demand but are too small to keep the solids in suspension. The net effect is for all heavy solids to settle out in the lagoon. Only the dispersed microbes are maintained in suspension with the soluble organics. The organic matter that settles to the bottom of the facultative lagoon undergoes anaerobic digestion and slowly releases organics into solution. Operating of four aerators (one row) or eight aerators (two alternative rows) was enough to reach the conditions of incomplete mixing. The DO was kept at the range of 1–3 mg/l. The polishing lagoon was designed with hydraulic retention time (HRT) of 5 days; its main function is to remove pathogens. The algae and surface air are responsible for increasing DO content in the lagoon. Effluent structures with control weirs are provided at the end of the aerated facultative and polishing lagoons to collect sewage from the surface layers and to maintain the water depth. The final effluent is discharged by gravity to El-Mahsama drain.

2.2. Physicochemical, Bacteriological, and Parasitological Analyzes

The study was carried out for a period of one year. All physical, chemical, and bacteriological analyses were determined according to APHA (2005) [8]. The sampling regime was conducted to included 24 h weighed composite and grab samples. The parasitological analysis was carried out according to Bailenger technique as recommended by WHO [9]. The microscopic identification was counted in a McMaster counting cell.

3. RESULTS AND DISCUSSION

3.1. Plant Operation

During the present study the influent raw sewage was an average of 100, 4,230 m^3/day, ranged from 79,607 to 135,644 m^3/day. The HRT was 8.5 day and the range was (7–11) days. During this period, two aerated lagoons (A1, A2), one aerated facultative lagoon (F1 or F2) and mostly one polishing lagoon (P1 or P2) were in service. The characteristics of raw sewage and the different effluents are shown in Table 2. The DO was increased through the system and the mean was 7.39 mg/l in the effluent. The total dissolved solids (TDS) contents were also increased in the effluent, due to the high evaporation rate along the lagoons surface (Table 3).

3.2. Wastewater Treatment Efficiency (BOD_5, COD, VSS, and TSS)

In the aerated lagoons (AL), organic load depletion rates were 22% of BOD_5, 3.3% of chemical oxygen demand (COD), 3.2% of volatile suspended solids (VSS) and 0.55% of TSS (Table 4). The present removal efficiency is lower than

Table 2. Physico-chemical characteristics of the raw sewage and effluents of the AL system in Ismailia (Mean ± SD and range)

Locations	Temp °C	D.O mg/l	pH	T. alkalinity mg/l	TDS mg/l	NO_3-N mg/l	PO_4-P mg/l	Chlorophyll A mg/l
RW	23 ± 5	0.4 ± 0.10	7.7 ± 0.13	274 ± 10	588 ±152	1.5 ± 1.2	2.9 ± 0.7	–
	(15 – 34.8)	(0.28 – 0.70)	(7.4 – 7.9)	(255 – 302)	(400 – 970)	(0 – 6)	(1.4 – 3.9)	
AL. eff.	25 ± 3.3	2.89 ± 0.47	8 ± 0.14	283 ± 17.3			3 ± 0.4	
	(19.7 – 30.8)	(2.21 – 4.11)	(7.5 – 8)	(245 – 306)			(2.4 – 4.3)	
A. FL. eff	25 ± 3.8	2.89 ± 1.02	8 ± 0.1	309 ± 17.9			5 ± 0.6	0.033 ± 0.035
	(18.1 – 31.3)	(1.4 – 4.87)	(7.5 – 7.9)	(284 – 336)			(3.8 – 6.4)	(0.005 – 0.133)
PL. eff	25 ± 4.1	4.47 ± 1.32	8 ± 0.12	309 ± 20.7			5 ± 0.6	0.11 ± 0.09
	(17.6 – 33.5)	(2.8 – 7.4)	(7.47 – 8)	(278 – 341)			(3.6 – 5.9)	(0.02 – 0.46)
Final	24.3 ± 4.87	7.44 ± 0.39	7.9 ± 0.1	305 ± 18	653 ± 161	13.4 ± 4.6	4.8 ± 0.5	0.14 ± 0.10
effluent	(15.2 – 33.5)	(6.69 – 8.1)	(7.6 – 8.1)	(277 – 331)	(505 – 1080)	(0.0 – 22)	(4.2 – 5.8)	(0.03 – 0.46)

RW: Raw Sewage, AL. eff.: Aerated Lagoon effluent, A. FL. eff.: Aerated Facultative Lagoon effluent, PL. eff.: Polishing Lagoon effluent

Table 3. Mineral salts in the raw sewage and final effluent of the AL system in Ismailia (Mean ± standard deviation and range)

Locations	Cl^- mg/l	CO_3^- mg/l	HCO_3^- mg/l	Mg^{++} mg/l	Ca^{++} mg/l	Na^+ mg/l	K^+ mg/l
Raw sewage	187 ± 43	0.0 ± 0.0	396 ± 28	38 ± 12	62 ±15	134 ± 32	22 ± 2
	(157 – 282)	(0.0 – 0.0)	(369 – 448)	(26 – 61)	(40 – 81)	(74 – 168)	(18 – 24)
Final effluent	195 ± 40	0.0 ± 0.0	377 ± 13	37 ± 14	53 ±13	130 ± 24	22 ± 2
	(166 – 285)	(0.0 – 0.0)	(360 – 397)	(25 – 66)	(24 – 74)	(96 ±154)	(19 – 26)

expected that may attributed to the insufficient oxygen supply by aerators and/or the short retention time (<1day). This situation could be improved by dividing the sewage influent between the two aerated lagoons equally and supplying the required oxygen demand by the aerators. The BOD_5 was removed faster in the hot seasons where the removal rate was 60%. In the aerated facultative lagoon, organic load depletion rates were 57% of BOD_5, 63% of COD, 78.8% of VSS and 78% of TSS (Table 4). The statistical analysis showed that there was a weak (–ve) correlation between BOD_5 and temperature ($r = -0.37$; $p < 0.05$). The organic load was removed mainly in the aerated facultative lagoon (AFL). In the polishing lagoon, organic load depletion rates were –9.5% of BOD_5, 19.5% of COD, 40.6% of VSS and 41.9% of TSS (Table 4). The increase of BOD_5 in the polishing lagoons (PL) is mainly due to the large algal bloom as explained by Mara [3, 5].

The overall removal efficiency of organic load in the system was 68.8% of BOD_5, 74.2% of COD, 87.4% of VSS, and 88.4% of TSS (Table 4). The statistical analysis showed that there was a very strong (+ve) correlation between TSS and VSS ($r = 0.97$; $p < 0.05$), indicating that the algae were responsible for increasing TSS in the effluent. The statistical analysis showed a (+ve) correlation between retention time and BOD_5, TSS, VSS, and chlorophyll A ($r = 0.45, 0.39, 0.41$ and 0.41; $p < 0.05$), respectively.

3.3. Nitrogen and Phosphorus Removal

Removal of ammonia varied from –19 to 73% and the mean was 26.5% (Table 4). Horan (1990) reported that the factors affecting the growth of nitrifying bacteria, are the substrate content, temperature, DO and pH [1]. The results showed that the increase in temperature was correlated with decrease in ammonia and alkalinity. Slight increase in the orthophosphate was recorded due to the acceleration of bio-degradation of polyphosphate by microorganisms as well as the increase in temperature. Meanwhile, the increase in pH and DO was correlated with a decrease in ammonia.

The overall removal of ammonia varied from –111 to 59% and the mean value was –6.02% (Table 4). This indicates that the effluent was mostly higher than the influent due to different reasons. First, the aeration capacity for carbonaceous and nitrogenous BOD removal in the aerated facultative lagoon was not enough. Secondly, the depth of polishing lagoons (3.5 m) and the presence of anaerobic zones in the bottom. Thirdly, the unsuitable media for nitrifyers to oxidize ammonia to nitrate [10]. For the phosphorus and orthophosphate, the overall removal efficiency was –33.4 and –39.5% respectively (Table 4). Li et al. [11] stated that the negative removal of orthophosphate was due to the

Table 4. Removal efficiency of physico-chemical characteristics in the AL system in Ismailia (Mean ± SD, range and percentage of removal)

Locations	TSS mg/l	VSS mg/l	BOD_5 mg/l	COD mg/l	TKN mg/l	NH_4-N mg/l	*T–P* mg/l	F. coliform N/100 ml
RW	207 ± 28 (136 – 254)	156 ± 20 (118 – 192)	188 ± 44 (120 – 291)	322 ± 204 (100 – 775)	44.9±13.1 (17.8 – 65)	18.4 ± 5.8 (9 – 33.7)	6.1 ± 1.85 (3.4 – 9.4)	$16.5 \times 10^6 \pm 11.3 \times 10^6$ ($5.4 \times 10^6 - 51.5 \times 10^6$)
AL. eff.	205 ± 37.1 (104–260) 0.55%	151 ± 28.3 (80 – 192) 3.2%	142 ± 38.9 (58 – 205) 22.1%	311 ± 137 (140–550) 3.3%	–	13 ± 4.6 (6–23.5) 26.5%	–	1.3×10^6 ($14 \times 10^4 - 4 \times 10^6$) 84.5%
A.FL. eff	43 ± 12.9 (17–72) 78.2%	32 ± 8.4 (14 – 46) 78.8%	64 ± 42.5 (18–140) 57%	114 ± 45 (71–195) 63.2%	–	18 ± 5.99 (9.9–28.4) –54.1%	–	3.1×10^5 ($2.1 \times 10^3 - 1.4 \times 10^6$) 84.9%
PL. eff	24 ± 7.3 (9 –37) 41.9%	19 ± 6.5 (8–29) –40.6%	62 ± 38.1 (7.3 – 129) –9.5%	92 ± 31 (40–158) 19.5%	–	18 ± 7.22 (2.2–29.5) –2.9 %	–	3.3×10^4 ($2.4 \times 10^3 - 1.1 \times 10^5$) 64%
Final effluent	23.9 ± 7.7 (7.5–37.3) 88.4%	19.7 ± 6.8 (7 – 32) 87.4%	61.3 ± 41 (17 – 145) 8.8%	83 ± 24.6 (59–158) 74.2%	32.4 ± 7.3 (6–51) 27.8%	18.5 ± 7.3 (6.1–30.1) –6.02%	8.1 ± 2.3 (3.3–11.2) –33.4 %	$5.1 \times 10^4 \pm 4.9 \times 10^4$ ($1.9 \times 10^3 - 2.03 \times 10^5$) 99.5762%

RW: Raw sewage; AL. eff.: Aerated lagoon effluent; A. FL. eff.: aerated facultative lagoon effluent; PL. eff.: PL effluent

Table 5. Heavy metals in the raw sewage and final effluent of the aerated lagoons system in Ismailia (Mean, range and percentage of removal)

Locations	Cd mg/l	Cr^- mg/l	Ni mg/l	Hg mg/l	Pb mg/l	Cu mg/l	Zn mg/l	Fe mg/l	Mn mg/l
Raw sewage	0.0003 (0–0.0009)	0.005 (0–0.01)	0.0057 (0–0.017)	0.0006 (0.0 – 0.0008)	0.023 (0 – 0.03)	0.01 (0 – 0.03)	0.065 (0.06–0.07)	1.05 (1.69–1.24)	0.007 (0–0.013)
Final effluent	0.0002 (0 – 0.0006) 33.3%	0.0 (0–0) 100%	0.0034 (0– 0.005) 41%	0.0002 (0.1 – 0.4) 66.7%	0.017 (0 – 0.002) 26%	0.0069 (0– 0.008) 31%	0.036 (0.002–0.055) 45%	0.45 (1.55–0.42) 57%	0.0052 (0.003–0.09) 26%

conversion of phosphorus by bacteria. They concluded that there is a dynamic equilibrium in the transition between organic and inorganic phosphorus and the precipitation in the lagoon.

3.4. Heavy Metals Removal

The mean elimination heavy metals namely; Cd, Cr, Hg, Pb, Fe, Cu, Mn, Zn, and Ni was 33.3, 100, 66.7, 25, 12, 31, 26, 45, and 41% respectively (Table 5). It is important to notice that the pH of final effluent (7.6 – 8.1) was slightly higher than the pH of the influent (7.4 – 7.9) as shown in Table 2. Therefore, little enhancing of metal precipitation may took place. It is well known that the solubility product of metals is governed by the pH value. Metals are generally associated with the suspended solids as well as the coprecipitation by microbial activity. However, the major part was concentrated in the sludge (Table 6). Therefore, such sludge should not be used for the agricultural purposes without adequate treatment to reduce the level of metals down to the permissible limits [12].

Table 6. Level of heavy metals in the sludge of the AL system in Ismailia (as mg/kg dry weight)

Locations	Cd	Cr	Ni	Hg	Pb	Cu	Zn	Fe	Mn
Dry Sledge	0.033 (0.024– 0.038)	1.88 (2.49– 1.72)	3.98 (3.02– 3.12)	0.003 (0.0– 0.0048)	1.77 .(1 02– 2.56)	133 (97– 149)	432 (389– 449)	9,572 (7,881– 11,071)	66 (54– 75)

3.5. Faecal Coliform Removal

Table 4 indicates that the results do not comply with the national regularity standards (< 5,000 cfu/100 ml). It is well documented that the retention time is the most important factor that allows pH, temperature, and other changes to affect bacterial die-off [1, 6]. Meanwhile, Pearson [6] found that increasing the depth of maturation pond or polishing lagoons in order to increase the retention time, leads to decrease the removal efficiency for faecal coliforms. The depth in the present polishing lagoon is 3.5 m, which is inconvenient to achieve a good removal. The shallower lagoons are known to be more effective in pathogen removal due mainly to greater light penetration. Deep lagoons are more likely to have anaerobic zones, which have been shown to enhance bacterial survival.

Almasi and Pescod [1, 3] reported that the reduction of faecal pathogens is achieved by a combination of many factors, such as organic load decrease, long retention time, high sedimentation rate, adsorption on the settleable solids, predation by microorganisms, starvation, high pH, high oxygen content, and sunlight penetration. Long retention time is always associated with an increase in the algal content of the polishing lagoon effluent [4]. The present results demonstrated that faecal coliform bacteria were quickly destroyed in presence of algae. Unfortunately, algae seemed to be responsible for the high BOD and TSS in the effluent.

3.6. Parasite Cysts and Eggs Removal

Almost all raw wastewater samples were found to contain eggs as follows: 91% of the contaminated samples contained nematode eggs, whilst 9% contained eggs of cestodes. Eggs of *Ascaris lumbricoides* were predominant (relative frequency 54.6%), followed by: hookworm, *Ancylostoma duodenale* (39.9%), *Hymenolepis nana* (7%), *Trichuris trichuria* (4%), *Hymenolepis diminuta* (3.8%); while *Enterobius vermicularis* and *Taenia saginata* were scarce. Concerning the protozoan cysts, most samples contained *Entamoeba histolytica* as a predominant (relative frequency 61.9%), followed by *Entamoeba coli* (24.5%) and *Giardia lamblia* (16.4%). Seasonal variation exhibited different number of eggs and cysts in the raw sewage. Larger numbers of helminth eggs were found during autumn (68 egg/l) and summer (40.7 egg/l) than in spring (22.5 egg/l) or winter (22.2 egg/l). Concerning the protozoan cysts, larger numbers of cysts were found during the months of winter (109 cyst/l) and summer (49.9 cyst/l) than in autumn (37.5 cyst/l) or spring (26.2 cyst/l). In the AL, the removal efficiency of intestinal parasite was (3.83%) for all helminthes eggs, (5.2%) for nematodes, and (53.9%) for protozoan cysts.

Remarkable removal of the parasite cysts and eggs was recorded in the aerated facultative lagoon. The results indicated that sedimentation is the removal mechanism of eggs and cysts, with high removal efficiency. In the AFL, compared to AL. In the polishing lagoon, the removal efficiency of parasites was 89.8% for all helminth eggs, 88.4% for nematodes, 100% for cestodes, and 95.7% for protozoan cysts. The statistical analysis showed that there was a strong positive correlation between helminth eggs numbers in AFL effluent and the PL ($r = 0.70$; $p < 0.05$).

The overall removal efficiency of intestinal parasites in the system was 99.2% for all helminth eggs, 98.4% for nematodes, 100% for the cestodes, and 98.1% for the protozoan cysts. The mean of helminth eggs in the final effluent was 0.5 eggs/l. Few samples were found to contain cysts, mainly of *Entamoeba coli.* The mean of protozoan cysts in the final effluent was 0.9 cyst/l. Table 6 shows the removal % of individual helminth eggs and protozoan cysts by the aerated lagoons system in Ismailia. The obtained results showed that the helminth eggs were reduced with the increase in retention time ($r = -0.21$; $p < 0.05$). The long retention times in many lagoons have produced very high efficiencies. Previous results showed that high temperature has a beneficial effect on nematode eggs removal. The present results show that the high temperature was correlated with a high degree of helminth eggs removal. It can be concluded that AL are not very effective in removing helminth eggs. Nevertheless, combination of AL followed by faculatative and PL can achieve high removals. The human health risks of the presence of *Ancylostoma duodenale* larvae and other helminths such as *Stronglyoides* spp. in the wastewater effluents raise special concern. The eggs of these nematodes could hatch to infective larvae in the aerobic medium of the PL.

Futhermore, the results of the parasite cysts and eggs indicated such sludge contained *Ascaris lumbricoides* eggs that were found to be viable. It is known that time treatment of sludge affects eggs severely. The eggs had been irrevocably damaged by the high pH of sludge and were unable to develop and cause infection (i.e. they were no longer pathogenic). In addition, the resulted high-pH precipitate all metals at variable degrees thus eliminates any hazard effect on soil or plants.

It is worth mentioning that part of the treated effluent is reused as follow: 3,000 m^3/day for irrigating fruit trees and palm trees on an area of 170 feddan (71.4 ha) and 15,000 m^3/day for irrigating lumber forest trees of an area of 1000 feddan (420 ha). All these land areas were desert.

4. CONCLUSIONS

The following foundings was observed during the operation of the AL as oxidation ponds for wastewater treatment in the sub-tropical climate.

1. TDS contents were increased in the effluent, due to the high evaporation rate along the lagoons surface.
2. BOD_5 increased in the polishing lagoons, which is mainly due to the large algal bloom. Meanwhile, the algae were responsible for increasing TSS in the effluent (i.e. algae are responsible for the high BOD and TSS in the effluent).
3. The increase in temperature was correlated with decrease in ammonia and alkalinity.
4. As the temperature, increase acceleration of biodegradation of polyphosphate by microorganisms was observed, and slight increase in the orthophosphate was recorded.

5. High temperature has a beneficial effect on nematode eggs removal. Results showed that the high temperature was correlated with a high degree of helminth eggs removal in the system.

Acknowledgment: The authors wish to express their deep appreciation to the facilities provided by the Serabium Wastewater Treatment Plant, Ismailia, Egypt, the European Community under MEDA WATER PROGRAMME initiative and the National Research Centre of Egypt. NATO provided financial support to present the paper at the ARW – Wastewater Reuse – Risk Assessment, Decision-making and Environmental Security held in Istanbul, Turkey, during October 12–16, 2006.

5. REFERENCES

1. Horan NY. (1990). Biological wastewater treatment systems, J. Wiley. NY. USA. 267.
2. Hills S et al.: (2001). Water recycling at the Millennium Dom, Wat. Sci. Tech. 43, 10, 287–294.
3. Mara DD. (1996). Water stabilization ponds: effluent quality requirements and implications for process design, Wat. Sci. Tech. 33, 7, 23–31.
4. EPA (1971). "Waste treatment lagoons - State of Art." Missouri Basin Engineering Health Council, Cheyenne, Wyoming, USA.
5. Mara DD. (2001). Appropriate wastewater collection, treatment and reuse. Proceeding of the Institution of CE & ME, 145 (4): 299–303.
6. Pearson HW. (1996). Expanding the horizons of pond technology and application in the world, Wat. Sci. Tech. 33, No. 7, 1–9.
7. WHO: World Health Organization, Health (1989) Guidelines for the use of wastewater in agriculture and aquaculture. Technical Report No. 778, Geneva.
8. Abdel-Shafy HI, Al-Kaff HA, Ali AA. (2004). "Risk reduction of sewage disposal by oxidation," J. Nat. Science, 8, 2, 315–320.
9. Li J, Wang J, Zhang J. (1991) Removal of nutrient salts in relation with algae in ponds. Wat. Sci.Tech. 24 (5): 75-83.
10. Abdel-Shafy HI, Hegmann W, Guldner C. (1996). "Fate of heavy metals via Chemical-biological upgrading," J. Envi Manag & Health, 7(3) 28–33.
11. Almsi A, Pescod MB. (1996). "Pathogens removal mechanism in anoxic waste stabilization ponds". Wat. Sci. Tech. 33 (7), 144–140.

COMBINING WASTEWATER IRRIGATION WITH DESALINATION: A MULTIDISCIPLINARY APPROACH

Nava Haruvy
Netanya Academic College, Netanya, Israel

Corresponding author: navaharu@netvision.net.il

Abstract:

A multidisciplinary water management model was constructed that combines hydrological, technological, economic, and regional planning aspects, and applied it to a case study of two regions in Israel. The model enables to forecast the chloride concentration in the aquifer over time, estimate the cost of different desalination technologies, and plan the supply and treatment of different water resources.

Keywords: Wastewater, desalination, hydrological model, chlorine, salinity.

1. INTRODUCTION

In this research, a water management model was constructed based on a multidisciplinary approach that combines hydrological, technological, economic, and regional planning aspects [1–4]. The model enables to plan the supply and treatment of different water sources, including groundwater, national carrier water, wastewater, and seawater. The model starts with a given initial level of water salinity by source, and a specific permitted water quality threshold for water supply by use—town, agriculture, and aquifer [5–10]. Desalination is initiated when the salinity level exceeds the permitted threshold. The desalinated water is mixed with other water sources in order to reach the threshold level [11–12].

2. METHOD

The model was applied for a selection of hydrological cells in two different regions in Israel. A hydrological cell is rectangular shaped, and stretches from the soil surface to the base of the aquifer. It is composed of two layers: the upper layer is the unsaturated zone, from the soil surface to the aquifer surface, and the lower layer is the saturated zone, from the aquifer surface to its base. The water supply in each hydrological cell area includes freshwater and wastewater, which is consumed by agricultural and urban sectors, and in each cell, the demand

M.K. Zaidi (ed.), Wastewater Reuse–Risk Assessment, Decision-Making and Environmental Security, 185–190.

equals the supply. The hydrological model predicted the water levels and salinity over time, based on regional planning data inputs, which included water sources by hydrological cell and water uses by sector. The technological model examined different desalination technologies and their costs. The outputs of these two models were used as an input for the economic model, which evaluated the scope and costs of desalination for each scenario.

Each scenario included policies of water pumping, supply and allocation. The water supply policy was based on the agricultural and urban water quality thresholds. The water allocation policy allows a pumped water supply priority for either urban or agricultural use; and the pumping policy can be based on capacity or cost considerations.

Several scenarios were examined. In the basic scenario (Scenario A), the threshold for urban water quality is 250 mg/l of chlorides. Water from local sources (pumping) has the first priority, based on pumping capacity. The rest of the water demand is filled by the Israeli National Water Company (Mekorot), which is the sole source of water supply for these areas. Agricultural users have priority in access to pumped water, and the policy includes wastewater irrigation.

Five additional scenarios were created which differed by threshold levels, allocation and pumping policies. Scenarios B, C, and D include thresholds for agricultural as well as urban use of 250, 150, and 50 mg/l Cl; Scenarios E and F include irrigation with or without wastewater (Table 1).

Table 1. Water policy scenarios

Scenario	Urban water salinity threshold (mg/l Cl)	Agricultural water salinity threshold (mg/l Cl)	Irrigation with wastewater
A	250	–	yes
B	250	250	yes
C	150	150	yes
D	50	50	yes
E	250	–	no
F	250	–	yes; high salinity

3. RESULTS

3.1. Results of the Hydrological Model

The water allocation was calculated for each of the scenarios A–F. The water allocation according to the basic scenario (Scenario A) is shown in Table 2. The model predicts that the salinity levels will increase gradually over time during the next 100 years in all the hydrological cells, except for those who are not pumped. Table 3 shows the model's predictions of the salinity level in each area after 100 years, under each scenario shown in Table 1.

The Table shows that the salinity levels in the 100th year decrease as the scenarios progress from Scenario A (threshold of 250, only for urban use), through Scenario B (introduction of a threshold of 250 for agricultural use as well), Scenario C (thresholds decreased to 150), and to Scenario D (thresholds decreased to 50 mg/l Cl). Irrigation without wastewater (Scenario E) decreases the salinity level as compared with irrigation with wastewater (Scenario F).

Table 2. Water allocation according to Scenario A (MCM)

Water use	Water supply	Region I	Region II
Urban	Wells – pumped	2.55	14.93
	National carrier	0.02	9.77
Agricultural	Wells – pumped	9.41	21.38
	National carrier	0.22	9.96
Total water use	Wells – pumped	11.96	36.32
	National carrier	0.24	19.74

Table 3. Concentration of chlorides in groundwater in the 100th year (mg/l Cl)

Scenario	Scenario characteristics	Region I	Region II
A	Urban threshold 250 mg/l Cl	716	158
B	General high threshold* 250 mg/l Cl	453	159
C	General medium threshold* 150 mg/l Cl	357	130
D	General low threshold* 50 mg/l Cl	182	84
E	Irrigation without wastewater	716	157
F	Irrigation with wastewater-high salinity level	907	174

* The term "general threshold" refers to a common urban and agricultural threshold level

3.2. Results of the Technological Model

The model incorporated different desalination technologies, in order to evaluate the relevant alternatives and derived cost. The cost of desalination are influenced by different physical parameters such as the size and location of the desalination plant and the quality of the water supplied to the plant. The cost is also influenced by planning parameters such as return on investment, cost structure, availability of operations, and the cost of energy, membranes, chemicals, labor, maintenance, and overhead. The input data include the water quantity and quality, product quantity, rate of discharge, drilling depth, distance of the pipe system, height of the desalination plant, and volume of impounded water.

The desalination alternatives using the reverse osmosis method include desalination of brackish water, system water, wastewater—including pretreatment such as tertiary and premembrane treatments, and seawater (mixed with national carrier water), which are meant to be used mainly as an additional water supply. The average cost of desalination in Region I, as a function of the initial conditions, is shown in Table 4. The costs of desalinating brackish water in Region I is 36 cents/mc; the cost of desalinating national carrier water is 29.4 cents/CM (depending on the plant size); the cost of desalinating wastewater is 41.6 cents/mc, and the costs of seawater desalination is 54.2 cents/mc (for a plant capacity of 50 MCM/year).

Table 4. Average desalination costs (cents/mc)

Cost structure	Brackish water	National system water	Wastewater	Seawater
Infrastructure	13.0	14.6	3.3	32.5
Desalination	23.0	14.8	38.3	21.7
Total	36.0	29.4	41.6	54.2

3.3. Results of the Economic Model

For each scenario, we evaluated the total cost of the water supply to a region which includes an urban and an agricultural area. The input included the combination of water sources derived from the planning model, the forecast of chloride concentration by scenario and by year from the hydrological model, and the cost of the various water sources from the technological cost budgets. For each scenario, we calculated the relevant cost of the water supply for the combined hydrological cells of Region I.

In Scenario A (urban threshold 250 mg/l Cl, no wastewater desalination), the annual cost varied from \$4.75 million for groundwater desalination up to \$5 million for seawater desalination. In Scenario B (urban and agricultural thresholds of 250 mg/l Cl), the annual cost varied from \$5.28 million for groundwater desalination to \$8.53 million for wastewater desalination and up to \$14.78 million for seawater desalination. In Scenario C (urban and agricultural thresholds of 150 mg/l Cl), the annual cost varied from \$7.19 million for groundwater desalination to \$9.9 million for wastewater desalination and up to \$16.21 million for seawater desalination. In Scenario D (urban and agricultural thresholds of 50 mg/l Cl), the annual cost varied from \$10.12 million for groundwater desalination to \$11.27 million for wastewater desalination and up to \$17.93 million for seawater desalination. It is evident that in every scenario, the lowest desalination cost is that of national carrier water, followed by groundwater, wastewater, and seawater desalination.

It should be mentioned that the seawater is desalinated primarily in order to increase the total volume of the water sources. Therefore, the costs of their desalination to improve the water quality are only the marginal costs. Maintaining a threshold of 250 mg/l Cl for the aquifer water involves an additional annual cost of \$2.53–\$4.68 million.

4. CONCLUSIONS

A database was developed and operated a hydrological model for planning the water sources and forecasting the chloride concentration in the aquifer. The cost estimation of different desalination technologies under the regional conditions was researched. From these calculations, it was concluded that desalinating brackish water involves the lowest cost, desalinating national carrier water is effective if the volume is large, desalinating wastewater is significant for the agricultural threshold, and desalinating seawater is worthwhile when their contribution is important for the national water balance.

The model developed and applied can be used to examine regional planning, economic aspects of supply, desalination of different water resources, and to estimate the impact on the aquifer and the environment. The model's advantage lies in its multidisciplinary character and its practical applicability. It can also evaluate and direct scenarios of supply and treatment of different water resources.

Acknowledgment: The presentation of this paper at the workshop was enabled through a NATO Advanced Research Workshop grant.

5. REFERENCES

1. Draper AJ, Jenkins MW, Kirby KW, Lund JR, Howitt RE. (2003). Economic-engineering optimization for California water management. Water Resources Planning and Management, 129(3), 155–164.
2. Rainwater K, Stovall J, Frailey S. (2005). Transboundary impacts on regional ground water modeling in Texas. Ground Water, 43(5), 706–716.
3. Xevi E, Khan S. (2005). A multi-objective optimisation approach to water management. J. Environmental Management, 77(4), 269–277.
4. Shepherd KA, Ellis PA, Rivett MO. (2006). Integrated understanding of urban land, groundwater, baseflow and surface-water quality—Birmingham, UK. Science of the Total Environment, 360(1–3), 180–195.
5. Haruvy N, Bachmat Y, Shalhevet S, Yaron D. (2000). Effect of Urban Development on Water Quality-Environmental Concerns. Proceedings, Regional Science Association, 40th European Cong. Barcelona, Spain.
6. Fernandez, L. (1997). Estimation of wastewater treatment objectives through maximum entropy. J. Env. Econ. Management, 32(3), 293–308.
7. Haruvy N, Shalhevet S, Ravina I. (2004). Irrigation with Treated Wastewater in Israel. J. Financial Management and Analysis, 16 (2), 142–146.
8. Haruvy N. (2004). Irrigation with Treated Wastewater in Israel–Assessment of Environmental Aspects. pp. 375–384, In: Linkov I. and Ramadan A. (eds.), Comparative Risk Assessment and Environmental Decision Making. Kluwer Academic Press.
9. Haruvy N. (2006). Reuse of Wastewater in Agriculture-Economic Assessment of Treatment and Supply Alternatives as Affecting Aquifer Pollution. pp. 257–263, Morel, B. and Linkov, I (eds.). Environmental Security & Environ. Management: The Role of Risk Assessment. Springer.
10. Haruvy N, Shalhevet S. (2005). Land use and water management in Israel- Economic and Environmental analysis of sustainable reuse of wastewater. Proc. 45th Cong. Euro. Reg. Sc. Assoc, Vrije Univ. Amsterdam, the Netherlands.
11. Haruvy N, Shalhevet S. (2006). Wastewater modeling to reduce disaster risk. Proc. International Disaster Reduction Conference, Davos, Switzerland.
12. Thomas A, Tellam J. (2006). Modelling of recharge and pollutant fluxes to urban groundwaters. Science of the Total Environ. 360(1–3), 158–179.

SESSION 6. RISK ASSESSMENT: ECONOMIC AND MANAGERIAL ASPECTS

WASTEWATER REUSE, RISK ASSESSMENT, DECISION MAKING—A THREE-ENDED NARRATIVE SUBJECT

Sureyya Meric and Despo Fatta
Department of Civil Engineering,
University of Salerno, 84084 Fisciano (SA), Italy
Laboratory of Environmental Engineering, University of Cyprus, 75 Kallipoleos, 1678 Nicosia, Cyprus

Corresponding author: msureyya@unisa.it

Abstract:

The current demographic scenario of the world shows that the population of developing countries is growing at an unprecedented rate with the increasing scarcity of freshwater resources. Therefore there is a great need to propose appropriate wastewater treatment in such countries to reuse wastewater. Wastewater treatment should not only target pathogens removal but also many compounds named endocrine disruptors (EDCs) which have not been included yet in reuse wastewater legislations. This paper provides information on the fate of EDCs in wastewater treatment, current wastewater legislations and it also summarizes a recently developed decision support tool for decision makers (wastewater technologists and risk analyzers) to improve their decision processes in respect to wastewater treatment and reuse schemes.

Keywords: Wastewater reuse, risk assessment, decision making.

1. INTRODUCTION

At present, water shortage is one of the main problems in the Mediterranean area but also elsewhere in the world and represents one of the main key factors for future economic and social development. Apart from the scarcity of freshwater in many regions, the quality of available resources is deteriorated due to pollution attributed to the wastewater treatment facilities and control mechanisms/laws. At present, in many developing or third countries such as Ghana, Senegal, Pakistan, etc., wastewater is used without any treatment for crop, fodder, and green space irrigation with a high risk of easy transmission of waterborne diseases and groundwater pollution. Therefore there is a great need to propose appropriate wastewater treatment in such countries. In the light of sustainable development, wastewater reuse is a resource of growing global importance which must be carefully managed in order to obtain substantial benefits and to minimize serious risks. Risk is one of the key areas of concern for stakeholders involved in water

M.K. Zaidi (ed.), Wastewater Reuse–Risk Assessment, Decision-Making and Environmental Security, 193–204.

management while experts tend to define risk in terms of the high value placed by the public on health issues, the public viewpoint on what risk is and the value they attribute to their definition of risk is more complex. People's risk perception and the potential severity of that risk are not based solely on numerical data. If the said risk is seen to be more under the direct control of an individual then it seems to be more acceptable than if the risk is controlled by others. Water resource management including wastewater reuse issue is typically perceived as the latter type of activity. One of the key parameters in order to assess the suitability of effluent wastewater for reuse is the occurrence of pathogens which are usually removed at high percentages in conventional sewage treatment. Table 1 summarizes the bacteria removal in each treatment phase. However, secondary treatment should be accompanied with a final disinfection process to remove pathogens more effectively [1]. Table 2 compares common disinfection methods [2].

Table 1. Removal/destruction of bacteria by different treatment processes [1]

Process	Percent removal
Coarse screens	0–5
Fine screens	10–20
Grit chambers	10–25
Plain sedimentation	25–75
Chemical precipitation	40–80
Trickling filters	90–95
Activated sludge	90–98
Chlorination of treated wastewater	98–99.999

On the other hand, it is well known that the conventional disinfection processes (chlorine use) produce unwanted chlorinated by-products (DBPs) such as trihalomethanes (THMs), many chlorinated compounds etc. [3]. Some of those chlorinated compounds are known as carcinogenic and many ongoing studies are being carried out toward this direction. Meanwhile, there are many chemical compounds entering the water cycle exhausted from all motor vehicles, cigarette smoke, ordinary household products (breakdowns products of detergents and associated surfactants). Several studies have shown that the presence of some chemical substances, the so-called endocrine disrupting chemicals (EDCs) must be removed before the assessment of the potential wastewater reuse for nonindustrial applications, such as irrigation of crops and aquaculture, since they could constitute a contamination source of food chain threatening both humans and environment [4] (Table 3).The effects related to EDCs have been reported in mollusks, crustacea, fish, reptiles, birds, and mammals in various parts of the world as early as in the 1930s [4]. The priority list of the Water Frame Directive

Table 2. Comparison of different disinfection treatments [2] (L: Low, M: Middle; H: High; N: none)

Disinfection/criteria	Cl_2/Dechlorination	UV	Ozone	Microfiltration	Ultrafiltration	Nanofiltration
Safety	L	H	M	H	H	H
Bacterial removal	M	M	M	M	H	H
Virus removal	L	L	M	L	H	H
Protozoa removal	Ne	N	M	H	H	H
Bacterial regrowth	L	L	L	N	N	
7Residual toxicity	H	N	L	N	N	
By-products	H	N	L	N	N	
Operating/investment costs	L/M	L/M	M/H	H/H	H/H	H/H

Table 3. List of endocrine disruptor compounds (EDCs) [4]

Organohalogens	Dioxins and furans, PBBs, PCBs, hexachlorobenzene, octachlorostyrene, pentachlorophenol
Pesticides:	2,4,5-T, 2,4-D, alachlor, aldicarb, *d-trans* allethrin, amitrole, atrazine, benomyl, β-HCH, carbaryl, chlordane, chlozolinate, cyhalothrin, *cis*-nonachlor, cypermethrin, DBCP, DDT, DDT metabolites, dicofol, dieldrin, endosulfan, esfenvalerate, ethylparathion, fenvalerate, glyphosate (the active ingredient in Roundup), hepoxide, heptachlor, iprodione, kelthane, kepone, ketoconazole, lindane, linurone, malathion, mancozeb, maneb, methomyl, methoxychlor, metiram, metribuzin, mirex, nitrofen, oxychlordane, permethrin, procymidone, sumithrin, synthetic pyrethroids, toxaphene, *trans*-nonachlor, tributyltin oxide, trifluralin, vinclozolin, zineb, ziram
Phthalates:	Diethylhexyl phthalate (DEHP), butyl benzyl phthalate (BBP), Di-*n*-butyl phthalate (DBP), Di-*n*-pentyl phthalate (DPP), Dihexyl phthalate (DHP), Dipropyl phthalate (DprP), Dicyclohexyl phthalate, (DCHP), Diethyl phthalate (DEP)
Others	Penta- to nonyl-phenols, bisphenol A, bisphenol F, styrene dimers and trimers, benzo(a)pyrene, ethane dimethane, sulphonate, tris-4-(chlorophenyl), methane, tris-4-(chlorophenyl), methanol, *b*enzophenone, N-butyl benzene, 4-nitrotoluene, 2,4-dichlorophenol, cyanazine, diethylhexyl adipate, (DES) diethylstilbestrol
Metals:	Arsenic, cadmium, depleted uranium, lead, mercury
Pharmaceuticals:	drug estrogens—birth control pills, cimetidine

[5] includes 33 substances while several of them (e.g., polyaromatic hydrocarbons (PAHs), organotins (OTs), brominated flame retardants) have been already proven to be or are potential EDCs [6–7]. In a monitoring campaign in Italy, France, Greece, and Sweden, more than 20 individual pharmaceuticals belonging to different therapeutic classes and classified among EDCs, were found in the effluent of the sewage treatment plants STPs [8].

1.1. Wastewater Treatment for the Removal of Endocrine Disrupting Compounds before Reuse

The removal of EDCs in wastewater treatment processes depends on the inherent physicochemical properties of the pollutants and on the nature of the treatment process involved. It is generally recognized that there are four main removal pathways for organic compounds during the conventional wastewater treatment: (i) adsorption onto suspended solids or association with fats and oils, (ii) aerobic and anaerobic degradation, (iii) chemical (abiotic) degradation by processes such as hydrolysis, and (iv) volatilization. The physicochemical data of a compound can be used to predict physical processes, such as sorption, volatilization, and dissolution. The important properties to be considered are octanol–water partition coefficient (K_{ow}), aqueous solubility, acid dissociation constant, and Henry's Law constant (H_c).

Knowledge of the chemical partitioning between the aqueous and solid phase is needed to assess pathways of EDCs transport and transformation [9]. However, many of those EDCs, and other emerging pollutants are partially or at a low grade removed from wastewater in conventional treatment systems. Several studies confirmed that in activated sludge treatment plants the EDCs are relatively removed. Sorption and biodegradation are considered the principal mechanisms for steroid estrogens removal, influenced by hydraulic retention time (HRT) and high sludge retention time used in sewage treatment plants [10, 11]. Regarding the physical treatment methods, the nonpolar and hydrophobic nature of many EDCs cause them to adsorb onto particulates. This suggests that the general effect of wastewater treatment processes would be to concentrate organic pollutants, including EDCs, in the sewage sludge while mechanical separation techniques, such as sedimentation, would result in significant removal from the aqueous phase to primary and secondary sludge.

The result of this is treated wastewater discharged relatively free of EDCs and a sewage sludge which could constitute a new source of pollution if applied as fertilizer on agricultural fields. The use of advanced oxidation processes (AOPs), including photocatalysis as economic and environmentally friendly process has been investigated in respect to their ability to remove EDCs from the effluents [12]. Photocatalysis is a chemical oxidation process in which a metal oxide semiconductor immersed in water and irradiated by near UV light (λ <385 nm) results in the formation of free hydroxyl •OH radicals. Although several semiconductors exist, TiO_2 is the most widely used catalyst, mainly because of its photostability, nontoxicity, low cost, and water insolubility under most environmental conditions. This process has been found to be effective for the degradation of many EDCs in wastewater as seen in Table 4 [13] as well as for microbial removal also, as the tertiary treatment for urban wastewater treatment plants [14]. However these studies should be evaluated on the basis of complex environmental matrices at full scale plants. Because of the reactivation of bacteria, disinfection before reuse is absolutely required. In this case the interaction among the by-products and the chlorine should be considered.

1.2. Wastewater Reuse Limits

Wastewater reuse is a resource of growing global importance which must be carefully managed in order to obtain substantial benefits and to minimize serious risks. Table 5 revises the wastewater reuse limits applied in the world [15]. However these limits do not contain most of the EDCs. In 1996, the EPA formed the Endocrine Disruptor Screening and Testing Advisory Committee (EDSTAC) to recommend a conceptual framework, priorities, screening, and testing methodologies for EDCs. In 2001, the EPA formed the Endocrine Disruptor Methods Validation Subcommittee (EDMVS) to evaluate the battery of test suggested by EDSTAC. California is considering establishing regulations based on the potential impacts of EDCs and pharmaceuticals and personal care products (PPCPs), where municipal wastewater effluent is recycled for indirect potable reuse [16]. Because California's criteria often establish precedents for programs throughout the world, other regulatory agencies will likely adopt similar choices in their own water recycling programs [17]. The Water Framework Directive [18] leads the member states toward the promotion of (1) wastewater reuse through e.g., reuse for irrigation, (2) lower water consumption, and (3) recycling of processed wastewater in industries, no European regulations on water reuse have been set yet. This is expected to promote to removal of EDCs effectively naturally before discharge, either for reuse. In Italy, very strict limits for municipal wastewater reuse are set (Table 6) [19]; however, the limits are not categorized according to the reuse application.

Table 4. Removal of some EDCs during TiO_2 based photocatalytic suspended systems [13] (a) compound (b) TOC))

Compound	Concen. (mg/l)	Removal (%)	pH	Reaction time	Added use	References
17β-Estradiol	0.82	98 (a)	5.5	3.5 h	1.0 g/l of TiO_2	Coleman et al. (2000)
Atrazine	25	99 (a)	6.94	60 min	1.0 g/l of TiO_2	Parra et al. (2004)
Atrazine	25	86.1 (a)	2.03	60 min	1.0 g/l of TiO_2	Parra et al. (2004)
Atrazine	25	67.6 (a)	9.99	60 min	1.0 g/l of TiO_2	Parra et al. (2004)
2,4-Dichlorophenol	22.5	96 (a)	6.6	2 h	0.2 g/l of TiO_2	Malato et al. (2002)
Methomyl	18	67 (b)	7	400 min	0.2 g/l of TiO_2	Malato et al. (2003)
Bisphenol A	100	63 (a)	6	1 h	10 g/l of TiO_2	Kaneco et al. (2004)
Bisphenol A	20	95 (a)	3	1 h	2 g/l of TiO_2 (pH = 3)	Chiang et al. (2004)
Bisphenol A	20	99 (a)	10	1 h	2 g/l of TiO_2 (pH = 3)	Chiang et al. (2004)
Bisphenol A	20	62.5 (b)	3	1 h	2 g/l of TiO_2 (pH = 3)	Chiang et al. (2004)
Bisphenol A	20	12.5 (b)	10	1 h	2 g/l of TiO_2 (pH = 3)	Chiang et al. (2004)
Aldrin	5	71 (a)	–	8 h	0.2 g/l of TiO_2	Bandala et al. (2002)
Aldrin	5	68 (a)	–	8 h	0.2 g/l of TiO_2 + 3 g/l H_2O_2	Bandala et al. (2002)
Aldrin	5	90 (a)	–	8 h	0.2 g/l of TiO_2 + 6 g/l H_2O_2	Bandala et al. (2002)
Malathion	70	50 (a)	–	1 h	0.1 g/l of TiO_2	Muszkat et al. (1995)
Atrazine	90	81 (a)	–	4 h	0.1 g/l of TiO_2	Muszkat et al. (1995)
Lindane	0.1	99 (a)	7.3	30 min	0.5 g/l of TiO_2	Vidal et al. (1999)
3-Amino-2-chloropyridine	64.2	98 (a)	4.6	1.5 h	2 g/l of TiO_2	Abramovic et al. (2004)
3-Amino-2-chloropyridine	321.2	98 (a)	4.6	7 h	2 g/l of TiO_2	Abramovic et al. (2004)
Malathion	10	94 (a)	7	250 min	1 g/l of TiO_2	Doong et al. (1997)
Diazinon	10	99 (a)	7	220 min	1 g/l of TiO_2	Doong et al. (1997)
Malathion	10	99 (a)	7	3 h	1 g/l of TiO_2 + 20 mg/l H_2O_2	Doong et al. (1997)
Diazinon	10	99 (a)	7	160 min	1 g/l of TiO_2 + 20 mg/l H_2O_2	Doong et al. (1997)

Table 5. Summary of water recycling guidelines and mandatory standards in the United States and other countries

Country/Region	Fecal coli. (CFU/100ml)	Total coli. (cfu/100ml)	Helminth eggs (#/l)	BOD_5 (ppm)	Turbid. (NTU)	TSS (ppm)	DO (%)	pH	Cl_2 res. (ppm)
Australia (New South W)	<1	<2/50	–	>20	<2	–	–	–	–
Arizona	<1	–	–	–	1	–	–	4.5–9	–
California	–	2.2	–	–	2	–	–	–	–
Cyprus	50	–	–	10	–	10	–	–	–
France	<1000	–	<1	–	–	–	–	–	–
Florida (m)	25 for any sample for 75%	–	–	20	–	5	–	–	1
Germany (g)	100(g)	500(g)	–	20(g)	1–2 (m)	30	80–120	6–9	–
Japan (m)	10	10	–	10	5	–	–	6–9	–
Israel	–	2.2 (50%); 12 (80%)	–	15	–	15	0.5	–	0.5
Kuwait, crops cooked Raw	–	10,000	–	10	–	10	–	–	1
Kuwait crops eaten raw		100		10		10			1
Oman 11A	<200	–	–	15	–	15	–	6–9	–
Oman 11B	<1,000			20		30		6–9	
South Africa	0 (g)	–	–	–	–	–	–	–	–
Spain (Canary Islands)	–	2.2	–	10	2	3	–	6.5–8.4	1
Texas (USA) (m)	75(m)	–	–	5	3	–	–	–	–
Tunisia	–	–	<1	30		30	7	6.5–8.5	–
US EPA (g)	14 for any sample, 0 for 90%	–	–	10	2	–	–	6–9	1
WHO (lawn irrigation)	200(g); 1000(m)	–	–	–	–	–	–	–	–

(g) signifies that the standard is a guideline and (m) signifies that the standard is a mandatory regulation

Table 6. Limit values for reuse of wastewater in Italy [19]

pH	–	6–9.5	Thallium	mg/l	0.001
SAR	–	10	Vanadium	mg/l	0.1
Total suspended solid	mg/l	10	Zinc	mg/l	0.5
BOD_5	mg/l	20	Total cyanides (as CN)	mg/l	0.05
COD	mg/l	100	Sulfides	mg/l	0.5
Total phosphorus	mg-P/l	2	Sulfites	mg/l	0.5
Total nitrogen	mg-N/l	15	Sulfates	mg/l	500
Ammonium	mg/l	2	Active chlorine	mg/l	0.2
Conductivity	μS/cm	3,000	Chlorides	mg-Cl/l	250
Aluminium	mg/l	1	Fluoride	mg-F/l	1.5
Arsenic	mg/l	0.02	Grease, animal and vegetable oils	mg/l	10
Barium	mg/l	10	Mineral oils	mg/l	0.05
Beryllium	mg/l	0.1	Total phenols	mg/l	0.1
Boron	mg/l	1	Pentachlorophenol	mg/l	0.003
Cadmium	mg/l	0.005	Total aldehydes	mg/l	0.5
Cobalt	mg/l	0.05	Tetrachloroethylene + trichloroethylene	mg/l	0.01
Total chromium	mg/l	0.1	Total chlorinated solvents	mg/l	0.04
Chromium VI	mg/l	0.005	THMs	mg/l	0.03
Iron	mg/l	2	Total aromatic organic solvent	mg/l	0.01
Manganese	mg/l	0.2	Benzene	mg/l	0.001
Mercury	mg/l	0.001	Benzo(a)pyrene	mg/l	0.00001
Nickel	mg/l	0.2	Total organo-nitrogen solvent	mg/l	0.01
Lead	mg/l	0.1	Total surfactants	mg/l	0.5
Branch	mg/l	1	Chlorinated pesticides (each)	mg/l	0.0001
Selenium	mg/l	0.01	Phosphorated pesticides	mg/l	0.0001
Tin	mg/l	3	Other pesticides (total)	mg/l	0.05
Escherichia coli	UFC/100ml		10 for the l'80% of samples, 100 maximum value		

2. DECISION MAKING—DECISION SUPPORT SYSTEMS

Decision making in environmental projects can be complex principally due to the inherent existence of trade-offs between sociopolitical, environmental, and economic factors. This is why the MEDAWARE project has focused its work on the development of a software tool able to apply a scoring system for existing wastewater facilities based on the potential safe wastewater reuse [20, 21]. The tool developed is able to guide the responsible authorities to the most efficient solutions in terms of health and safety for the agricultural reuse of the produced effluent as well as to help them undertake actions that can be sustainable [22]. The tool components are classified in inputs, water quality requirements, treatment technologies, legislative base, etc., to optimize also the cost needs of the treatment alternatives (Figure 1). The input data for the specific model are quite simple and can be easily collected (e.g., data concerning the population served by the facility, the possibilities for agricultural reuse of the water in the area, specific requirements or preferences on cultural, economical, technological, or social issues), while the outcome is the ranking of the alternative scenarios and the suggestion of specific processes and treatment systems (Figure 2) [22]. The risk due to wastewater reuse is considered according to the effects on human and environment as well. However, here, the risk due to the existing legislations does not include any toxicity or EDCs, including DBPs parameters.

This software calculates the design of each unit process based on the influent to the system and the quality requirements for a number of possible final uses, then applies a cost to the designs and compares them in order to show a ranking of possible solutions to a specific problem. The tool carries out a systematic evaluation of the different alternative designs with respect to a set of criteria. The first step of the design methodology is to define objectives. The following step is to select the alternatives to solve each issue and the criteria against which the design alternatives will be evaluated. These criteria are specific to each decision. Once the alternatives and criteria are selected, the evaluation of alternatives can be carried out. The output of the system is, then, a list of treatment methods (consisting of several unit processes), satisfying the objectives of effluent quality previously set. This list of treatment methods can be ranked by cost. The alternative that meets the quality objectives while at the same time is less costly will be the one recommended by the system. However the final decision will be taken by the process designer [22].

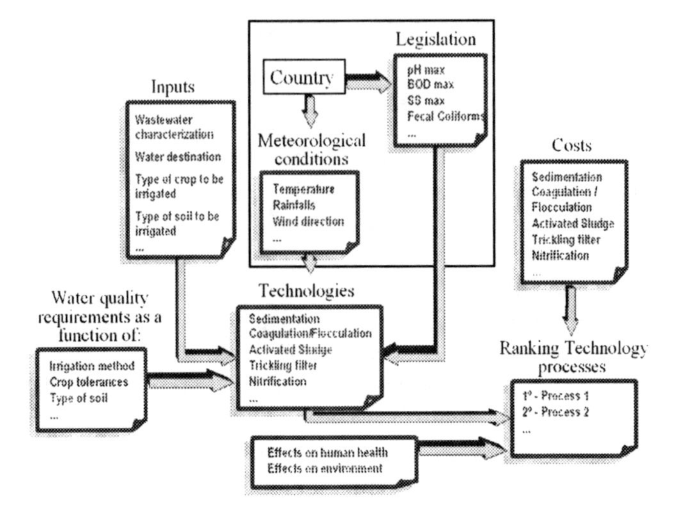

Figure 1. Flow-Information system

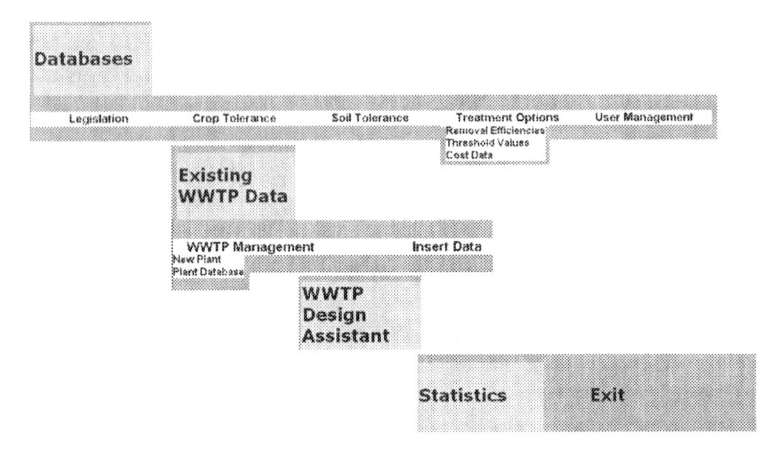

Figure 2. Screen options for '*Admin*' user

3. CONCLUSIONS

Decision making in environmental projects can be complex principally due to the inherent existence of trade-offs between sociopolitical, environmental, and economic factors.The developed tool in the framework of the MEDAWARE project has been based on the existing legislations to provide schemes in compliance with wastewater treatment and reuse related to regional, national or international regulations. Wastewater reuse issue should be evaluated as a triangular-narrative subject among the wastewater reuse legislations, risk analyzers, and decision makers.

Acknowledgments: This paper has been inspired by the MEDAWARE project (# ME8/AIDCO/2001/0515/59341-P033) which is funded by the Euro-Mediterranean partnership and more specifically by its Regional Program for Local Water Management and within the scope of the EU COST Action 636 "Xenobiotics in the Urban Water Cycle." The financial support by the NATO Advanced Research Workshop organizing committee on Water Reuse, Decision-making and Environmental Security is acknowledged.

4. REFERENCES

1. Tchobanoglous G, Burton FL. (2003). Wastewater Engineering Treatment and Reuse, fourth edition, Metcalf & Eddy Inc, M-Hill, New York.
2. Urkiaga A, Fuentes L. (2004). Best Available Technologies for Water Reuse and Recycling. Needed Steps to Obtain the General Implementation of Water Reuse, Fundación Gaiker, Parque Tecnológico De Bizkaia, Spain.
3. Rook JJ. (1974). Formation of haloforms during the chlorination of natural water. Wat. Treat. Exam. 23, 234–243.
4. WHO (World Health Organization). (2002). Global Assessment of the state-of-the-science of Endocrine disruptors. www.who.int/ipcs/publications/newissues/ endocrine_disruptors/en/.
5. Proposal for a DIRECTIVE OF THE EUROPEAN PARLIAMENT AND OF THE COUNCIL on environmental quality standards in the field of water policy and amending Directive 2000/60/EC (presented by the Commission){COM(2006) 398 final}{SEC(2006) 947}Brussels, 17.7.2006, COM(2006) 397 final.
6. Ying GG, Williams B, Kookana R. (2002). Environmental fate of alkylphenols and alkylphenol ethoxylates-a review. Environ. Int. 28, 215–226.
7. Stasinakis AS, Thomaidis NS, Nikolaou A, Kantifes A. (2005). Aerobic biodegradation of organotin compounds in activated sludge batch reactors, Environmental Poll. 134, 431–438.
8. Andreozzi R, Raffele M, Nicklas P. (2003). Pharmaceuticals in STP effluents and solar photodegration in aquatic environment. Chemosphere 50, 1319–1330.
9. Birkett, JW, and Lester, JN. (2003). Endocrine disrupters in wastewater and sludge treatment processeses, Lewis Publishers, Chelsea, MI, USA.
10. Korner W, Bolz U, Submuth B, Hiller G, Schuller W, Hanf V, et al. (2000). Input/output balance of estrogenic active compounds in a major municipal sewage plant in Germany. Chemosphere 40, 1131–1142.
11. Johnson AC, Sumpter JP. (2001). Removal of endocrine disrupting chemicals in activated sludge treatment works. Environ. Sci. Tech. 35, 4697–4703.

12. Auirol M, Filali-Meknassi Y, Tyagi RD, Adams CD, Surampalli RY. (2006). Endocrine distrupting compounds removal from wastewater: a new challenge. Process Biochem. 41, 525–539.
13. Belgiorno V, Rizzo L, Fatta D, Della Rocca C, Lofrano G, Nikolaou A, Naddeo V, Meric S. (2007). Review on endocrine disrupting-emerging compounds in urban wastewater: occurence, removal by photocatalysis and ultrasonic irradiation for reuse. Desalination special issue, 215, 166–176.
14. Bekbölet M. (1997). Photocatalytic bactericidal activity of TiO_2 in aqueous suspensions of *E. coli*. Wat. Sci. Tech. 35(11–12), 95–100.
15. WHO (World Health Org. (2006). WHO Guidelines for the Safe Use of Wastewater Excreta and Greywater. Vol II: Wastewater Use in Agriculture.
16. USEPA (2004.) Guidelines for Water Reuse. EPA/625/R-04/108.
17. Boyd GR, Reemtsma H, Grimm DA, Mitra S. (2003). Pharmaceuticals and personal care products (PPCPs) in surface and treated waster in USA and Ontario, Canada. Sci. Tot. Environ. 311, 135–149.
18. European Commision (EC). (2000). Directive of the European Parliament and of the council 2000/60/EC establishing a framework for community action in the field of water policy, Official Journal C513, 23/10/2000.
19. DECRETO LEGISLATIVO (DLgs). (2006). 3 April 2006, No. 152, published in Gazzetta Ufficiale, n 88, 14 04 2006, Suppl Ordinario n 96, Italy.
20. Achilleos A, Kythreotou N, Fatta D. (2005). Development of Tools and Guidelines for the Promotion of the Sustainable Urban Wastewater Treatment and Reuse in the Agricultural Production in the Mediterranean Countries (MEDAWARE) Task 5: Technical Guidelines on Wastewater Utilisation. http://www.uest.gr/medaware/
21. Fatta D, and Anayiotou S. (2007). MEDAWARE Project for wastewater reuse in the Mediterranean countries: An innovative compact biological wastewater treatment system for promoting wastewater reclamation in Cyprus. Desalination special issue, 211, 34–47.
22. Hidalgo D, Irusta R, Martinez L, Fatta D, Papadopoulos A. (2007). Development of a multi-function software decision support tool for the promotion of the safe reuse of treated urban wastewater. Desalination special issue, 215, 90–103.

GUIDELINES FOR IRRIGATION MANAGEMENT OF SALINE WATERS ARE OVERLY CONSERVATIVE

John Letey
Distinguished Professor of Soil Science, Emeritus
University of California, Riverside, CA USA

Corresponding author: John.letey@ucr.edu

Abstract:

A general trend exists toward increasing the salt concentration of surface and ground waters in the semiarid regions of the world. Quantitative knowledge on the interaction between irrigation management and crop yield is becoming increasingly important as water salinities increase. Guidelines for irrigation management with saline waters which have been based on steady-state analyses overpredict the amount of irrigation required, and/or underestimate the yield that can be achieved when irrigating with saline waters.

Keywords: Wastewater, salinity, leaching fraction, irrigation, irrigation guidelines.

1. INTRODUCTION

A general trend exists toward increasing the salt concentration of surface and ground waters in the semiarid regions of the world thus jeopardizing their utility. Irrigation of crops contributes to this trend by concentrating root-zone soil water through evapotranspiration of pure water. The urban sector increases the water salinity by adding salts to the water as it is used and passed through the city. Salts are not removed by the typical sewage treatment facilities.

Quantitative knowledge on the interaction between irrigation management and crop yield is becoming increasingly important as water salinities increase. This information is important to the farmer, water resource managers, and to regulatory agencies that might be considering imposing saline water quality standards for streams specifically designed to be protective of agricultural production. The initial guidelines for managing saline irrigation waters, which are still extensively used, were based on steady-state analyses. Subsequently, with the increased knowledge of the physical–chemical–biological interactions that occur in the soil–water–plant matrix, and the advent of high-speed computers, models have been developed that take into account the dynamic interactions.

M.K. Zaidi (ed.), Wastewater Reuse–Risk Assessment, Decision-Making and Environmental Security, 205–218.

Steady-state analyses are simpler than transient-state analyses. The common assumption is that with time, a transient system will converge into a steady-state case and provide justification for the steady-state analyses. However, in a soil–water system, steady-state implies that water is continuously flowing at a constant rate. This condition never exists in the upper soil profile on lands which are not submerged under water. Thus the validity of conclusions drawn from steady-state analysis is questionable. The dynamic versus steady-state approaches to evaluate irrigation management of saline waters were studied. They concluded that the steady-state analysis overpredict the amount of irrigation required, and/or under estimate the yield that can be achieved when irrigating with saline waters [1].

1.1. Basic Steady-State Analyses Considerations

Plants extract and transpire pure water. Thus the salt concentration in the soil-water increases proportionately to the amount of extracted water. To avoid the accumulation of salts in the root zone to levels that decrease yield, excess salts must be periodically leached from the root zone by irrigating with more water than is needed for evapotranspiration (ET). The leaching requirement (LR) was defined as the minimum fraction of the total amount of applied water that must pass through the soil root zone to prevent reduction in crop yield from the excessive accumulation of salts [2]. A major focus of the steady-state analysis has been to quantify the LR based on crop tolerance to salinity and the irrigation water salinity. Typically a mass balance approach was taken assuming no dissolution or precipitation of salts within the root zone. Under these conditions, mass balance dictates that the concentration of the water leaving the root zone is equal to the irrigation water salinity divided by the leaching fraction (LF). LF is the fraction of the applied water (AW) that percolates below the root zone (DP), or LF = (AW − ET)/AW. Hereafter, the water salinity will be characterized and quantified by the electroconductivity (EC) of the water. The EC of 1 dS/m is approximately equal to a salt concentration of 640 mg/l. The EC of the drainage water (EC_d) is equal to the EC of the irrigation water (EC_i) divided by LF.

As the water flows through the root zone, water is extracted by the plant. Therefore, the salt concentration increases with increasing depth. The distribution of the soil-water concentration with depth depends upon the water extraction pattern within the root system. The resulting concentration also depends upon LF. A low LF results in a high soil-water concentration because most of the water is extracted before it leaves the root zone. A relationship between the salt concentration in the root zone and crop growth must be known to interpret the effects on crop production.

Crop tolerances to salinity data were reported by Maas and Hoffman [3] for numerous crops. They quantified salt tolerance by two parameters: the threshold and the rate of yield decline as salinity increases beyond the threshold. These parameters have been referred to as the Maas and Hoffman coefficients. Most of the data on tolerance of crops to salinity were obtained from artificially salinized plots that were managed by means of a high LF to obtain a nearly uniform salt distribution throughout the root zone. Under these conditions, all roots were exposed to a similar level of salinity.

Although the plant would be expected to react to the salt concentration in the soil-water surrounding the root (EC_{sw}), the Maas and Hoffman coefficients are reported in units of EC of the saturated soil extract (EC_e). For the experimental conditions from which the data were derived, EC_e is approximately $EC_{sw}/2$. Furthermore, since the experimental conditions created fairly uniform salt distribution through the root zone, the threshold EC_e (EC_e*) is assumed to represent the average root zone salinity. These two assumptions: EC_e equals $EC_{sw}/2$, and the EC_e* represents the average root zone salinity have been made in numerous reports.

Ayers and Westcot [4] illustrated the salt concentration with depth for various values of LF. Assuming that the water extraction pattern was 40-30-20-10 whereby the number represents the fraction of roots from the upper through lower quarter of the root zone. They then took the linear average of the salt concentration and reported the concentrating factor for each value of LF.

The assumption that the plant response to the linear average ECSW in the root zone is not supported by experimental evidence for crops. This assumption indicates that the salinity in the least dense root zone is equally effective in affecting water uptake as the salinity in the most dense root zone. Gardner [5] reviewed the water extraction patterns with depth from several experiments and concluded that water is initially extracted from the regions nearest the surface where the roots are more prolific with the zone of extraction progressing downward in the profile as the water becomes limiting in the upper zones. Furthermore, there is evidence that plants will extract more water from portions of the root zone where the water status is good to compensate for reduced water uptake for parts of the root zone where the water status is poor. For example, van Schilfgaarde et al. [6] analyzed data from an experiment with alfalfa and concluded "apparently as long as the roots have access to water of low salinity, they are able to utilize some water of higher salinity without adverse effects." They found that the roots extracted by far the largest amount of water from the upper less saline zone. Hoffman and van Genuchten [7] calculated the root-weighted average salinity rather than the linear average. The root-weighted average is lower than the linear average concentrating factor. Rhoades [8] proposed the following relationship:

$$LR = EC_i/(5EC_e^* - EC_i) \tag{1}$$

A comparison between the LR calculated using the linear salt concentration, the root-weighted average salinity concentration, and Rhoades equation results in the highest LR for the linear average, the lowest LR for the root-weighted average, and an intermediate LR calculated from Equation 1.

Irrigation to achieve maximum yield, consistent with the LR requirement concept, may not be the economically efficient practice. The economic goal is to maximize profits, and maximum yield may not produce the highest economic benefit. Furthermore the matric water stress, associated with soil dryness; as well as the osmotic stress, associated with salinity, must be considered in assessing the consequences of irrigation management practices. Letey et al. [9] developed a crop-water production function model for saline irrigation waters, and the simulated functions for several crops using this model were reported by Letey and Dinar [10]. The model, which was based on steady-state considerations, allows the computation of relative crop yield, amount of drainage water, and the EC of the drainage water for various amounts of applied irrigation water of given salinity for crops based on the crop tolerance to salinity. These functions can be used for the economic analyses of irrigation management.

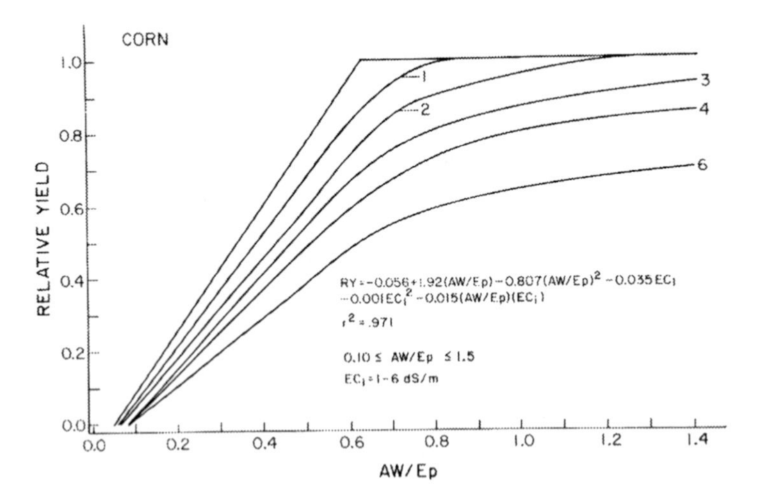

Figure 1. Each curve is for given EC of irrigation water (dS/m) [10]

The computed amount of applied water from the model that produces maximum yield would be the amount that produces the LR. However, the same shortcomings identified above for the steady-state approach are inherent in the model. Specifically the relationship between the average EC_e, EC_i, and LF is the relationship proposed by Raats [11] based on steady-state conditions. In fact, the concentrating factor related to LF from the Raats relationship are greater the ones reported above from other sources. Although the model computations of LR are comparable to the other steady-state analyses, the consequences of applying less water on crop yield differs significantly from the other steady-state approaches. A linear relationship between dry matter production and ET, consistent with the reported results of numerous experiments, is part of the model. A reduction in yield reduces ET so that for the same amount of water applied, a part of it will cause leaching and reduce the soil salinity. As a consequence, the water application asymptotically approaches the amount that provides maximum yield [9, 10]. The relationships between the amount of applied water and EC_i to relative crop yield are illustrated in Figure 1 for corn (*Zea mays* L.) and in Figure 2 for cotton (*Gossypium hirsutum* L.). Corn is more sensitive to salinity than cotton. A fairly large reduction in water application, below that which produces maximum yield, results in only a small (often less than 5%) reduction in computed yield. Therefore, although the model overpredicts the LR, it provides more reasonable relationships between applied water and yield at lower water application rates.

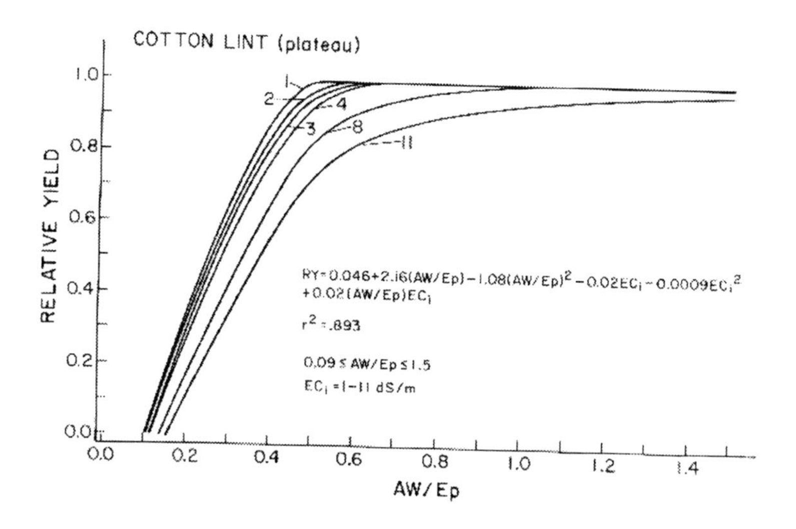

Figure 2. Computed relative yields of cotton for various quantities of applied water which are scaled to pan evaporation (Ep). Each curve is for given EC of irrigation water (dS/m) [10]

Clearly the steady-state approach that has been extensively used has many deficiencies. Very high concentrating factors, which differ depending on the assumptions used, are computed at the lower values of LF. As such low LFs were not prescribed, even for water relatively low in salinity. A "true" steady-state condition is never achieved in irrigated agriculture, even if identical management is imposed for a long time. However, steady-state was the best approach available until the advent of high-speed computers facilitated the use of transient models.

Most analyses of salinity, including the material above, have assumed that the salt concentration was conservative, i.e., there was no precipitation or dissolution of salts. However, depending on the chemical composition of the irrigation water and the minerals in the soil, salts in the soil-water may precipitate or soil minerals dissolve and modify the soil-water concentration. The most probable effect in irrigated lands in the semiarid regions of the world is for salt precipitation. Low LF contributed to increased soil-water concentration and increases the probability for salt precipitation. Thus failure to account for precipitation leads to an overestimate of the LR required to achieve high yield. WATSUIT is a model that can be used to compute the expected precipitation of salt from irrigation of various chemical compositions as related to LF [12, 13].

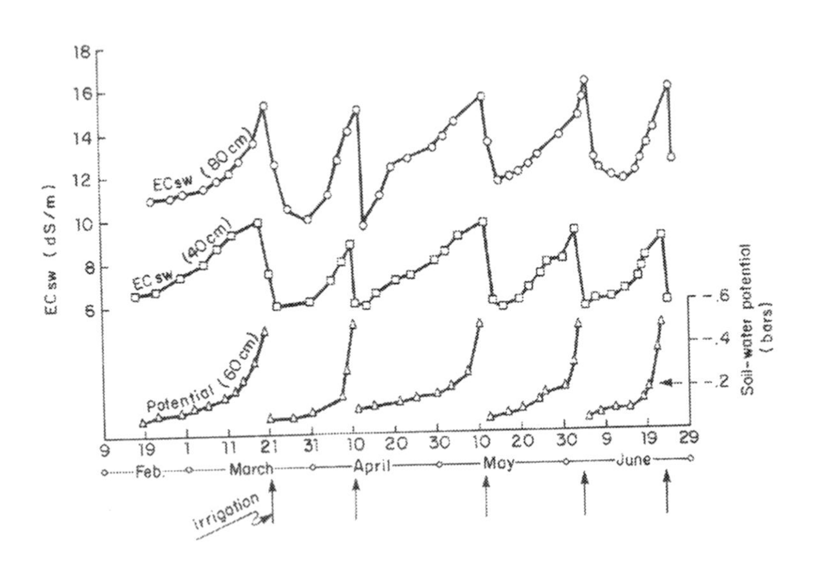

Figure 3. Change in salinity of soil-water (EC_{sw}) between irrigations of alfalfa due to ET use of stored water [14]

One condition of steady-state is that the water content or salt concentration at a given point remains constant with time, which can be achieved with constant steady water flow. This condition never exists in the root zone of a crop. Both water and EC_{sw} are continually changing with time in the root zone as water is extracted. With irrigation, water flushes the salts downward and replenishes the water in the soil. This phenomenon is illustrated in Figure 3 from measurements of salinity and water as a function of depth and time as reported by Rhoades [14]. If the same management pattern is used over a period of time, the pattern of water and EC_{sw} with time will repeat, but this is not truly steady-state.

2. TRANSIENT-STATE MODELS

Transient-state models are required to account for all of the time dependent variables encountered in the field. These variables include switching crops with different salinity tolerance, variable irrigation water salinity including rainfall that is pure, timing and amount of irrigation, initial soil salinity conditions, etc. Some basic concepts concerning transient-state models are as follows.

The water flow and salt transport equations are the basic components of transient-state models. The water flow, which takes into account the water uptake by plant roots, is quantified using the Darcy-Richards equation:

$$\frac{\partial \theta}{\partial t} = \frac{\partial}{\partial z}\left[K(\theta)\frac{\partial h}{\partial z} + K(\theta) \right] - S \tag{2}$$

where θ is the volumetric soil-water content, t is time, z is soil depth taken positive downward, K is the hydraulic conductivity, h is soil-water pressure head, and S is the root water uptake term. Salt transport is calculated using the convection–dispersion equation for a nonreactive, noninteracting solute:

$$\frac{\partial (c\theta)}{\partial t} = \frac{\partial}{\partial z}\left[\theta D\frac{\partial c}{\partial z} - qc \right] \tag{3}$$

where c is the concentration of salt, D is the dispersion coefficient, and q is the volumetric water flux. Salt uptake by the plant is assumed to be negligible. The nonlinearity of Equations 2 and 3 requires solution by numerical methods which is facilitated by high-speed computers. The appropriate characterization of the plant root uptake function, S, is necessary for the analyses to accurately depict behavior in the field. Cardon and Letey [15] evaluated two major types of uptake functions that had been reported in the literature. They concluded that the best term had the form:

$$S = \alpha(h, \pi) S_{max} \quad (4)$$

where α is a dimensionless empirical stress response function, π is the osmotic pressure head, and S_{max} is the maximum water uptake by the plant (potential transpiration) that is not stressed. They further used the relationship proposed by van Genuchten [16] for α so that Equation 4 becomes

$$S = \frac{S_{max}}{1 + \left(\frac{ah + \pi}{\pi_{50}} \right)^3} \quad (5)$$

where a accounts for the differential response of the crop to matric and osmotic pressure head influences and is equal to the ratio of π_{50} and h_{50} where 50 represents the values at which S_{max} is reduced by 50%.

Note that these models do not compute the actual yield of a crop. They compute the yield relative to that which would be achieved if there was not any water stress causing a decreased yield.

Pang and Letey [17] programmed the model to allow the potential amount of water uptake until α (h, π) was equal to the threshold value. They further programmed the model to allow extra water from nonstressed root zone to compensate for reduced water uptake from stressed zone of the roots. The latter feature is consistent with the experimentally determined behavior of plants.

The total ET for a crop season is usually assumed to be only dependent on the crop and climate, i.e., the plant size is not considered to be a factor affecting ET. However, the scientific literature consistently reports that plant transpiration is linearly related to plant dry matter production. This finding significantly affects soil-water flow and chemical transport through soil when plants are growing. Models should, therefore, include a feedback mechanism between the soil–water–chemical status and plant growth, which affects water flow and chemical transport. For example, assume that the soil-water salinity is sufficiently high to stress the plant and reduce plant growth. The smaller plant will transpire less water than a nonstressed plant. With the reduced transpiration, more water will flow through the soil for a given amount of applied water. The increased water flow through the soil will increase the transport (leaching) of salt. Cardon and Letey [15] accounted for this phenomenon by adjusting the value of S_{max} with time to account for reductions in plant size.

2.1. Comparison of Steady-State and Transient-State Simulations

Computations were done using Equation 1, the crop-water production function model [9], and the ENVIRO-GRO model [17] for the production of corn when irrigated with water of 1 or 2 dS/m. The Maas and Hoffman coefficients for corn are a threshold of 1.7 dS/m and a slope of 12 beyond the threshold. The initial soil salinity selected influences the crop yield in transient-state models for the first and possibly more years. Therefore the ENVIRO-GRO simulations were conducted for successive years until the initial soil condition was no longer a factor. The results that are reported are for this time because they would most closely approximate the steady-state conditions. Comparisons between the three models were made for 85, 90, 95, and 100% production and presented in Table 1. The irrigation amount is normally reported in terms of applied water (AW) rather than in LF. Therefore the three comparisons were done on the comparable basis of the ratio of AW to the potential evapotranspiration (PET).

Table 1. Comparison of the calculated ratio of applied water to potential evaportranspiration that produced RY of corn using Equation 1, the crop-water production function model (WPF), and the ENVIRO-GRO model (E-G)

Specified relative yield (RY)	Equation 1	WPF	E-G
	AW/PET	AW/PET	AW/PET
	Irrigation water salinity of 1 dS/m		
1.00	1.15	1.28	<1.05
0.95	1.11	1.14	<1.05
0.90	1.09	1.05	<1.05
0.85	1.08	1.00	<1.05
	Irrigation water salinity of 2 dS/m		
1.00	1.44	2.04	1.17
0.95	1.31	1.56	1.09
0.90	1.24	1.28	<1.05
0.85	1.19	1.14	<1.05

The results from the steady-state approaches are dependent on the assumptions made in computing the relationship between LF and the salt distribution in the rooting zone. Different concentrating factor were reported above for the various approaches. The water production function model had the highest concentrating factor of the reported cases. Thus the irrigation requirement to achieve maximum yield was higher for the production function model than Equation 1 (Table 1). However, because the production function model accounts for extra leaching as a result of decreased yield, a lower AW is specified from the water production function model than Equation 1 for the lower yields. Significantly the transient-state analysis simulates a very much lower water requirement than the steady-state approaches under all conditions. The lowest AW/ET used in the transient-state analysis was 1.05.

The reasons that the transient-state analysis simulated a much lower irrigation amount than the steady-state approaches for a given yield are as follows: The steady-state approaches assumed that the plant responded to the average root zone salinity that increased greatly as the LF decreased. However the major amount of water is extracted from the upper part of the root zone. Furthermore, the salt concentration at a given depth in the field does not remain constant with time, but is continually changing as illustrated in Figure 3. The salts become concentrated by water extraction, but the irrigation water "flushes" the salts downward thus reduces the concentration to a lower value at a given depth after irrigation. The concentration immediately after irrigation near the surface would be close to the concentration in the irrigation water. For most soils, the volumetric soil-water content would be reduced by less than half between irrigations. Thus the salts would concentrate by less than two between irrigations. Note that the salts concentrated by a factor less than two in the experiment reported by Rhoades [14] as illustrated in Figure 3. Therefore as a general guideline, a water with a salt concentration equal to the Maas and Hoffman threshold value can be used and irrigated with a relatively low LF. This conclusion is based on the fact that the Maas and Hoffman coefficients are on the basis of EC_e which is about $EC_{sw}/2$. The soil-water can therefore be concentrated by a factor of two without exceeding the threshold value. The threshold value for corn is 1.7 dS/m. Note in Table 1 that the ENVIRO-GRO model simulated maximum yield for irrigation water of 2.0 dS/m with an AW/PET of 1.17 (LF of 0.15). Maximum yield was simulated for irrigation water salinity of 1.7 when AW/PET = 1.12.

None of the models used to simulate the results presented in Table 1 account for salt precipitation. Thus, the results are conservative because most irrigation waters would have some salt precipitation at the low LF and reduce the hazard. However, the effect may not be too great because most of the salts would be precipitated at the lower part of the root zone where they are most concentrated, and the salinity at the lower part of the root zone does not greatly impact the plant response.

2.2. Other Irrigation Management Factors

Uniformity of irrigation, rainfall effects, and frequency of irrigation are factors which must be considered in designing optimum irrigation strategies. All of the analyses reported here are for one-dimensional flow and assume that the irrigation is uniform, i.e., the same amount of water infiltrates in all parts of the field. This does not occur in the field except for well designed and managed pressurized irrigation systems like drip. Optimal irrigation management must consider the uniformity of irrigation. Application of water in the amount specified for uniform irrigation causes parts of the field to be over irrigated and other parts of the field to be under irrigated. Thus, there is an unavoidable trade-off between irrigation to maximize yield and irrigation to minimize deep percolation. Letey et al. [18] proposed a method of computing the average field yield to average water application if the distribution of irrigation uniformity is known. A major problem is in quantitatively characterizing and measuring irrigation uniformity.

The steady-state approaches cannot account for rainfall effects. The dynamic approaches must consider rainfall because the initial soil conditions must be specified and then the amount, timing, and salinity of the irrigation water must be programmed. Rainfall, even if it does not occur during the growing season decreases the impact of irrigating with saline waters. This is another reason why the steady-state analyses overestimate the LR or overestimate the negative consequences of irrigating with saline waters. Indeed, if off-season precipitation leaches most of the salts from the root zone, the application of larger quantities of saline water for the supposed intent of having a high leaching fraction add more salt to the soil and causes the root zone salinity to increase more rapidly than applying less water. Conversely if the soil has a high salinity level at the beginning of the season, the extra water for leaching would have a positive effect. All of these possibilities are included in the transient-state analyses but neglected in the steady-state analyses.

Based on the transient-state analysis by Bradford and Letey [19], the effect of rainfall can be approximated by calculating the weighted average of the water salinity including rainfall and the salinity of the irrigation water. The rain during the off-season can be as effective as during the season because it reduces the salinity of the soil at the time of seeding. So if the amount of rainwater that infiltrates the soil is equal to the amount of irrigation, the irrigation water would behave as if its salinity was halved.

More frequent irrigation was considered to be a good strategy when irrigating with saline waters. The rationale for this conclusion is that the soil-water salinity increases in direct proportion to the water that is removed through the plant roots and that the salinity does not reach the higher concentrations if irrigation is frequent and the soil-water content is not low. This seems logical, but experimental findings do not generally support the benefits of more frequent irrigation [20]. The steady-state models have no basis for addressing this matter. Conversely, the transient-state models incorporate the time-dependent soil-water and soil-salinity levels on a root spatial basis, and the effects of irrigation frequency can be evaluated. Considering the fact that the water is initially extracted near the surface and then progressively downward with time, frequent irrigations prescribe small amounts of water that do not penetrate deeply. Thus the salts tend to accumulate in the upper part of the root zone which is the zone of maximum water uptake. With less frequent irrigations, water is removed from a larger portion of the root zone. Subsequently, irrigation with a higher amount to recharge the soil leaches salts deeper into the root zone. Thus the initial salinity in the upper part of the root zone immediately after irrigation is less than occurs with more frequent irrigation.

2.3. Implication on the Transport of Other Chemicals through Soil

This report has focused on salinity and the transport of salts through the soil. Some of the findings of this study however have significant implications to the transport of other chemicals that are transported by water flowing through the soil. The amount of transport is very dependent on the water flow. One factor that is always neglected in the transport analysis is the implication of ET as it affects water flow. The ET is generally considered to be a fixed number for a given crop and climate. The literature is replete with experimental evidence that transpiration is generally linearly related to total plant mass production. In other words, larger plants transpire more water than smaller plants under the same climatic conditions. A reduction in transpiration translates into more water flow for a given amount of water application through irrigation or precipitation. Therefore the effects on plant growth should not be ignored in evaluating the transport of soluble chemicals through soil.

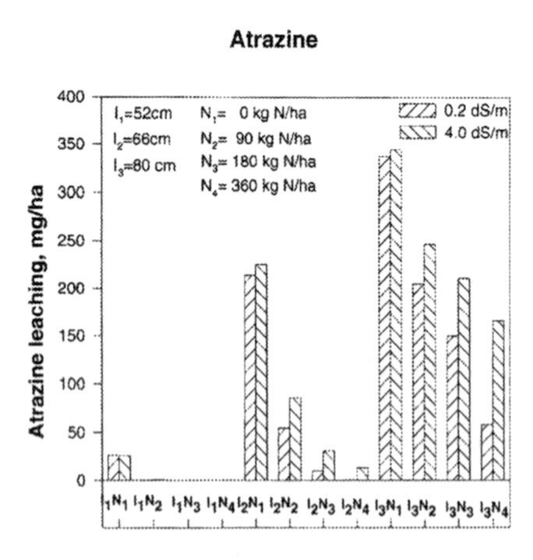

Figure 4. The effects of irrigation (I), nitrogen application (N), and irrigation water salinity (dS/m) on the simulated leaching of atrazine [21]

Groundwater degradation by nitrate is a common occurrence in irrigated agriculture. An obvious factor in the degree of degradation would appear the amount of nitrogen applied. However, some unexpected consequences may happen. If soil-nitrogen is inadequate to sustain maximum plant growth, the reduced plant growth causes less transpiration which increases the transport of nitrate through the soil resulting in even less nitrogen to support plant growth. Pang and Letey [17] used an ENVIRO-GRO model and illustrated the significant interaction between nitrogen supply, plant growth, and nitrogen leaching.

Attempts to reduce nitrate leaching by reducing nitrogen application may be counterproductive under some circumstances. Pesticide leaching was also demonstrated to be affected by the amount of irrigation, salinity, and nitrogen application [21]. Excessive water application, excessive salinity, or deficient nitrogen all contributed to increased leaching of pesticides below the root zone as illustrated in Figure 4. The main conclusion from these reports is that plant growth can significantly impact the transport of chemicals through soil and this factor must be included in the evaluation of any chemical transport.

3. CONCLUSIONS

Transient-state models, such as ENVIRO-GRO, can be used to track crop yield and soil salinity as a function of time for a management sequence. The management sequence can include initial soil salinity, shifting crops, rain, irrigation water quality, etc. Plant growth can significantly impact water flow which transports chemicals through the soil and this effect can not be ignored in a quantitative analysis of chemical transport.

Acknowledgment: I am thankful to the NATO ARW organizing committee to provide financial support to make this presentation possible.

4. REFERENCES

1. Letey J, Feng GL. (2007). Dynamic versus steady-state approaches to evaluate irrigation management of saline waters. Ag. Water Manag. (under review).
2. U.S. Salinity Laboratory Staff. (1954). Diagnosis and improvement of saline and alkali soils. Ag. Handbook. No. 60, USDA, Washington, D.C.
3. Maas EV, Hoffman GJ. (1997). Crop salt tolerance—Current assessment. J. Irrig. And Drainage Div., ASCE, 103, No. IR2, Proc. 12993, 115–134.
4. Ayers RS, Westcot DW. (1985). Water Quality for Agriculture. FAO Irrigation and Drainage Paper 29. FAO, United Nations. Rome. 174.

5. Gardner WR. (1983). Soil properties and efficient water use: An overview. In: Taylor HM, Jordan WR, Sinclair TR. (Eds.) Limitations to Efficient Water Use in Crop Production. Amer. Society of Agronomy. Madison, WI. 45–64.
6. Van Schilfgaarde J, et al. (1974). Irrigation management for salt control. J. Irrig. Drain. Div., ASCE. 100 (No. IR3). Proc. Paper 10822, 321–338.
7. Hoffman GJ, Van Genuchten MT. (1983). Soil properties and efficient water use: Water management for salinity control. In: Taylor HM, Jordan WR, Sinclair TR. (Eds.) Limitations to Efficient Water Use in Crop Production. Am. Soc. Agron. 73–85.
8. Rhoades JD. (1974). Drainage for salinity control. In: van Schilfgaarde, J. (Ed.) Drainage for Agriculture. Am. Soc. Agron. Monograph No. 17, 433–462.
9. Letey J, Dinar A, Knapp KC. (1985). Crop-water production function model for saline irrigation waters. Soil Sci. Soc. Am. J. 49, 1005–1009.
10. Letey J, Dinar A. (1986). Simulated crop-water production functions for several crops when irrigated with saline waters. Hilgardia 54, 1–32.
11. Raats PAC. (1974). Steady flow of water and salt in uniform soil profiles with plant roots. Soil Sci. Soc. Am. Proc. 38, 717–722.
12. Rhoades JC, Oster JD, Ingvalson RD, Lucker JM, Clark M. (1974). Minimizing the salt burdens of drainage waters. J. Env. Qual. 3, 311–316.
13. Oster JD, Rhoades JD. (1990). Steady state root zone salt balance. In: Tanji, KK. (Ed.) Agricultural Salinity Assessment and Management. ASCE No. 71 Manuals and Reports on Engineering Practice. Am. Soc. Civil Engr. 469–481.
14. Rhoades JD. (1972). Quality of water for irrigation. Soil Sci. 113, 277–284.
15. Cardon GE, Letey J. (1992). Plant water uptake terms evaluated for soil-water and solute movement models. Soil Sci. Soc. Am. J. 32, 1876–1880.
16. Van Genuchten MT. (1987). A numerical model for water and solute movement. Res. Rep. 121 USDA-ARS, U.S. Salinity Lab. Riverside, CA.
17. Pang XP, Letey J. (1998). Development and evaluation of ENVIRO-GRO, an integrated water, salinity, nitrogen model. Soil Sciences Society of America (SSSA) Journal 62, 1418–1427.
18. Letey J, Vaux HJ, Feinerman E. (1984). Optimum crop-water application as affected by uniformity of water infiltration. Agron. J. 76, 435–441.
19. Bradford S, Letey J. (1992). Cyclic and blending strategies for using nonsaline and saline waters for irrigation. Irrig. Sci. 13, 123–128.
20. Shalhavet J. (1994). Using water of marginal quality for crop production: major issues. Ag. Water Manage. 25, 233–269.
21. Pang XP, Letey J. (1999). Pesticide leaching sensitivity to irrigation, salinity and N application: model simulation. Soil Sci. 164, 922–929.

PRICING OF WATER AND EFFLUENT IN A SUSTAINABLE SALT REGIME IN ISRAEL

Orr Goldfarb and Yoav Kislev
The Water Commission, Tel Aviv 61203, Israel
Hebrew University, Rehovot 76100, Israel

Corresponding author: kislev@agri.huji.ac.il

Abstract:

Water withdrawal and irrigation in arid zones increase salt concentration in aquifers. The utilization of effluent further augments the concentration by adding salt from households and industry. A sustainable salt regime can be maintained if salt is removed from at least some of the water sources. The paper analyzes theoretically the pricing of water and effluent in a sustainable regime for the coastal aquifer in Israel.

Keywords: Irrigation, salt concentration, effluent, pricing, sustainability.

1. INTRODUCTION

A substantial proportion of water used in the urban sector finds its way to the sewerage and is treated and recycled as effluent. Urban users add salt to the water and are carried in the effluent. Consequently irrigation with recycled water adds salt to the soil and the water beneath its surface; the accumulated salt is detrimental to soil structure and plants. A sustainable salt regime is a set of policies maintaining salt concentration at a constant level and preventing its accumulation. The analysis of a sustainable salt regime cannot be limited to effect effluent and a comprehensive approach is will cover both quantities and prices. Prices are part and parcel of any water policy: they affect supply and use of the resources. An economic analysis of a sustainable salt regime in the coastal region, hydrology of water and salt in the region are presented, alternative policies are compared, and associated prices are derived [1].

1.1. Background

Israel has three major water reservoirs: Lake Kinneret (the Sea of Galilee, the coastal aquifer along the shore of the Mediterranean Sea, and the Mountain aquifer, partly under the hills of the West Bank [2]. The coastal region, receives water from all three reservoirs. The water is supplied from here to the region's urban centers and agriculture [3].

M.K. Zaidi (ed.), Wastewater Reuse–Risk Assessment, Decision-Making and Environmental Security, 219–225.

2. METHODS

2.1. An Illustration

In the illustration (Figure 1) precipitation is added to the groundwater (replenishment), part of the water is withdrawn for irrigation and the rest is outflow to the sea. Irrigated water evaporates from the surface of the land and plants and part of it reaches the groundwater as irrigation return flow. In parallel to the water flow, salts are recorded in the diagram: concentration in ppm chlorides in parentheses and a quantity in tons. These magnitudes will be explained below.

2.2. The Algebraic Model, Water and Salt [4]

Quantities in the illustrative model are flows per year. The variables are

	Water in MCM	Salts, chlorides in tons/y
Replenishment	R	
Autonomous salts		Δ
Irrigation (freshwater)	H	M_H
Irrigation return flow	Z	M_Z
Outflow to sea	Y	M_Y
Evapotranspiration	E	

As indicated above, salts are added to aquifers from ocean spray, underground brines, and seawater intrusion. These sources are termed here autonomous since the amount of salt added to the aquifer in this way is not a function of the quantity of water used (in the coastal aquifer, in reality, there is also entry of salt from noncoastal sources in quantities proportional to the water used). Hence, we treat the replenishment as if it did not carry any salt and write the autonomous amounts separately. The balancing equations for a reservoir in the steady state are

Condition	Balancing equation	Equation number
Water balance	$R + Z = H + Y$	Equation 1
Salt balance	$\Delta + M_Z = M_H + M_Y$	Equation 2
Irrigation return flow	$Z = 0.17H$	Equation 3
Irrigation return flow	$H = E + Z$	Equation 4

Equations 1 and 2 describe the entry of water and salts into the reservoir and exit away from it. The water supply is augmented with the irrigation return flow. Salt comes from autonomous sources. Equation 3 defines that the return flow is 17% of the quantity of water in irrigation (an assessment received from hydrologists). Equation 4, completes the picture, it separates irrigation water to the part evaporated and the part returning to the reservoir.

For the sake of the illustration, let us assume (water in MCM/Y; salt, chlorides, in tons per year): replenishment $R = 90$, autonomous salt $\Delta = 4{,}500$. Groundwater outflow to the Mediterranean Sea $Y = 30$. As indicated, in the steady state, entering quantities are identical to quantities leaving the aquifer. By Equation 1, the balance for water is $90 + 0.17H = H + 30$

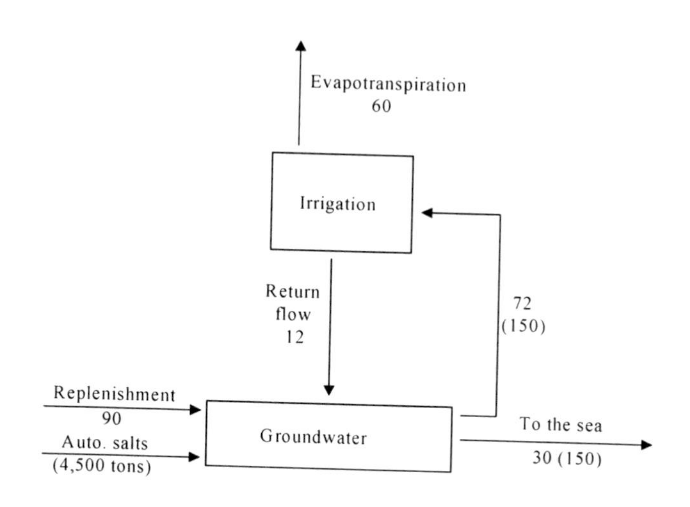

Figure 1. Conceptual illustration, water flows

Hence irrigation $H = 72$, return flow $Z = 12$, evapotranspiration $E = 60$ (Figure 1).

To calculate the concentration of salt in the flow of water in the model, in the steady state, salt added to the aquifer leaves it. Also, salt deposited by irrigation move to the aquifer with the return flow; hence $M_Z = M_Y$ and these variables may be dropped from Equation 2. Accordingly, define the concentration of salt in the outflow to the sea as P and write the equality $\Delta = M_Y$

$$4500 = 30P \tag{1}$$

The concentration of salts in the outflow to the sea is 150 ppm chlorides. Also, since groundwater is the source of the outflow, the concentration of chlorides in the groundwater will be 150 ppm as well.

To sharpen the intuitive grasp the concept of the steady state notice that the endogenous variable in the model is the concentration of salt in the groundwater and consider a slight modification of the basic data: assume that exit of water to the sea is not 30 MCM/Y but 25 MCM/Y. The solution of Equation 5 will now be 180 ppm chlorides (not 150). With a smaller quantity flowing to the sea, the concentration of salt in the aquifer is larger; the larger concentration ensures that, even with a smaller outflow, all the salt added from whatever sources are flushed to the sea (180 × 25 = 4,500).

2.3. The Coastal Aquifer

Only salt from autonomous sources entered the aquifer in the above illustration, and they all left the reservoir with water drained to the sea. Table 1 lists water and salt as forecasted for the coastal region for the year 2020. These are the data incorporated into our empirical analysis to be described below. The natural sources supplying to the coastal region, 640 MCM, are the coastal aquifer, Lake Kinneret, and the mountain aquifer, but only the salt carried by the water of the two last sources are listed in the table—the salts contained in water withdrawn from the coastal aquifer and used for irrigation above it are just recycled, they do not add to the quantity of salt in the reservoir.

As explained above, autonomous sources add salts that are not carried in water. The forecast is that 393 MCM of effluent will be used above the coastal aquifer in 2020; the salts in the effluent line are the salts added by urban users of water. Salts in the water supplied to households and other consumers in town are listed in other lines in the Table 1. Water is expected to leave the aquifer in 2020 through two channels, 30 MCM each, one is water withdrawn from the aquifer and exported to other regions, and the other is drainage to the sea. Repeating the calculation of the above illustration for the data of Table 1, one finds that a steady state could be maintained in the aquifer's water with salt concentration of 500 ppm. This concentration is however too high for households and agriculture; the sustainability of the aquifer cannot rely solely on natural processes and salt must also be removed actively.

Table 1. Quantities of water and salt in the coastal aquifer, 2020 forecast

	Water (MCM)	Salts (tons chlorides)
Natural sources	640	70,200
Autonomous		12,600
Des. seawater	225	4,500
Effluent	393	39,300
Export, freshwater	–30	
Drainage to the sea	–30	
Return flow	50	
Total	1,268	126,600

2.4. Prices and Extraction Levies

We set price equal to marginal cost. The economic rational behind marginal cost price is that they deliver the relevant information; they present the individual users of water, in any of the sectors, with the cost of the resource to society at large. With this policy, individuals can act freely, following their own private interests but, when doing so, directed to take into account the correct effects of their actions on others. Among the marginal cost items we include the scarcity value of water; its corresponding price is the extraction levy. Scarcity values arise when sources of water are utilized up to capacity—up to the safe yield. The scarcity value is a marginal cost since, when all available water is utilized; an additional unit supplied to one individual is taken away from another. The loss where the supply was reduced is the cost. Unlike conventional prices paid to the providers of water, the extraction levy is collected by the government; the government functions here as the representative of society, of the public at large, since the public is the owner of the resource.

The marginal cost is determined theoretically in a mathematical programming model presented in the Appendix. The model is both broader, in some aspects, and narrower (in others) than the framework of the discussion in the article. It incorporates agricultural production, a feature that is not included explicitly in the paper, but, for simplicity, import of mountain water and exit of water and effluent from the region, either to the sea or to other regions is disregarded in the formulation of the Appendix. The objective function of the programming model is the value of agricultural output *minus* the cost of the water economy. The sources of freshwater are the coastal aquifer, Lake Kinneret, and seawater desalination; effluent is used in agriculture. There are two consuming sectors in the model: urban and agriculture. The urban sector receives a predetermined quantity of water. A given ratio of the water used in this sector is collected as sewage and, after treatment, provided as effluent. Irrigation deposits salt on the surface of the land and identical quantities are added to the water in the aquifer. Additional salt come from autonomous sources (ocean spray and underground brines). Freshwater desalination is used to remove the salt. By assumption, prices are set equal to marginal cost.

In the low demand case, the coastal region is supplied with freshwater from the local aquifer and from Lake Kinneret; seawater is not desalinated. Salt is removed by desalination of natural water. The price of freshwater is determined by the marginal product of water in agriculture and it is set, in equilibrium (at maximum net income), to equal the cost of moving water from Lake Kinneret *plus* the cost per CM of removing from the water of the coastal aquifer the salts imported with the lake's water. The price farmers pay for the effluent is a fraction of the price of freshwater, the fraction representing the comparative productivity of the recycled wastewater.

In this low demand case, only the coastal water is scarce and has, in the model, a scarcity value. This value, and hence the extraction levy of coastal water, is equal to the price of freshwater *minus* the cost of its withdrawal from the aquifer. No scarcity value is attributed to the water of Lake Kinneret. The urban sector is seen in the program as if selling the effluent to agriculture; hence the net price urban dwellers pay for water equals the opportunity cost, the marginal productivity of water in agriculture, *plus* the cost of treating the sewage *minus* the price farmers pay for the effluent (recall that only part of the water used in town ends as effluent).

In the high demand case, seawater desalination is activated and the marginal productivity of water in agriculture is equal to the cost of desalination *plus* the cost of the removal of the (small amount) of salts left in the desalinated water. Desalinated water is supplied when the other water sources cannot satisfy the demand. Hence, in this case, the withdrawal constraint in Lake Kinneret is binding and the scarcity value of its water is positive; it is equal to the marginal productivity of water at the cost *minus* the cost of moving the water from the lake, and *minus* the cost of removing the salt carried in the Kinneret water. It is interesting to examine the difference in the cost of the lake's water in the two demand cases. In the low demand case, the users of water pay for the removal of the imported salts; the higher salt concentration in the water from Lake Kinneret, the higher the price of water. In the high demand case, on the other hand, water users pay the same price whatever the salt concentration in the lake's water is.

3. CONCLUSIONS

The utilization of effluent further augments the concentration by adding salts from households and industry. A sustainable salt regime can be maintained if slats are removed from the water resources. A theoretical analysis of pricing of water and effluent in a sustainable regime for the coastal aquifer in Israel was presented.

Acknowlegment: The financial support from the NATO Advisory committee is respectfully acknowledged.

4. REFERENCES

1. Kislev Yoav. 2006, The water economy of Israel, in: K. David Hambright, F (eds.) Water in the Middle East, Oklahoma U. Press, USA, pp. 127–150.
2. Goldfarb Orr and Kislev Yoav. 2002, A sustainable salt regime in the coastal aquifer, The Water Commission and Economic Research (Hebrew).

3. Fattal B, Lampert Y, Shuval H. 2006, A fresh look at Wastewater Irrigation in Agriculture, a risk assessment & cost effectiveness. IDRC Pub. pp. 1–14.
4. Simon CP, and Blume L. 1994, Mathematics for Economists, W. W. Norton, New York and London.

REGIONAL PLANNING OF WASTEWATER REUSE FOR AGRICULTURAL IRRIGATION AND RIVER REHABILITATION

G. Axelrad and E. Feinerman
Department of Agricultural Economics and Management,
The Hebrew University of Jerusalem, Rehovot 76100, Israel

Corresponding author: axelrad@agri.huji.ac.il

Abstract:

A single-year planning model being developed for a region in Israel which consists of a city and three potential wastewater consumers. The model incorporates, in one endogenous system, the economic, physical, and biological relationships in the water–soil–plant–environment system and its objective is to maximize the regional social welfare composed of the sum of agricultural and environmental net benefits. The model determines the optimal crop mix and the optimal allocation of the limited water and land resources among all potential users.

Keywords: Wastewater, allocation, linear programming, transferable utility games.

1. INTRODUCTION

In Israel, despite its modest role in the national product (less than 2% of the GDP), agriculture consumes about 60% of the nation's limited freshwater supply. There is a consensus among policy-makers and water experts that the supply of potable water (i.e. urban consumption) should receive top priority [1]. Therefore, the supply of reclaimed sewage and other alternative sources is expected to grow substantially due to increases in water consumption in the growing domestic and industrial sectors, and the expansion of irrigation with recycled effluents [2, 3]. Indeed, a large-scale transition in agricultural water use, from good-quality freshwater to treated wastewater, is expected within the next few years [4, 5]. This shift requires the development of many more environmentally safe water treatment plants, reservoirs, and conveyance systems. In addition, under certain conditions, the use of municipal wastewater for river rehabilitation is also an effective means of wastewater removal combined with significant economic advantages [6, 7].

The current research is divided into two linked parts. First, we develop a planning model which determines the optimal crop mix and the optimal allocation of the limited water and land resources among all potential users [8–10]. The model's

M.K. Zaidi (ed.), Wastewater Reuse–Risk Assessment, Decision-Making and Environmental Security, 227–238.

objective is to maximize regional social welfare, which is composed of the sum of the agricultural and environmental net benefits (gross benefits minus relevant costs) in the examined region, while taking into consideration the impacts of salinity and nitrogen on the commercial yields of the various crops [11, 12] and the environmental damage associated with irrigation with recycled water over an aquifer. The model incorporates, in one endogenous system, the economic, physical, and biological relationships in the water–soil–plant–environment system while taking into consideration the possibility of using recycled wastewater for river restoration. In this part, we expand the relevant optimization models [13, 14].

The result obtained from the planning model determines the total net benefits in the region that should be allocated among the various players. Obviously, each player would like to have the largest possible share. Here we assume the following allocation procedure: the players reach a cooperative agreement about the basic principles (such as anonymous, symmetrical, efficient, and rational) of allocation, and then they nominate an objective middleman who determines the actual allocation, subject to the agreed-upon principles. Specifically, we assume that these principles can be satisfied if the decisions of the middleman are based on an allocation scheme from the concept of transferable utility (TU) games [15–17]. The different approaches are referred to the allocation of the additional net benefit (gross benefit minus the parties' stand-alone values) obtained in the examined area, among the parties. The models are applied to the Sharon region in central Israel, which includes two cities (operating one wastewater treatment plant [WWTP]), the Yarqon River Authority (RA), and two groups of farmers.

2. A REGIONAL PLANNING MODEL

The objective of the model is to maximize the sum of net benefits in the examined region, subject to a given supply of wastewater, environmental and health regulations, and the farmers' capability and willingness to utilize the recycled wastewater for crop irrigation. The impacts of salinity and nitrogen on the commercial yields of the various crops and the environmental damage associated with irrigation with recycled water over an aquifer are taken into consideration.

The following economic entities or "players" are involved: (1) the city that owns a WWTP is "the effluent producer"; (2) a river authority responsible for the river's restoration: (3) a group of farmers located (geographically) close to the WWTP, "the nearby farmers" and (4) a group of farmers located far from the WWTP, "the distant farmers". The two groups of farmers and the RA are the potential consumers of the city's recycled effluent.

The motivation for regional cooperation among these players is concealed in the economic and environmental advantages related to recycled wastewater reuse for irrigation and river rehabilitation. Utilization of recycled wastewater may increase the net benefits to the farmers and the utility of the RA. The fixed costs associated with the establishment of recycled wastewater—conveyance structures and upgrading of the WWTP as required by health regulations, combined with the treatment and operation costs, are significant.

The economic analysis refers to a period of one year, with all long-term revenues and costs expressed on an annual basis. However, the model takes into account the possible effects of present recycled wastewater reuse decisions on the future of salt accumulation in the aquifer. We assume a central planner (CP) who operates under certainty conditions. The decisions of the CP can be divided into two major groups: (1) choosing the optimal organizational structure that yields the maximum benefit in the region (Figure 1) determining the recycled wastewater and freshwater allocation to the involved players in the region. The decisions regarding those issues depend upon the following factors: (a) the relevant costs and the cost-sharing scheme associated with each of the examined organizational alternatives, (b) the crop mix and land area cultivated by each group of farmers, (c) the geographical distribution of the players in the region, and (d) the level of predetermined quotas of freshwater assigned to the farmers.

The objective of the CP is to allocate the water (recycled wastewater and freshwater) among the potential consumers (i.e., the RA and farmers) such that total benefits are maximized in the region. The CP does not determine the allocation of the net benefits among the four players. For the sake of clarity, the possible water allocations in the examined region (Table 1).

Table 1: Water allocation schemes in the examined region

j/i	1	2	3	4	Description
1	–	W^{12}	W^{13}	–	The amount of wastewater that the city will purify to secondary (*j*=2) and tertiary (*j*=3) levels
2	–	–	$W^{23} = W^{R}$	–	The amount of tertiary wastewater that will be allocated to the RA
3	W^{31}	W^{32}	W^{33}	W^{34}	The amount of water from the various sources that will be allocated to the nearby farmers
4	W^{41}	W^{42}	W^{43}	W^{44}	The amount of water from the various sources that will be allocated to the distant farmers
5	–	–	$W^{53} = W^{M}$	–	The amount of tertiary wastewater that will be disposed of by release into the sea

i - The economic entity in the region (i = 1 represents the city; i = 2 represents the RA; i = 3 represents the nearby farmers; i = 4 represents the distant farmers and i = 5 represents the sea).
j - The water sources in the region (j = 1 represents fresh water available to the groups of farmers; j = 2 and j = 3 represent secondary and tertiary recycled wastewater, respectively, and j = 4 represents the tertiary recycled wastewater that is allocated to the river.
W^{ij} – Water allocation (in cubic meters) from source j to economic entity i.

A schematic description of the possible organizational alternatives faced by the CP is shown in Figure 1.

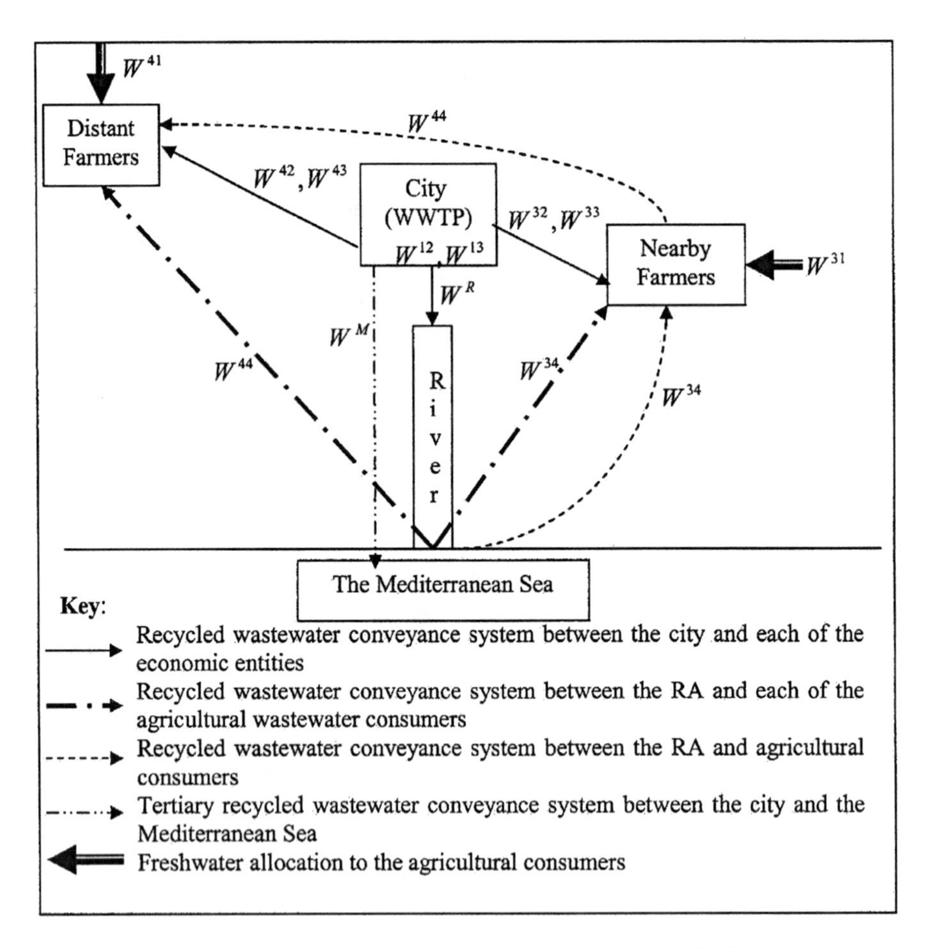

Figure 1: The Possible Organizational Alternatives in the Region

Specifically, we analyze the following organizational alternatives:

1. No cooperation: each party acts independently
2. Cooperation between the city and one of the parties
3. Cooperation between the city and two of the parties
4. Cooperation between all parties within the region (the grand coalition)

We assume that, in the alternatives where the city cooperates with the RA, the possibility exists of there being no need to build a conveyance system between the city and the sea (the recycled wastewater will be disposed of at sea, via the river). Let us now describe in relative length the characteristics and assumptions associated with the assumed economic entities.

2.1. The City and the Wastewater Treatment Plant

The city is obligated by law to purify and discharge the wastewater from its jurisdiction while satisfying several environmental and health requirements, such as threshold levels of biochemical oxygen demand (BOD), chemical oxygen demand (COD) and total suspended solids (TSS) (Water Law, 1959; The Ministry of Health, 1999; Water Commission, 1999). We assume that the city is obligated by law to dispose of its entire exogenously given amount of wastewater ($\overline{W}$). In other words, at the end of the allocation process, no wastewater is left in the city's jurisdiction. Formally,

$$\overline{W} = \sum_{i=3}^{4} \sum_{j=2}^{3} W^{ij} + W^{R} + W^{M} \qquad (1)$$

We assume that the effluent can be purified to two quality levels: (a) secondary level—the current level at the WWTP and (b) tertiary level—the upgraded level. According to the conclusions of the Halperin Committee [18], the tertiary level is the minimal quality of wastewater purification that is safe for ecological uses (i.e., river rehabilitation), for disposal in the sea and for all crops without toxic accumulation in the soil and leaching into the aquifer. Irrigation with secondary-level recycled wastewater is not allowed over an aquifer in order to avoid toxic percolation. In addition, this low-quality water cannot be used for irrigating some crops, such as vegetables, deciduous plants, artichoke, etc.

Due to the agricultural (i.e., irrigation) and ecological (i.e., river rehabilitation) potential of reusing wastewater that has been purified to the tertiary level, we examine the possibility of upgrading the current WWTP to be able to purify to this level (in addition to the current, secondary level), implying,

$$\overline{W} = W^{13} + W^{12} \tag{2}$$

The operating cost (in dollars) of conveying 1 m^3 of recycled wastewater of quality j ($j=2, 3$) to the ith customer $(i \neq 1)$ is denoted by e^{ij}, and it depends, among others things, on the location of the potential users (distance from the city and topography), pipe pressures, pumping costs, energy costs, and more. The purification cost (in dollars per m^3) is α^j ($j=2, 3$), and it includes the operational and maintenance costs, energy costs, and costs of materials and activated-sludge disposal.

The annual equivalent investment cost (in dollars) of building a new conveyance system from the WWTP to the potential users is F_1^i, $i \neq 1$, taking into consideration the distance of the potential users from the city, the pipe length, and the topography. The annual equivalent cost (in dollars) of upgrading the current purification level of the WWTP from secondary to tertiary is denoted by E and it includes the costs of planning, designing and building, purchasing materials, etc.

The city's effluent causes health and environmental damage. Without loss of generality, we assume that the additional net benefit to the city (in monetary terms), EU^1, from wastewater discharge is equal to the cost of conveying it to the sea, namely,

$$\text{EU}^1 = e^{53} W^M \tag{3}$$

The total cost associated with purifying and discharging effluent from the city excluding the additional profit to the city, EU^1, is given by:

$$\text{TC}^1 = \sum_{i=2}^{4} \sum_{j=2}^{3} \left[\left(e^{ij} + \alpha^j \right) \cdot \left(W^{ij} + W^R + W^M \right) \right] + \sum_{i=2}^{5} F_1^i + E \tag{4}$$

Note that in the case in which the city does not allocate wastewater to the other economic entities, $W^{ij} = W^R = 0$ for $j = 2, 3$ and $i = 3, 4$, all the recycled effluent (tertiary level) will be disposed of in the sea (see Equation 1), yielding:

$$\text{TC}_0^1 = \alpha^3 \cdot W^M + F_1^5 + E \tag{5}$$

where TC_0^1 can be viewed as the city's "threat alternative" or "zero alternative" since the entire purification and discharge costs are born by the city.

2.2. The River Authority

The RA has potential ecological benefits to gain from utilizing the city's recycled wastewater. As already mentioned, the RA can use only tertiary wastewater and it also conveys recycled wastewater to the agricultural groups. This activity requires investing in conveyance structures from the RA to the two groups of farmers. Equation 6 presents the RA's benefit from the flow of recycled wastewater in the river:

$$\mathrm{EU}^2 = uW^R \tag{6}$$

where u is an estimation of the benefit (in dollars) from a flow of 1 m^3 of tertiary recycled wastewater in the river. Adding the RA into the regional cooperation between the city and the farmers imposes the following additional major costs: e^{i4}, the operating cost (in dollars) of conveying 1 m^3 of tertiary recycled wastewater to the ith agricultural consumer, ($i = 3, 4$). This cost depends on the farmers' distance from the river's downstream, on pipe pressures, on pumping and energy costs, etc. The annual equivalent investment costs (in dollars) for building a new conveyance system from the RA to the nearby farmers and to the distant farmers are denoted by F_2^3 and F_2^4, respectively.

The total cost, TC^R, associated with building a new conveyance system and transferring the tertiary recycled wastewater from the RA to the potential users is

$$\mathrm{TC}^R = c^3 W^R + \sum_{i=3}^{4}\left(e^{i4}W^{i4} + F_2^i\right) \tag{7}$$

By subtracting Equation 7 from Equation 6, we get the RA's benefit from utilizing the city's tertiary recycled wastewater

$$\pi^2 = \sum_{i=3}^{4}\left[u \cdot W^R + e^{i4} \cdot W^{i4} - F_2^i\right]. \tag{8}$$

2.3. The Farmers

The two assumed groups of farmers in the region differ in their location, land area, freshwater quotas, and crop mixes, among other things.

Each farmer has a freshwater quota (in m^3) of $\overline{Q}^i$ ($i = 3, 4$). He is endowed with $\overline{X}_g^i$ hectares (ha) of agricultural land, some of which is located above a groundwater aquifer (indexed by $g = 1$) and some of which is not (indexed by $g = 2$), and he can choose to grow up to N crops. The farmer's net benefit from utilizing recycled wastewater for irrigation is

$$\pi^{i} = \sum_{n=1}^{8}\sum_{g=1}^{2}\sum_{j=1}^{4} p_{n}^{ij} x_{ng}^{ij} - \sum_{n=1}^{3}\sum_{g=1}^{2} h_{n}^{i} \cdot \widetilde{x}_{ng}^{i} - c_{W} W^{i1},\ i = 3,4 \tag{9}$$

where

p_{n}^{ij} – net income (in dollars) from 1 ha of crop n ($n = 1,\ldots, N$), irrigated with water from source j minus the fixed and variable costs, not including the expenses associated with recycled wastewater use (see Appendix I).

x_{ng}^{ij} – the area (ha) of crop n grown by the ith farmer irrigated with water from source j and located above ($g = 1$) or not ($g = 2$) the relevant aquifer (a decision variable).

h_{n}^{i} – the cost of uprooting orchard n ($n = 1, 2, 3$) grown by farmer i (in dollars per ha).

$\widetilde{x}_{ng}^{i}$ – the uprooted area (ha) of the nth orchard ($n = 1, 2, 3$) grown in area g by farmer i (a decision variable).

c_{W} – the cost of freshwater diverted for agricultural use (in dollars per m^3).

3. RESULTS

Regional cooperation can be established among the city and some or all of the wastewater consumers. Therefore, the number of a priori feasible coalitions in the region is 11 (less than the potential number of $2^4 – 1$):

Non-cooperative coalitions (i.e., stand alone) (1), (2), (3), (4)
Partial coalitions (1, 2); (1, 3); (1, 4); (1, 2, 3); (1, 2, 4); (1, 3, 4)
Grand coalition (1, 2, 3, 4)

The empirical analysis focuses on the monetary incentive for cooperation between the city and some or all of the wastewater consumers. It is shown that acting alone (i.e., noncooperatively) is not a desirable solution. Under the grand coalition, the highest additional net profit is achieved—3.38 million dollars (Table 2), and therefore a grand coalition will be preferred by the CP. Moreover, the optimal cooperative solution enables the farmer (or each group of farmers) to efficiently reallocate his freshwater quota, to cultivate new land areas without uprooting orchards, and to expand the area planted for crops which can be irrigated only with freshwater or with wastewater purified to a tertiary level (the total cultivated area in the region is increased by 23% relative to the non-cooperative situation, see Table 3). In addition, the use of tertiary wastewater by the farmers reduces the amount of irrigation with freshwater by 1.1 M^3, compared to the noncooperative scheme, as illustrated in Table 4.

Table 2. Total value of any feasible coalition in the optimal solution, the sum of players' stand-alone values in the coalition and the net benefit values of the coalitions (in million dollars)

	Column 1	Column 2	Column 3
Coalitions (i), (s)	The coalition value Π^S	The summation of players' stand-alone income in coalition s $\sum_1^4 \pi(i)$	The additional net benefit of a coalition $v(S)$
(1)	–3.31	–3.31	0.00
(2)	0.00	0.00	0.00
(3)	0.42	0.42	0.00
(4)	0.61	0.61	0.00
(1, 2)	–0.58	–3.31	2.73
(1, 3)	–2.52	–2.88	0.36
(1, 4)	–2.16	–2.7	0.54
(1, 2, 3)	0.06	–2.88	2.94
(1, 2, 4)	0.32	–2.7	3.02
(1, 3, 4)	–1.37	–2.27	0.9
(1, 2, 3, 4)	1.1	–2.27	3.38

Column 1 is the value of coalition(s) obtained from the optimization model
Column 2 is the sum of the "stand-alone" values of each coalition member
Column 3 is the difference between column 1 and column 2. In other words, $v(s)$ is the additional net profit of any feasible coalition s

Table 3. Cultivated area allocation under the extreme alternatives (in ha)

Crops (N) Coalitions	Distant farmers		Nearby farmers	
	(1, 2, 3, 4)	(4)	(1, 2, 3, 4)	(3)
1. Citrus	93.9	70.4	83.5	62.7
2. Avocado	54.1	40.6	–	–
3. Deciduous	–	–	125.1	93.9
4. Watermelon	93.8	56.5	–	–
5. Artichoke	–	–	62.5	62.5
6. Green plants	58.8	35.3	–	–
7. Potatoes	90	54	66.3	39.8
8. Animal feed (corn)	98.3	137.5	63.3	70
Total orchard uprooting	0	37	0	51.9
Total area irrigated with wastewater	331.8	0	233.6	0
Total cultivated area	488.9	394.3	400.7	328.9

Tables 3 and 4 present the major changes for each group of farmers which are induced by the cooperation. The nearby farmers increase the amount of irrigation water by 0.88 M^3 of tertiary wastewater and their irrigated area increases from 328.9 to 400.7 ha. The usage of freshwater by these farmers decreases by 0.385 M^3. The distant farmers increase the amount of irrigation water by 1.55 M^3 of tertiary wastewater and their irrigated area increases from 394.3 to 488.9 ha. The usage of freshwater decreases by 0.71 M^3.

Table 4. Water allocation under the extreme alternatives (in M^3)

Data coalition	Distant farmers		Nearby farmers	
	(1, 2, 3, 4)	(4)	(1, 2, 3, 4)	(3)
Freshwater	0.98	1.69	1.25	1.635
Tertiary wastewater from the RA	1.55	0	0.88	0
Total	2.53	1.69	2.13	1.635

4. DISCUSSION

The RA gains the second largest additional net profit allotment (Table 5; about a third). The RA's double role in the examined region, as the largest tertiary wastewater consumer and as a tertiary wastewater supplier to the farmers, provides it with relatively high negotiation strength compared to the other wastewater consumers. The nearby and distant farmers serve only as wastewater consumers in the region. At first, the farmers' allocations of the additional net profits appear relatively small next to the city and RA allotments. However, comparing the solutions obtained to their initial status (column 2, Table 2) shows that the nearby and distant farmers improve their profits by 40 and 35.5%, respectively.

Table 5. Allocations of additional net benefits based on game theory principles (in million $)

Player	The nucleolus solution	Percentage of the total net benefit	The shapley value	Percentage of the total net benefit
1	1.74	%51.60	1.72	%50.90
2	1.24	%36.68	1.27	%37.56
3	0.18	%5.30	0.17	%4.96
4	0.22	%6.42	0.22	%6.58

5. CONCLUSION

A year planning model was developed, for a region in Israel which consists of a city and three potential wastewater consumers, to incorporates the economic, physical and biological relationships in the water–soil–plant–environment system and its objective is to maximize the regional social welfare composed of the sum of agricultural and environmental net benefits. The model determines the optimal crop mix and the optimal allocation of the limited water and land resources among all potential users.

Acknowledgment: I would like to express my thanks to the NATO program for security through science for the financial support.

6. REFERENCES

1. Zaslavski D. The Face of Water in Israel, Draft Manu., Haifa, Technion, 2001.
2. Di Pinto AC, Lopez A, et al. "Wastewater Reuse as an Alternative to Traditional Water Resources", Ann. Chimica, 89/9–10, 689–697, 1999.
3. Pereira LS, Oweis T, Zairi A. "Irrigation Management under Water Scarcity," Agricultural Water Management, 57(3), 175–206, 2002.
4. Yaron D. "The Israel Water Economy" in Decentralization and Coordination of Water Resource Management, D. Parker & Y. Tsur, Kluwer 9–22, 1997.
5. Kislev Y. "Urban Water in Israel," Paper. 6.02, Hebrew University, Oct. 2002
6. Fleischer A, Tsur Y. "Measuring the Recreational Value of Agricultural Landscape," Euro. Review Agricultural Economics, 27, No. 3, 385–398, 2000.
7. Loomis J, Kent P, Strange L, Faush K, Covich A. "Measuring the Total Economy Value of Restoring Ecosystem Services in an Impaired River Basin: Results from a Contingent Valuation Survey," Ecol. Econ. 33, 103–117, 2000.
8. Dinar A, Yaron D, Kannai Y. "Sharing Regional Cooperative Gains from Reusing Effluent for Irrigation," Water Res. Research, 22/3, 339–344, Mar. 1986.
9. Haruvy N. "Wastewater Reuse—Regional and Economic Considerations," Agriculture Ecosystems & Environment, 23, 57–66, 1998.
10. Haouari M, Azaiea MN. "Optimal Cropping Patterns Under Water Deficits," European Journal of Operational Research, 130, 133–146, 2001.
11. Mass EV, Hoffman GL. "Crop Salt Tolerance—Current Assessment," ASCE, Journal of the Irrigation and Drainage Division, IR2, 10, 115–134, 1977.
12. Feinerman E, Yaron D. "Economics of Irrigation Water Mixing Within a Farm Framework," Water Resources Research, 19/2, 337–345, Apr. 1983.
13. Haruvy N, et al. "Wastewater Irrigation—Economic Concerns Regarding Beneficiary and Haz. Effects of Nutrients," Water Res. Manag. 13, 303–314, 1999.
14. Dinar A, Yaron D. "Treatment Optimization of Municipal Wastewater and Reuse for Regional Irrigation," Water Resources Research, 22/3, 331–338, 1986.

15. Young HP. "Cost Allocation," in Handbook of Game Theory with Economic Applications, by Aumann, J.R. & Hurt, S., Elsevier Pub. 2, 1193–1236, 1994.
16. Lejano PR, Davos AC. "Cost Allocation of Multiagency Water Resource Project: Game Theoretic Approaches and Case Study," Water Res. Res. 31(5), 1387–93, 1995.
17. Loehman, E.T., "Cooperative Solution for Problems of Water Supply," in Water Quality/Quantity Management and Conflict Resolution Institutions, Processes and Economic Analysis, edited by Dinar, A. & Loehman, E.T., Praeger Publishers, Westport, Connecticut, London, pp. 301–319, 1995.
18. The Ministry of Health, "Principals for Wastewater Irrigation Permits," (in Hebrew) The Halperin Committee Report, Aug. 1999.

SESSION 7. RISK ASSESSMENT/NATIONAL POLICY MAKING INTERFACE

SEWERAGE INFRASTRUCTURE: FUZZY TECHNIQUES TO MANAGE FAILURES

Yehuda Kleiner, Balvant Rajani, and Rehan Sadiq
Urban Infrastructure Program, Institute for Research in Construction
National Research Council, Ottawa, Ontario, Canada K1A 0R6

Corresponding author: yehuda.kleiner@nrc-cnrc.gc.ca

Abstract:

An approach is presented to model the deterioration of buried, infrequently inspected infrastructure, using scarce data. The robustness of the process is combined with the flexibility of fuzzy mathematics to arrive at a decision framework that is tractable and realistic. In applying this approach to sewerage infrastructure, the scoring scheme was converted using current guidelines. A rule-based model is used to replicate and predict the possibility of failure. The model can be used to plan the renewal of the asset subject to maximum risk tolerance. The concepts are demonstrated using data obtained from Canadian municipalities.

Keywords: Sewerage, deterioration, fuzzy sets, risk management, inspection.

1. INTRODUCTION

Water Research Centre (WRc), UK initiated in 1978 a 5-year research project to investigate failures of sewer mains. Based on this research, sewerage, rehabilitation manual (SRM) was developed [1–4]. The manual has gone through several revisions and the latest is the 4th edition which includes a computerized grading system compatible with European defect coding systems, and new design methods for renovation techniques [4]. The National Research Council of Canada (NRC) protocol is known as Guidelines for Condition Assessment and Rehabilitation of Large Sewers [5].

Table 1 illustrates point score schemes of both protocols. Note that both protocols contain separate scoring schemes for structural and operational observed defects. In this paper we deal only with the structural aspects of pipe deterioration. It is clear that while WRc uses a five-grade rating, NRC uses six, from zero to five where 0 = excellent, 1 = good (G), 2 = fair (F), 3 = poor (P), 4 = bad (B), and 5 = imminent collapse (IC). Table 1 provides a summary of structural distress indicators (defects) and their associated point scores (deduct values) in the two protocols.

M.K. Zaidi (ed.), Wastewater Reuse–Risk Assessment, Decision-Making and Environmental Security, 241–252.

Table 1. Comparison of two point scoring protocols

Protocol	Structural conditions					
	O (E)	1 (G)	2 (F)	3 (P)	4 (B)	5 (IC)
WRc Scores [3]	n/a	<10	10–39	40–79	80–164	>165
NRC Scores [5]	0	1–4	5–9	10–14	15–19	20

The process of applying the protocol to real situations is inherently imprecise and subjective. Often, two different evaluators will provide different scores to the same observed defects. Moreover, on observing Table 2, one cannot help wonder whether a deduct value of 80 (WRc protocol) is in fact equivalent to 164, when both scores would translate to a condition state of 4.

Additionally, the current framework does not lend itself easily to the consideration of the time dimension in order to obtain rate of deterioration. Evaluating the rate of deterioration is important in order to facilitate educated decisions on scheduling pipe renewal or alternatively, scheduling the next inspection.

In this paper a framework is proposed, whereby the scores obtained from conventional scoring protocols are transformed into a fuzzy-based condition ratings. Subsequently this fuzzy condition rating is used in conjunction with a fuzzy Markov-based deterioration model (developed elsewhere).

The model is used to predict pipe future condition and make some risk-based decisions on whether to renew the pipe or schedule the next inspection. This framework is presented with the help of a case study, using real data on sewers obtained from a Canadian municipality. The rest of this paper is structured as follows:

2. METHOD

The Markovian deterioration process requires that the condition of the deteriorating asset be encoded as an ordinal condition state (e.g., excellent, good, fair, etc.). The condition assessment of buried pipes comprises two steps. The first step involves the inspection of the asset using direct observation (visual, video) and/or NDE techniques (radar, sonar, ultrasound, sound emissions, etc.), which reveal distress indicators. The second step involves the translation of these distress indicators to determine the condition rating of the asset. As described earlier, there are already well-accepted protocols to define the nature and weight of various distress indicators in sewers, therefore we shall use them as the basis from which to transform into a fuzzy-based rating method. For lack of space, it is assumed that the reader has basic notions of fuzzy sets, however for the uninitiated, [6] provides a brief description with an emphasis on the application of fuzzy sets to condition rating of assets.

Table 2. Distress indicators and their assigned scores (deduct values)

Distress indicator (defect)[a]	Distress level[b]	Unit	Scores	
			NRC	WRc
Longitudinal crack	• Light (up to 3 cracks, no leakage)	m	3	10
	• Moderate (>3 cracks, leakage)	m	5	40
Circumferential crack	• Light (up to 3 cracks, no leakage)	m	3	10
	• Moderate (>3 cracks, leakage)	m	5	40
Diagonal crack	• Light (up to 3 cracks, no leakage)	m	3	n/a
	• Moderate (>3 cracks, leakage)	m	5	n/a
	• Severe (multiple cracks, leakage)	m	10	40
Longitudinal fracture	• Light (<10 mm)	m	5	40
	• Moderate (10–25 mm/more than one)	m	10	80
	• Severe (>25 mm)	m	15	n/a
Circumferential fracture	• Light (<10 mm)	m	5	40
	• Moderate (10–25 mm/more than one)	m	10	80
	• Severe (>25 mm)	m	15	n/a
Diagonal fracture	• Light (<10 mm)	m	5	40
	• Moderate (10–25 mm or multiple)	m	10	80
	• Severe (>25 mm)	m	15	n/a
Deformation	• Light (<5% change in diameter)	m	5	20
	• Moderate (5–10% change in diameter)	m	10	80
	• Severe (11–25% change in diameter)	m	15	165
Surface damage (spalling)	• Light	m	3	5
	• Moderate	m	10	20
	• Severe	m	15	120
Joint displacement	• Light (<¼ pipe wall thickness)	each	3	n/a
	• Moderate (¼ –½ pipe wall thickness)	each	10	1
	• Severe (>½ pipe wall thickness)	each	15	2
Broken pipe		each	15	60
Collapse		each	20	165

[a] This nonexhaustive list is based on [3, 5]
[b] Definitions sometimes vary between the two protocols

We define a seven-grade condition scale: excellent, good, adequate, fair, poor, bad, and failing. Each grade (or state) is represented by a triangular fuzzy set. The WRc protocol has a range of scores from 0 to 165. This range will be mapped on the proposed seven-grade condition scale as described in Table 3 and illustrated in Figure 1. For example, an aggregate score of 100 (WRc protocol) will yield 40%

membership to the fair state and 60% membership to the poor state or symbolically its condition rating will be (0, 0, 0, 0.4, 0.6, 0, 0).

Table 3. Seven-grade fuzzy scale for condition rating

Order	Descriptor	Triangular fuzzy number (TFN) representation
1	Excellent	$(q_{1,1}; q_{1,2}; q_{1,3}) = (0, 0, 20)$
2	Good	$(q_{2,1}; q_{2,2}; q_{2,3}) = (0, 20, 50)$
3	Adequate	$(q_{3,1}; q_{3,2}; q_{3,3}) = (20, 50, 80)$
4	Fair	$(q_{4,1}; q_{4,2}; q_{4,3}) = (50, 80, 110)$
5	Poor	$(q_{5,1}; q_{5,2}; q_{5,3}) = (80, 110, 140)$
6	Bad	$(q_{6,1}; q_{6,2}; q_{6,3}) = (110, 140, 165)$
7	Failing	$(q_{7,1}; q_{7,2}; q_{7,3}) = (140, 165, 165)$

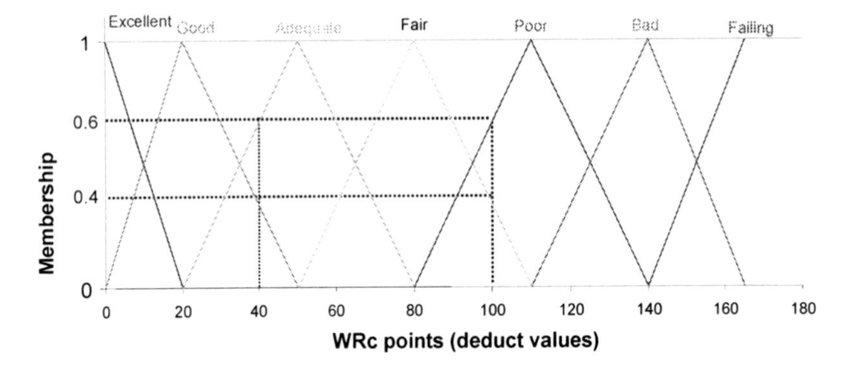

Figure 1. Seven-grade fuzzy scale for pipe condition rating

The first example is a 76 m long (manhole to manhole), 525 mm diameter vitrified clay (VC) pipe that was installed in 1962. The second is a 111 m long (manhole to manhole) 900 mm diameter concrete pipe installed in 1971. The observed distress indicators and their associated WRc scores (deduct values) are shown in Table 4. These distress indicators are assumed to have been observed on a single pipe segment (typically 2–4 m long). The scores translate to a fuzzy rating of (0, 0.4, 0.6, 0, 0, 0, 0) and (0, 0, 0, 0.4, 0.6, 0, 0), for examples 1 and 2, respectively (Figure 1). Both pipes were assumed to be mainly in excellent condition (0.9, 0.10, 0, 0, 0, 0, 0) upon installation. Unfortunately, we could not obtain information on whether these pipes have been inspected again since year 2000.

2.1. Rule-Based Fuzzy Markovian Deterioration Model

Markov process has frequently been used to modeling infrastructure asset deterioration [7–10] and a new fuzzy rule-based approach was introduced [6, 11]. It took advantage of the robustness of the Markov process and the flexibility of the fuzzy-based techniques, which seem to be particularly suited.

Table 4. Distress indicators observed in examples 1 and 2

Example	Insp. Year.	Description	WRc score
1	2000	Longitudinal crack (assumed)	40
2	2000	Longitudinal fracture (assumed)	100

The model yields a family of curves that represent the deterioration process (Figure 2). As the pipe ages, 20 to year 40, its memberships to good condition states diminish and memberships to worse condition states increase is a graphical representation of the condition rating in year 40. Using the threshold values concept [6], the model is formulated to mimic a reality in which a given asset at a given time cannot have significant membership values to more than two or three contiguous condition states. The deterioration model is calibrated (or trained) on observed condition ratings obtained from historical inspections and condition assessments. Once the deterioration model has been trained, it can be used to predict the future condition of the pipe.

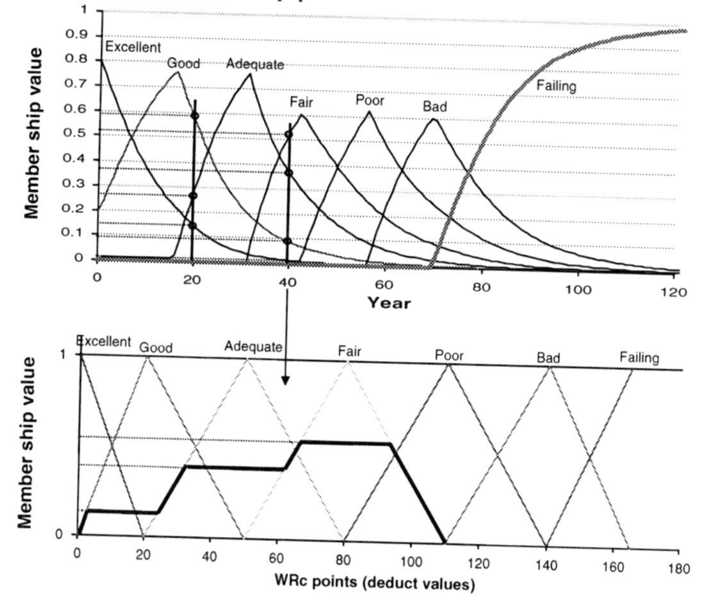

Figure 2. Rule-based fuzzy Markov deterioration model

Membership to the failing state can be transformed to a possibility mass function (as opposed to probability) of failure at every point along the life of the asset [11]. The possibility of failure can then coupled with failure consequence to obtain failure risk as a function of age, as is described later.

2.2. Training the Deterioration Model

Results of the training sessions on the Halton pipe data are illustrated in Figure 3. Note that example 1, which is an older pipe (installed in 1962) has a better condition rating than example 2 (installed in 1971), is anticipated to endure much longer. As mentioned earlier, data on subsequent inspections were not available; therefore we could not validate the forecasted condition ratings. However, even if data were available, they would correspond to inspection that would have taken place at most 5–6 years later than the previous inspection. This short gap is often insufficient to provide credible validation results, as deterioration processes are typically quite slow for this type of infrastructure assets.

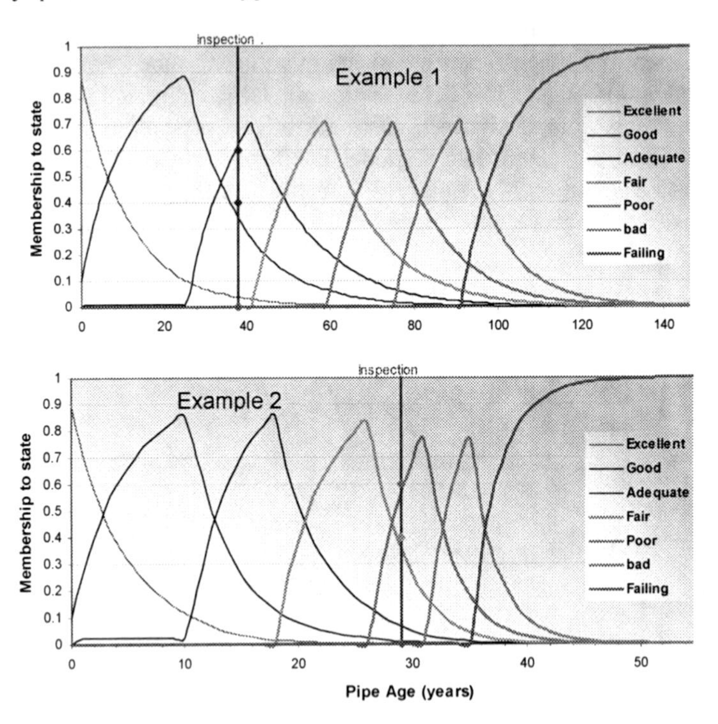

Figure 3. Examples: deterioration model training

2.3. Postrenewal Condition Improvement and Subsequent Deterioration

Renewal is defined as an action that improves the condition of the pipe and possibly impacts its deterioration rate as well. Typically, several pipe renewal technologies are available, each of which is assumed to have three specific attributes [6]. The first is a condition improvement matrix, which determines how much the condition of the pipe will improve immediately after renewal. The second is a postrenewal deterioration rate matrix, which determines how fast the pipe will continue to deteriorate after renewal. The third is the cost associated with the renewal alternative.

The condition improvement matrix can be populated based on hard field data. However until these types of data become available, this matrix would be established from expert opinion, as illustrated in Table 5. Similarly, the postrenewal deterioration rate matrix is also estimated from experience and expert opinion, as illustrated in Table 6. Renewal costs can usually be obtained from manufacturers/ contractors. Once the condition improvement and the postrenewal deterioration rate matrices are established, a new fuzzy Markov-based deterioration process can be modeled, where the pipe continues to deteriorate from its postrenewal condition. Figure 4 illustrates postrenewal deterioration for the pipe in example 1. If, after renewal it takes 29 years for the pipe to deteriorate to a condition rating similar to its prerenewal condition, it can be said that the renewal action "bought" 29 years of additional life.

2.4. Fuzzy Risk of Failure

2.4.1. Likelihood, consequences, and risk of failure

The risk of failure is determined jointly by the likelihood (possibility) and the consequences of a failure. Both WRc and NRC protocols do not explicitly address consequences of failure. However, both consider criticality of sewers in making decisions about prioritization for renewal and future inspections. This criticality is said to depend on location, importance, collateral damage, and replacement cost. Consequently, it can be safely concluded that criticality and failure consequences are largely one and the same.

Accurate data on failure consequences including direct, indirect, and social costs are difficult to come by. The fuzzy approach is therefore well suited to exploit the qualitative understanding many practitioners have about the conditions that affect these costs. The encoding process of failure consequences into fuzzy sets was beyond the scope of the underlying research, it was therefore assumed that these consequences could be described as a nine-grade (extremely low, very low, quite low, moderately low, medium, moderately severe, quite severe, very severe, and extremely severe) possibility mass function. It was further assumed that this mass function is largely stationary over the periods of time (planning horizons).

Table 5. Expert input to construct condition improvement matrix

Expression of confidence to get condition shift	To condition						
Condition	Excellent	Good	Adequate	Fair	Poor	Bad	Failing
Excellent	Highest						
Good	Highest	Lowest					
Adequate	Medium	Highest	Lowest				
Fair	Medium	Highest	Medium				
Poor		Lowest	Highest	Medium			
Bad		Medium	Highest	Lowest			
Failing			Lowest	Highest	Medium		

Table 6. Expert input for evaluating the postrenewal deterioration rate

Expression of confidence about the postintervention deterioration rate relative to the current (observed) deterioration rate				
Much lower	Lower	Same	Higher	Much higher
	Medium	Highest	Lowest	

Using a set of fuzzy rules, the fuzzy consequence mass function is coupled with the mass function that defines the possibility of failure to obtain another mass function that describes the risk of failure as a nine-grade (extremely low, very low, quite low, moderately low, medium, moderately high, quite high, very high, and extremely high) fuzzy set [11]. The risk mass function (which is a fuzzy number) is calculated for each time to obtain the fuzzy risk of failure throughout the life of the pipe. The result is a fuzzy risk curve as illustrated in Figure 5. The gray levels represent membership values to risk levels (darker gray for a higher membership).

3. MAKING DECISIONS

3.1. Maximum Risk Tolerance as a Decision Criterion

The owners of the sewer can, through a consensus-building process like Delphi, define their maximum risk tolerance (MRT), while considering both the possibility of failure and the failure consequences. Further, it can be assumed that any decision to renew or rehabilitate a pipe segment or section will always be preceded by an inspection and condition assessment. Thus, if the deterioration model predicts that MRT is going to be reached at a given time, it follows that an inspection/condition assessment will be scheduled around that time [11]. This inspection/condition assessment can have one of two outcomes:

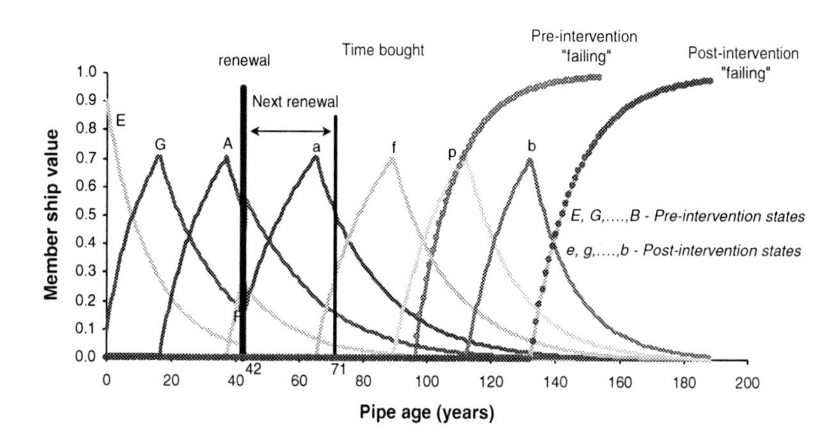

Figure 4. Deterioration curves before and after renewal (E = excellent, G = good, A = adequate, f = fair, p = poor, b = bad)

- The observed condition of the pipe is better than predicted (the model overestimated the deterioration rate) and MRT has not yet been reached. In this case, the deterioration model is recalibrated to include the newly acquired data, then reapplied and the next condition assessment is scheduled for the next time at which MRT is predicted to be reached.
- The observed condition of the pipe is the same or worse than the model predicted and current risk is equal to or exceeds MRT. In this case renewal work has to be planned immediately and implemented as soon as possible.

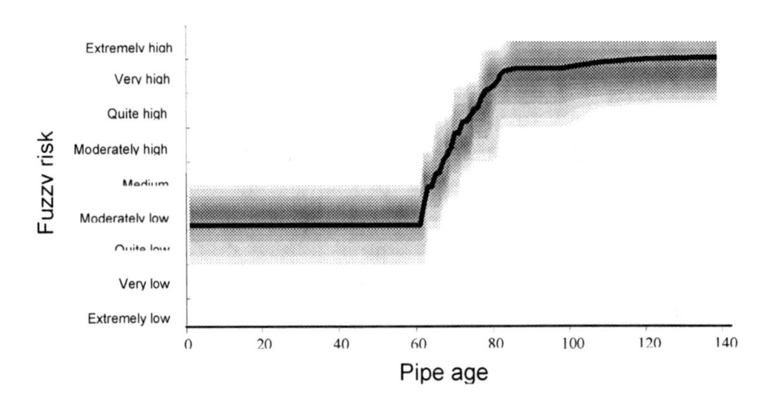

Figure 5. Lifetime fuzzy risk (dark line represents defuzzified values)

Several alternative renewal technologies available in the market need to be evaluated to select the most appropriate one when pipe renewal is required. In the selection, the owner has to consider both the improvement that the renewal action will affect and the postrenewal deterioration rate. The owner may resort to the time "bought" concept explained earlier to make this selection. If, for example, a renewal technology that costs $100,000 buys 20 years of additional life (i.e., postpones subsequent renewal by 20 years until the time at which MRT is reached again), the normalized cost of this technology can be thought of as $5,000 per year of extra life. The owner will usually select the technology with the lowest cost per year of extra life. It is noted that two assumptions are implicit in this decision approach: (a) cost of asset renewal is independent of the condition of the renewed asset, and (b) asset owner will always try to reach MRT and will not consider any risk-cost trade-off, i.e., lowering risk exposure at a higher cost. An alternative approach to decision making was outlined [12], where these two implicit assumptions are relaxed and the trade-off between cost and risk exposure is explicitly considered, but this is beyond the scope of this paper.

3.2. Fuzzy Risk of Failure

Continuing with the pipe in example 1, we assume the following: fuzzy consequence of failure is given by (0, 0, 0.5, 0.5, 0, 0, 0, 0, 0), i.e., 0.5 memberships to quite low and moderately low consequence grade; MRT = moderately high; analysis year is 2006 (pipe is 44 years old). The next inspection should be scheduled about 56 years into the future (Figure 6).

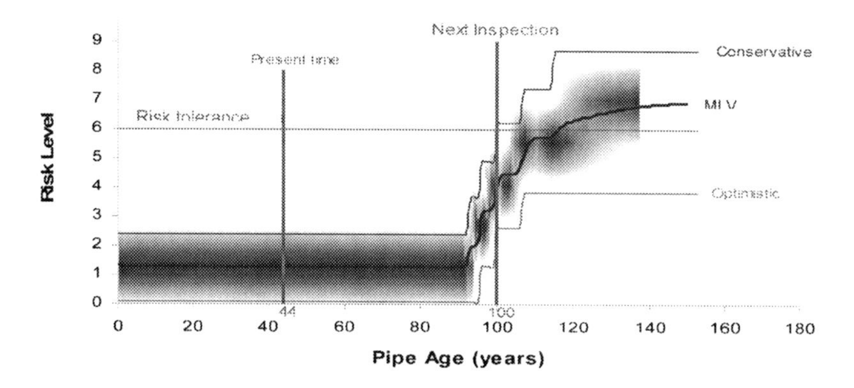

Figure 6. Example: decision (MLV = most likely value)

4. CONCLUSIONS

It is difficult to ensure that inspection and condition assessment carried out many years apart in a consistent manner. This is a limitation inherent in this type of activity and it requires the application of engineering judgement, perhaps to use variable weights for inspections that are deemed to be less reliable.

Acknowledgment: This research was cofunded by the American Water Works Association Research Foundation (AwwaRF) and the Institute for Research in Construction of the National Research Council of Canada (IRC/NRC), with special thanks to my colleague Dr. Dana Vanier who helped to obtain and process data and provided useful comments. NATO for providing the stage and financial support for presenting this research paper.

5. REFERENCES

1. WRc. (1986). Sewer Rehabilitation Manual, 2nd Ed. Water Research Centre plc, UK (WRC UK).
2. WRc (1993). Manual of Sewer Condition Classification, 3rd Ed. WRC UK.
3. WRc (1994). Sewerage Rehabilitation Manual, 3rd Ed. WRC UK.
4. WRc (2001). Sewerage Rehabilitation Manual, 4th Ed. WRC UK
5. Zhao JQ, McDonald SE and Kleiner Y. (2001). Guidelines for Condition Assessment and Rehabilitation of Large Sewers. Institute for Research in Construction, National Research Council of Canada, Ottawa.

6. Kleiner Y, Sadiq R and Rajani B. (2006). Modelling the deterioration of buried infrastructure as a fuzzy Markov process, J. Water Supp. Res. Tech: Aqua, 55(2): 67–80.
7. Madanat SM, Karlaftis MG and McCarthy PS. (1997). Probabilistic infrastructure deterioration models with panel data, J. Infrastructure Systems, ASCE, 3(1): 4–9.
8. Wirahadikusumah R, Abraham D and Isely T. (2001). Challenging issues in modelling deterioration of combined sewers, J. Infrastructure Systems, ASCE, 7(2): 77–84.
9. Mishalani RG and Madanat SM. (2002). Computation of Infrastructure Transition Probabilities Using Stochastic Duration Models, J. Infrastructure Systems, ASCE, 8(4): 139–148.
10. Kleiner Y. (2001). Scheduling Inspection and Renewal of Large Infrastructure Assets, J. Infrastructure Systems, ASCE, 7(4): 136–143.
11. Kleiner Y, Rajani BB and Sadiq R. (2006). Failure risk management of buried infrastructure using fuzzy-based techniques, J. Water Supp. Res. Tech: Aqua, 55(2): 81–94.
12. Kleiner Y. (2005). Risk approach to examine strategies for extending the residual life of large pipes, Middle East Water 2005, Proceedings of the 3rd International Conference for Water Technology, Manama, Bahrain.

GUIDELINES FOR GOOD PRACTICE OF WATER REUSE FOR IRRIGATION: PORTUGUESE STANDARD NP 4434

Maria Helena F. Marecos do Monte
Professor at Instituto Superior de Engenharia de Lisboa (ISEL)
University of Évora, Lisbon, Portugal

Corresponding author: hmarecos@dec.isel.ipl.pt

Abstract:

The growing number of municipal wastewater treatment plants in Portugal delivers about 500 million m^3 of treated wastewater that is discharged in river and coastal waters, representing a pollutant load to receiving waters and yet the waste of water resources which could be successfully used for irrigation in agriculture, landscape, golf courses, water reservoirs for fire protection. Because water reuse can contribute to the economic development it must be stimulated by central, regional, and local authorities, but within the framework of good practices and their monitoring. This paper presents the new Portuguese standard NP 4434 that presents guidelines on water quality, irrigation practice, management of environmental impacts, protection of public and animal health and, aspects of control and monitoring.

Keywords: Agriculture, wastewater, irrigation, Portuguese standard, reuse.

1. INTRODUCTION

In Portugal the mean annual hydrological balance in Portugal is about 740 mm. Taking into account the population (9 million), the country cannot apparently be classified as short in water resources compared with the mean values of Europe or the world. However, the actually available water resources are not as high as they potentially could be, due to the Mediterranean feature of the Portuguese climate: about 66% of the annual rainfall occurs during half of the year, and in some cases about 30% falls in one month. In addition to the uneven time distribution of rainfall there is a clear spatial heterogeneity: in general terms the half of the country located north of the River Tagus basin receives about 1065 mm of rainfall per year, while the southern part receives 641 mm. Other factors of the hydrological cycle—evapotranspiration, infiltration runoff—show a similar distribution. The result is that under a natural regime 57.5% of the country mainland suffers a water deficit. These areas are located mainly in the southern part of the

M.K. Zaidi (ed.), Wastewater Reuse–Risk Assessment, Decision-Making and Environmental Security, 253–265.

country and in northeast. Water shortage due to the simultaneous occurrence of low precipitation, high evaporation, and increased demands for irrigation and tourism is definitively found to be a factor hindering the social and economic development of a large part of the country that tends to be deserted by the active population. As a result, small communities are disappearing and the Portuguese population is concentrating along the cities near the coast.

Water conservation is the hydrological answer to the problem and certainly water reuse is an important component of water conservation strategies [1]. Other solutions can be implemented such as water savings (e.g., suppressing the leakage of supply networks, using more efficient irrigation techniques such as drip irrigation and small flush systems), tapping other resources (e.g., desalinating seawater or brackish water), [2]. Reducing demand through pricing is also a possible option but it raises many political difficulties. Water can be reused for one or more beneficial purposes: irrigation for agriculture or landscape purposes, industrial supply, nonpotable urban applications (such as street washing, fire protection), groundwater recharge, recreational purposes, and direct or undirected water supply.

Water reuse can have two important benefits. The most obvious is the provision of an additional dependable water resource. The second is the reduction of environmental impacts by reducing or eliminating wastewater disposal, which results in the preservation of water quality downstream. Therefore, when considered in the framework of an integrated water management strategy at catchments scale, the benefits of wastewater reuse should always be assessed taking into account that it contributes to both enhancing a region's water resource and minimizing its wastewater outflow.

Agricultural irrigation is the major application for water reuse in both developed and developing countries because it is not so difficult to match the quality of treated wastewater with the required quality of water for irrigation. Agriculture is the economic sector of human activity which, by far, consumes the largest volume of water. Basically, the water needed for agriculture is the amount required for irrigation since the water needed for livestock is a very small part (<1%) of the total.

In developed countries landscape irrigation is the second application of water reuse [3]. The water used or reused for landscape irrigation includes residential, commercial, and municipal applications. Therefore, the use of treated wastewater for irrigation means a positive contribution for appropriate water resources management, particularly in arid and semiarid regions. In addition to augmenting water resources, using reclaimed wastewater for irrigation can reduce the need for fertilizers thanks to the nutrients it contains.

2. THE BENEFIT OF WATER REUSE FOR IRRIGATION

In Portugal the volume of water for agriculture tends to increase, although its percentage will decrease in comparison with other water uses, such as industry, energy production, and drinking water supply, due to the development of industry and the use of more efficient irrigation technologies (allowing for less water losses in irrigation).

The volume of treated urban wastewater available in Portugal has been increasing significantly along the last decade in order to comply with the requirements imposed by the Directive 91/271/EEC of the Council of the European Communities regarding urban wastewater treatment (UWWT) Directive. Basically, Directive 91/272/EEC states that by the year 2005 in the European Union every community above 2000 Population equivalent (PE) must be served with secondary wastewater treatment or tertiary treatment where the effluent is to be discharged into a sensitive area; in addition, it is required that communities under 2000 PE must be served with some appropriate means of wastewater treatment. It is important to mention that this directive recommends the reuse of treated effluent wherever possible (article No.12).

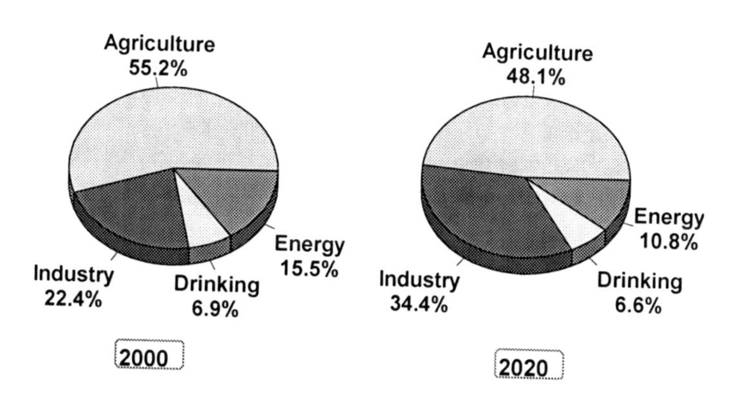

Figure 1. Water use per sector in 2000 and 2020

As a result of the heavy effort carried out by Portugal for the compliance to the UWWT Directive the annual volume of treated wastewater available in Portugal by the year 2006 is about 450 million m^3. The estimated volume of treated wastewater discharged in 2006 will be enough to cover about 10% of the water needs for irrigation in a dry year without seasonal storage. Seasonal storage can increase significantly the potential volume of available water, depending on the retention time. Major benefits will possibly be experimented in the southern region of Algarve (where water is badly needed for landscape irrigation and golf courses) and Alentejo (traditionally suffering of drought). It is very difficult to estimate the area that could be irrigated with treated wastewater, because this depends on a number of factors which have to be evaluated case by case, such as available flow, storage, distance to farms, local orography, etc. However, a rough estimate would range between 35,000 and 100,000 ha, depending on storage retention time. The interest of the use of treated wastewater for irrigation in Portugal is quite clear, if we take into account that the current irrigated area in the country is 900,000 ha and this is about 22% of the total area used for agricultural [4]. Secondary treatment is most commonly used in Portugal. Usually, secondary effluents are not disinfected. During the 1990s the tendency seemed to be towards increasing the number of waste stabilisation pond (WSP) systems, especially in the southern area of the country and a slight increase of fixed film reactors. In the first half decade of the new millennium, activated sludge treatment is getting more popular. Many recent wastewater treatment plant (WWTP) include ultraviolet (UV) disinfection aiming principally at the protection of bathing waters but enabling water reuse for irrigation too.

2.1. The Need for Water Reuse Guidelines in Portugal

Public health and environmental protection require that water reuse is carried out according to appropriate guidelines. Water reuse for irrigation is a growing practice in Portugal due to the pressure of water needs. Therefore, guidelines for water reuse for irrigation were found to be a priority among other purposes of water reuse. A committee was appointed to produce Portuguese guidelines, which were published in January 2006 as Portuguese Standard NP 4434.

2.2. Basis for Establishment of Portuguese Guidelines on Water Reuse for Irrigation

Although irrigation with wastewater is in itself an effective treatment (a sort of low-rate land treatment), some treatment must be performed previously for the protection of public health, the prevention of nuisances during storage and prevention of damage to the crops. In many circumstances, the level of wastewater treatment licensed for discharging the treated wastewater in a receiving water body is appropriate to enable water reuse for irrigation. However, treated wastewater contains chemical and microbiological components which may pose a potential public health risk.

It is generally assumed that only the soil–plant biosystem is significantly affected by the chemical characteristics of irrigation water by means of the absorption of phytotoxic substances through roots and leaves, although it is known that some elements tend to accumulate within parts of plants, sometimes reaching levels dangerous to animal health. On the other hand, there are several guidelines on the microbiological quality of reclaimed wastewater used for irrigation. The State of California was the pioneer and its wastewater reuse guidelines, first issued in 1918, have been adopted with more or less modifications in several North American states as well as in other countries.

Most of the existing regulations can be considered quite strict, in the sense that they require a high-grade water quality, namely concerning the microbiological characteristics, only achievable by means of expensive technology requiring sophisticated operation. Such strict regulations based on a "no risk" philosophy could not be adopted by many countries badly needing to reuse wastewater for irrigation. The only options left in such countries were both undesirable: either no wastewater reuse for irrigation or wastewater reuse without any respect by unbearable regulations, meaning in practice without any treatment or other sort of restriction, with all the consequent public health risks.

Several studies carried out by some international organisations (World Bank, UNEP, UNDP, FAO, IRCWD, and IDRC) identified some contradiction in the requirements of quality for water for irrigation and for other uses like bathing and lead to the conclusion that there was no justification for the stringency of most of the existing regulations on wastewater reuse (IRCWD, 1985). The consequence was that specialists were divided in what can be called two "schools": the less stringent epidemiological evidence school (WHO) and the nil risk school, represented by the US [5].

The establishment of guidelines for the use of reclaimed wastewater for irrigation in Portugal must consider the existing guidelines in other countries, principally in Europe (France) on one hand, and on the other to take into account the experience from other countries with similar level of development, namely in the Mediterranean region. The guidelines should cover two main areas: (a) the agronomic aspects, related to the maximization of crop yields and soil and groundwater preservation; and (b) the sanitary aspects, related to public health protection. The document should have the following general characteristics: (a) simplicity, to prevent unnecessary nonencouraging of wastewater reuse; (b) robustness and reliability, to assure public health protection, good crop production, and to prevent adverse impact in the environment. Finally, the document must be flexible enough to allow for further improvement steps in the light of new scientific and technologic developments and acquired experience.

Guidelines can be defined in great detail or in a broad manner, such as a simple list of chemical and microbiological parameters whose values should range within a certain interval. This simple approach, however, presents a rather serious shortcoming: in most of the southern European countries, where wastewater reuse for irrigation is more necessary, the monitoring of effluent quality from wastewater treatment plants is not completely assured, due to several factors, including lack of trained staff, available laboratories, and also financial reasons.

A second simple approach would consist of specifications for minimum treatment level required for certain irrigation purposes. For instance, not less than primary treatment should be authorized in any circumstances; only disinfected effluent would be allowed to irrigate raw eaten vegetables and sport lawns. This methodology is not satisfactory, however, because the operation of treatment plants is frequently unreliable in many areas and the effluent quality is unlike to correspond to the presumed level. A combination of both approaches, consisting of a reduction of the number of parameters in the list with some specification concerning the minimum treatment requested for certain irrigation purposes seems to be a better methodology. The selection of parameters and the requests for treatment specifications are now the question.

As mentioned previously the list of parameters for the monitoring of reclaimed wastewater quality for irrigation should include both chemical parameters with major agronomic impact and microbiological parameters affecting public health. However, these impacts depend on the agriculture practices, such as irrigation methods and relevant crops in the region, for example. Therefore, the guidelines should be the interception of three factors: specification of limits for some chemical and microbiological quality parameters in combination with treatment specifications and as a function of the field practices (crops and irrigation methods).

3. THE PORTUGUESE STANDARD ON THE USE OF TREATED URBAN URBAN WASTEWATER FOR IRRIGATION: NP 4434

3.1. Scope

The objectives defined for NP 4434—the Portuguese Standard on Reuse of Treated Urban Wastewater for Irrigation [1] are the following:
(a) specification of quality criteria for treated urban wastewater for irrigation; (b) guidelines for selection of irrigation equipment and methods; guidelines for environmental protection and environmental impact monitoring in areas irrigated with treated urban wastewater.

NP 4434 applies to agricultural irrigation (crops, forest, plant nurseries) and landscape irrigation (parks, gardens, sport lawns such as golf courses).

3.2. Quality Criteria of Treated Urban Wastewater for Irrigation

The Portuguese law No. 236/98 states in its Annex XVI the chemical quality criteria for water used for irrigation as maximum recommended value (MRV), which were adopted in NP 4434 and are presented in Tables 1 and 2.

Table 1. Standards on salinity of irrigation water

Parameter		Unit	MRL
Salinity	EC	dS/m at 25°C	1
	TDS	mg/l	640
Sodium adsorption ratio (RAS)		–	8
Total suspended solids (SST)		mg/l	60
pH		Escala de Sorensen	6.5–8.4

Concerning microbiological quality of irrigation water the Decree law n° 236/98 indicates that the water should contain no more than 10^3 faecal coliform/100 ml and less than 1 helminth egg/l. The rational for NP 4434 is based upon the principle that the required microbiological characteristics of irrigation water should be established taking into consideration the use of the irrigated plant (e.g., for food to be eaten raw by humans and animals, forest, industrial crop, sport lawns, etc.) together with the type of irrigation method.

Although every irrigation method was found acceptable for treated urban wastewater exception for overflow, NP 4434 recommends that preference is given to irrigation methods that limit contact between irrigation water and the plant, especially with the edible parts of the plant, and reduce the risk of runoff and spray generation and transportation by the wind.

3.3. Site Characterisation

In order to prevent adverse environmental impact the following characteristics of the irrigation site were given consideration in NP 4434: chemical properties, especially the soil heavy metal content; topography; hydrogeological vulnerability; distance to dwellings.

3.3.1. Soil chemistry

Some soil chemical characteristics are important to enable irrigation mainly from the fertility point of view: pH, salinity, cation exchange capacity, nutrients, organic matter, and percentage of exchangeable sodium. Usual agricultural techniques are adequate to control these characteristics. Therefore, only some soils with high heavy metal content may be excluded from irrigation with treated urban wastewater. In NP 4434 soils that present heavy metal content greater than allowed to soils for sludge disposal are not acceptable for irrigation with treated urban effluents (Table 3).

Table 2. Standard on chemical quality of water for irrigation

Element or ion	Maimum recommended value (mg/l)	Maximum permissible value (mg/l)
Aluminum (Al)	5.0	20
Arsenic (As)	0.10	10
Barium (Ba)	1.0	*
Beryllium (Be)	0.5	1.0
Boron (B)	0.3	3.75
Cadmium (Cd)	0.01	0.05
Lead (Pb)	5.0	20
Chloride (Cl^-)	70	–
Cobalt (Co)	0.05	10
Copper (Cu)	0.20	5.0
Chromium (Cr)	0.10	20
Tin (Sn)	2.0	*
Iron (Fe)	5.0	*
Fluoride (F)	1.0	15
Lithium (Li)	2.5	5.8
Manganese (Mn)	0.20	10
Molybdenum (Mo)	0.005	0.05
Nickel (Ni)	0.5	2.0
Nitrate (NO_3^-)	50	*
Selenium (Se)	0.02	0.05
Sulfate (SO_4^{2-}	575	*
Vanadium (V)	0.10	1.0
Zinc (Zn)	2.0	10.0

Table 3. Maximum permissible heavy metals in soils (mg/kg dry soil)

Metal	Soil pH		
	<5.5	5.5–7.0	>7.0*
Cadmium (Cd)	1	3	4
Lead (Pb)	50	300	450
Copper (Cu)	50	100	200
Chromium (Cr)	50	200	300
Mercury (Hg)	1	1,5	2
Nickel (Ni)	30	75	110
Zinc (Zn)	150	300	450

* Not applicable to crops for human consumption and pasture. For such crops values for soils with pH ranging from 5.5 to 7.0 are applicable.

3.3.2. Topography

The irrigation site should not present more than 20% sloper to prevent soil erosion and runoff and subsequent risk of surface water contamination. Some irrigation methods require soil slope as low as 3%, as recommended in NP 4434.

3.3.3. Hydrogeological vulnerability

Water reuse for irrigation should not be carried out over soil classified as hydrologically vulnerable such as very permeable soil or carsic rock. The water table should be always deeper than 1–4 m during irrigation, depending on the type of irrigation method: drip irrigation can be used when the water table is 1 m, but furrow irrigation requires that the water table is around 4 m.

3.3.4. Distance to dwellings

In NP 4434 distance from irrigated site to dwellings depends on the irrigation method and the wastewater treatment. No distance is required for drip irrigation. Near sources for drinking water a distance of 100 m should be observed. In any other situations NP 4434 recommends the distance shown in Table 4.

Table 4. Minimum distance from irrigated site to dwellings

Irrigation method	Type of dwelling	Faecal coliforms in treated wastewater (FC/100 ml)		
		$\leq 2\times10^2$	$2\times10^2-10^3$	$>10^3$
Sprinkler irrigation	Single house	30 m	60 m	70 m
	Group of houses	50 m	80 m	100 m
Other irrigation method	Single house	10 m	20 m	30 m
	Group of houses	30 m	60 m	70 m

3.4. Classes of Crops to be Irrigated with Treated Urban Wastewater

In NP 4434n crops are classified in four classes according to the level of risk of microbiological contamination generated by irrigation with treated urban wastewater:

Class A - vegetables to be eaten raw.
Class B - public parks and gardens, sport lawns, forest with public easy access.
Class C - vegetables to be cooked, forage crops, vineyards, orchards.
Class D - cereals (except rice), vegetables for industrial process prior to consumption, crops for textile industry, crops for oil extraction, forest and lawns located in places of difficult or controlled public access.

Vegetables whose edible parts are in close contact with the irrigated soil are not included in Class A. Irrigation of such crops with treated wastewater is not permitted in NP 4434. Crops of class A can only be drip irrigated. The irrigation of other classes of crops with treated urban wastewater depends on the treatment level according to Table 5.

3.5. Control of Environmental Impact and Public Health Risk

The safe water reuse for irrigation requires not only appropriate wastewater treatment, crop selection according to water quality and irrigation method, good site evaluation, as mentioned in previous items, but also measures aiming at reducing to a minimum the risks of contamination of groundwater and surface water, the contact of people and animals with the irrigation water, the transportation of droplets by wind, the inhalation of aerosols. NP 4434 establishes procedures to minimize such risks, concerning the irrigation installation and the irrigation site.

NP 4434 presents guidance on signaling the irrigation installation lay out of piping, time schedule of irrigation sessions, protection equipment for irrigation operators, animal access to irrigation field, wind speed for spray irrigation. Procedures for field drainage and protection with a tree curtain are also described.

3.6. Monitoring

The use of treated wastewater for irrigation brings water to the soil–plant biosystem but nutrients and other substances that may be beneficial and hazardous, depending on plant needs and soil buffer capacity. Therefore, the sustainable long term of water reuse for irrigation requires monitoring the amount of applied nutrients and heavy

Table 5. Microbiological quality of treated urban wastewater for irrigation

Classes	Type of crop	Faecal coliforms (NMP or cfu/100 ml)	Helminth eggs (egg/l)	Appropriate treatment	Notes
A	Vegetables to be eaten raw	100	1	Secondary⇒filtration⇒disinfection or Tertiary⇒ filtration⇒disinfection	UV disinfection (self-cleaning lamps) or O_3 preferable to chlorination
B	Public parks and gardens, sport lawns, forest with public easy access	200	1	Secondary⇒filtration⇒disinfection or Tertiary⇒ filtration⇒disinfection	UV disinfection (self-cleaning lamps) or O_3 preferable to chlorination. Irrigation must avoid contact with people
C	Vegetables to be cooked, forage crops, vineyards, orchards	10^3	1	Secondary⇒filtration⇒disinfection or Tertiary⇒ filtration⇒disinfection or Waste stabilization ponds (System with ≥3 ponds and $t_R \geq 25$ d)	UV disinfection (self-cleaning lamps) or O_3 preferable to chlorinaton. Irrigation of vinyards and orchards must avoid contact with fruit. Fruit fallen on the soil should not be collected
D	Cereals (except rice), vegetables for industrial process, crops for textile industry, crops for oil extraction, forest and lawns located in places of difficult or controlled public access	10^4	1	Secondary ⇒ maturation ponds ($t_R \geq 10$ d) or Secondary ⇒filtration⇒disinfection	UV disinfection (self-cleaning lamps) or O_3 preferable to chlorinaton. Irrigation must avoid contact with people

metals. NP 4434 includes a table where the irrigation operator records the volume of treated wastewater applied during every irrigation session and based upon the water analysis calculates the amount of nutrients (N, P_2O_5, and K_2O) and heavy metals (Cd, Cu, Cr, Pb, Hg, Ni, and Zn).

Usually nutrients applied to the biosystem soil–plant together with treated urban wastewater do not match completely the needs of crops and the addition of artificial fertilizers is necessary. The Portuguese standard NP 4434 presents a table with a fertilization program, which guides the irrigation operator to calculate the amount of fertilizers to be added in order to complement the fertilization carried by the treated wastewater. The table is based on a balance between crop needs, soil chemical analysis, and the estimated nutrient amounts to be applied together with the estimated irrigation volumes.

Guidance on the frequency of soil analysis is included. The real amount of applied fertilizers to both the irrigation water and the complementary artificial fertilizers is recorded by the irrigation operator in another table.

Monitoring of the impact of the use of treated urban wastewater for irrigation on groundwater quality is important and NP 4434 provides a table to record the results of analysis of samples of groundwater taken from piezometers. Monitoring details are given in the standard.

4. CONCLUSIONS

NP 4434 is an important tool in Portugal as it guides on the use of treated urban wastewater agricultural irrigation and landscape irrigation. It is the first regulation in the country to ensure safe practice, e.g., for selection of irrigation equipment and methods, guidelines for environmental protection, and includes environmental impact monitoring in areas irrigated with treated urban wastewater.

Acknowledgment: We were pleased to have financial support from NATO to attend the NATO ARW in Istanbul, Turkey.

5. REFERENCES

1. Instituto Portugues da Qualidade. 2006. Norma Portuguesa sobre Reutilização de Águas Residuais Urbanas Tratadas na Rega. NP 4434, IPQ, Caparica.
2. Marecos do Monte MH, Angelaqkis AN, Asano, T. 1996. Necessity for Establishment of European Guidelines for Reclaimed Wastewater in the Mediterranean Region. Water Science Tech. 33/10-11: 306–316.

3. Lazarova et al. 2000. Role of Water Reuse in Enhancement of Integrated Water Management and Catchment Scale. In: 1st IWA World Water Congress, 3–7 July, 2000, Paris, France, 8:33–40.
4. Asano T. Wastewater Reclamation and Reuse. Water Quality Management Library, 10, Technomic Publishing Inc., USA.
5. Marecos do Monte MH. 2006. Agricultural Irrigation with Treated Wastewater in Portugal. Chapter 18 in Asano, T. (Ed.) Wastewater Reclamation and Reuse. Water Quality Management Library, 10, Technomic Publishing. USA.

FARMERS' DEMAND FOR RECYCLED WATER IN CYPRUS: A CONTINGENT VALUATION APPROACH

Ekin Birol[1], Phoebe Koundouri[2], and Yiannis Kountouris[2]
[1] Department of Land Economy and Homerton College, University of Cambridge, UK
[2] Department of International and European Economic Studies, Athens University of Economics and Business, Greece

Corresponding author: pkoundouri@aueb.gr

Abstract:

The aims of this paper are twofold. First, to investigate farmers' opinion of adoption of a new program, which involves utilization of recycled water to replenish an aquifer. Second, to evaluate the economic viability of this new program. A contingent valuation study is undertaken with 97 farmers located near the Akrotiri aquifer in Cyprus, a common-pool water resource with rapidly deteriorating water quality and quantity. The results reveal that farmers are willing to adopt the new technology, and they derive higher values from a recycled wastewater use program, which provides high quality water, and high water quantity in the aquifer.

Keywords: Wastewater, quantity, quality, aquifer recharge.

1. INTRODUCTION

The Millennium Development Goals of the United Nations lists water scarcity among the most important global challenges. Scarcity of water resources can be a result of reduced water availability emerging from increased demand and/or of reduced water quality arising from increased pollution. It has been argued that both in developed and developing countries, the main cause of water quality and quantity deterioration is the increasing volume and intensity of the agricultural sector [1]. Agricultural production, in return, is likely to become unsustainable in the long run as a result of reduced water quality and quantity. The main economic reason behind the inefficient management of water quality and quantity is the public good nature of many water resources, which implies that water resources are not traded in the markets as other goods are, and hence they do not have readily available market prices which would make their management efficient and sustainable. Moreover, several water resources which are used for irrigation, such as groundwater, are common-pool resources, which face problems of

M.K. Zaidi (ed.), Wastewater Reuse–Risk Assessment, Decision-Making and Environmental Security, 267–278.

overexploitation and pollution, resulting in significant costs to the local economy in the long run, as well as in the short run [2].

The magnitude and gravity of the water scarcity problem, coupled with the imminent food security issues, highlight the urgent need for development and implementation of new technologies and economic measures for efficient and sustainable management of the world's scarce water resources. This paper investigates farmers' stance to a recently proposed technology to tackle water resources scarcity, namely reuse of recycled water for irrigation, and evaluates the economic viability of this new measure. The case study is the Akrotiri aquifer (Figure 1), a common-pool resource and the third largest aquifer in Cyprus, a semiarid country, which faces chronic water shortages. The aquifer is extremely important for the local economy. Extending over 42 km^2, the aquifer not only provides local farmers with irrigation water, but also supplies a significant portion of the water needs of the city of Limassol and the nearby British sovereign bases. The aquifer is replenished with runoffs from the Kouris River; releases from the Kouris River dam; rainfall, and agricultural return flows [3]. Akrotiri aquifer faces important water quality and quantity problems, which are expected to have significant adverse effects on the welfare of the local farmers in the not too distant future. After the construction of the Kouris river dam, inflow in the aquifer has decreased significantly resulting in a lower water table [3]. This has lead to intrusion of saltwater into the aquifer to maintain the hydrological balance. Water quality in the aquifer is deteriorating further because of the intensive use of fertilizers and pesticides in agricultural production in the area. Recently the Cypriot government has designated the Akrotiri aquifer as a "nitrate vulnerable area" [4]. The quantity of water in the aquifer is also adversely affected by uncontrolled and excessive pumping in the area, an artifact of lack of property rights, i.e., the open access nature of the aquifer.

In order to mitigate the adverse effects of reduced water availability and deteriorating water quality in the aquifer, its replenishment with treated effluent

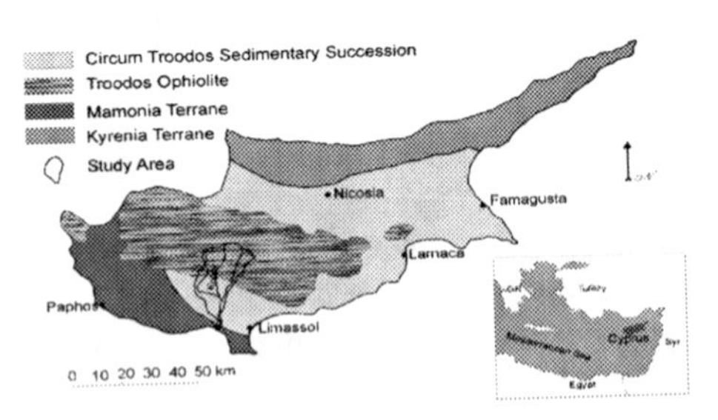

Figure 1. The Akrotiri aquifer

from Limassol and nearby villages has been proposed. Given that the public good and open access nature of the resource has resulted in its inefficient management, economic instruments, including water pricing, allocation of property rights, and rationalization of groundwater pumping have also been proposed to enable the efficient and sustainable management of the aquifer. In order to examine the feasibility of the two goals, namely adoption of the new technology, i.e., the use of recycled water for irrigation, and pricing of water pumped from the aquifer, a nonmarket valuation method, namely a contingent valuation study was undertaken with 97 randomly selected farmers in the Akrotiri aquifer area. The results of this study reveal that on average local farmers are willing to adopt the new technology, and that they also derive significant economic values from higher levels of water quantity in the aquifer, and replenishment of the aquifer with higher quality treated water, expressed in terms of willingness to pay. These results have important implications for adoption of this new source of water in Cyprus, as well as in determining the efficient price to charge for the water in the aquifer, once it is replenished with recycled water.

The next section presents the contingent valuation survey undertaken. The following section reports the results of this valuation exercise, and the final section concludes the paper and draws some policy implications.

2. CONTINGENT VALUATION STUDY ON FARMERS' VALUATION OF RECYCLED WATER

A contingent valuation (CV) survey was implemented in September and October 2006 to estimate farmers' valuation of recycled water use programs to be implemented in the Akrotiri aquifer. Following [5], four recycled water use programs were valued, where a respondent's valuation (i.e., total willingness to pay [WTP] for each program can be defined as the value of simultaneous change in the quantity (water level in the aquifer) and quality (treatment level of the recycled water used to replenish the aquifer) of water in the aquifer. This survey design, which is also known as scenario difference approach, enables estimation of the values of both the quantity and the quality of the recycled water used to replenish the aquifer.

More formally, the valuation exercise presented in this paper takes into account that a recycled water use program might have multidimensional impacts on the state, q, of the aquifer, affecting both its quantity (water level in the aquifer) and quality (treatment level of the recycled water used to replenish the aquifer). The definition of value used in this paper, therefore treats q as a vector [4].

Following [4], assume that q consists of two dimensions, the quantity and quality of the water in the aquifer, $q = (q_1\text{-}q_2)$, where the former is measured by the level of recycled water used to replenish the aquifer and the latter is measured by the quality of treated water used to replenish the aquifer. A farmer's preference function can be specified as $u = u(x(q_1, q_2))$ where x is the composite good, i.e., water for irrigation. For a multidimensional change in the program that results in the simultaneous change in both dimensions in q, the Hicksian compensating welfare measure is the amount of income paid or received that would leave the individual at the initial level of utility subsequent to the multiple impacts of policy. For the change from q^0 to q^1 a holistic measure of value is represented by:

$$\mathrm{WTP}(q^0, q^1) = e(p, q_1^0, q_2^0, u^0) - e(p, q_1^1, q_2^1, u^0) \tag{1}$$

where $e(\bullet)$ is the standard respondent expenditure function defined for market prices p and fixed utility u^0. Following [5], component values can be subsequently defined from Equation 1 by using a simultaneous valuation path that begins at $q^0 = (q_1^0, q_2^0)$ and ends at $q^1 = (q_1^1, q_2^1)$. The simultaneous valuation path estimates the effect of each element of q as the overall vector changes from q^0 to q^1. The disaggregated expression for 1 is given by:

$$WTP(q^0, q^1) = \int_{q^0}^{q^1} \left[\frac{\partial e(p, q_1, q_2, u^0)}{\partial q_1} \right] dq_1 + \int_{q^0}^{q^1} \left[\frac{\partial e(p, q_1, q_2, u^0)}{\partial q_2} \right] dq_2 \tag{2}$$

where each one of the two components of Equation 2 evaluates a derivative of the expenditure function $\partial e(p, q_1, q_2, u^0) / \partial q_i, i \in \{1,2\}$ as the overall recycled water use program shifts from its initial to postprogram level.

The CV survey consisted of three parts. In the first section, the respondents were informed of the serious water quality and quantity challenges faced by Cyprus. They were reminded of the irrigation water shortages in the Akrotiri area due to uncontrolled pumping from the aquifer. They were also explained that uncontrolled pumping lowers the groundwater level, causing seawater intrusion, and hence increasing water salinity, which makes the groundwater inappropriate for irrigating most crops. Farmers were further reminded that lower levels of water in the aquifer imply higher pumping costs. They were warned that ongoing groundwater overexploitation in the Akrotiri area will eventually result in the permanent desertification of presently fertile areas, thereby causing irreparable economic damage to local and national agriculture, and hence to the entire country's economy.

The farmers were then presented with the new technology, namely the use of recycled water for replenishment of the Akrotiri aquifer, which they were told would definitely provide long-term water security in the area. They were explained that under the recycled water use program, treated wastewater from Limassol and the nearby villages would be channelled into the aquifer to replenish its groundwater supplies. They were further explained in layman terms what recycled water is and how the programme would work. Finally, the farmers were told that if a recycled water use programme is implemented, they would be asked to pay a price to the government for each m^3 of water they pump from the aquifer, and the Ministry of Agriculture and Natural Resources would monitor the quantity of water pumped. They were explained that the quality of the recycled water used to replenish the aquifer, and the quantity of the water in the aquifer would depend on the price of each m^3 of water pumped from the aquifer.

In the CV survey, farmers were presented with four distinct recycled water use programs, characterized in terms of the quantity and quality of recycled water used to replenish the aquifer. The definitions of the recycled water use programs were based on the focus group discussions and personal interviews with the policy makers, ecologists, and hydrologists at the Cypriot Ministry of Agriculture, Natural Resources and the Environment. The farmers were explained that the government would be equally likely to undertake any one of the programs, or to not to undertake any recycled water use program at all, depending on the costs and benefits generated by each option. The four recycled water use programs and the present situation, i.e., status quo, were defined as follows:

- Status Quo: This is the present situation in which there are no recycled water use programs implemented to replenish the aquifer. In this case the quantity of water in the aquifer, which is currently at a medium level, will decrease rapidly to a low level within the next 10 years, implying that the pumping costs will double. In the present situation farmers are not expected to pay for the water they pump from the aquifer.
- Recycled Water Use Program A: In this program, low quality treated water is used to replenish the aquifer. Low quality treated water is appropriate for irrigating forestland, albeit in a controlled manner which ensures neither humans nor crops come in contact with the water. The quantity of water in the aquifer will stay at its current medium level and the pumping costs will remain the same during the next 10 years. If this

program is undertaken, then the farmers are expected to pay for each m^3 of water they extract from the aquifer.

- Recycled Water Use Program B: Under this program, medium quality treated water is used to replenish the aquifer. Medium quality treated water is appropriate for irrigation of trees, such as olive trees or vineyards, where water does not come in contact with the crops. The quantity of water in the aquifer will stay at its current medium level and the pumping costs will remain the same during the next 10 years. If this program is undertaken, then the farmers are expected to pay for each m^3 of water they extract from the aquifer.
- Recycled Water Use Program C: Under this program, medium quality treated water is used to replenish the aquifer. The quantity of water in the aquifer will increase to a high level, implying that the pumping costs will decrease to half or even quarter of what they are now during the next 10 years. If this program is undertaken, then the farmers are expected to pay for the m^3 of water they extract from the aquifer.
- Recycled Water Use Program D: Under this program, high quality treated water is used to replenish the aquifer. High quality treated water is appropriate for irrigation of crops whose edible parts do not come in contact with treated water. The quantity of water in the aquifer will increase to a high level, implying that the pumping costs will decrease to half or quarter of what they are now during the next 10 years. If this program is undertaken, then the farmers are expected to pay for the m^3 of water they extract from the aquifer.

An "advanced disclosure" approach was employed, where respondents were presented in advance with all four recycled water use programs and the status quo alternative [5]. The valuation questions consist of two parts: first the respondents were asked whether they would be WTP some amount of money for m^3 of water in order to move from the status quo to program A. In the case where the respondent was willing to participate in the recycled water use program, they were asked for their maximum WTP for m^3 of water, using a payment card with amounts ranging from Cyprus pounds (CYP) 0.01 (€0.018) to over CYP2 (€3.516). Similarly, the respondents were asked whether they would like to participate in recycled water use programs B, C, and D, and if they were, they were asked to state their maximum WTP to move from the status quo to each one of these programs. Before stating their WTP, the respondents were told to bear in mind how they think the programs described above would affect their production, current income, and other financial obligations, as well as their future agricultural income. They were also reminded that if the majority of farmers decline the recycled water use programs, other measures would have to be imposed, such as obligatory taxation, for water pumping. Debriefing questions were asked to identify between protest responses and true zero values. The second section of the survey collected information on the farm characteristics, farm management practices, as well as farmers' attitudes and perceptions of how they think consumers would perceive these program and what they think are the most important agricultural problems in Cyprus. The final section of the survey collected various socioeconomic data on the farmers and their families, including age, educational level, and household size.

Data collection took place during September and October 2006 in four villages located in the Akrotiri area. The sampling frame is comprised of all the farmers located in the area. The results reported in this paper, however, are from a pilot sample, which was envisaged to include randomly selected 100 farmers from the sampling frame. Overall, 97% of the pilot sample approached agreed to take part in the survey, and the results reported in the following section are representative of the Akrotiri area.

3. RESULTS

The sample statistics reveal that the main decision makers in the farm are all male and full-time farmers. Only 4.4% of them have part-time jobs in addition to full-time farming. Their average age is 46.1, which is slightly younger than the European Union average of 48 years [5]. The main decision-makers' average years of experience in farming is 20.44 years. 82.4% of the farmers have high school diplomas, whereas only 3.3% or less have primary school diplomas, and 4.4% have university degrees. The average farm household in the area comprises of 3.13 members, and the average number of children is 0.95. The total monthly expenditure of households (proxy for income) is CYP 1,598.7 (approximately €2,896). The average total area of land owned by the households (indicator of wealth) in the sample is 659.2 stremmas (one stremma equals 0.1 hectare); the average total area they cultivate is 649.6 hectares, of which an average of 347.6 stremmas is irrigated. 58.5% of farmers obtain their water for irrigation from a well located on their land, whereas 40.8% get their irrigation water from dams and reservoirs, and only 9.7% buy their irrigation water from other farmers in the area.

When asked how do the farmers think the consumers would react when they know that recycled water is used in agricultural production in the area, 13.2% of farmers think that consumers will stop buying agricultural products produced in the area, whereas 34% of them think that the consumers would decrease their consumption; 35.2% do not think that the consumers would change their behavior;

13.2% think that the consumers would slightly increase their consumption, and 4.4% of the farmers think that the consumers might increase their consumption significantly. That is the consensus among the farmers on consumers reaction is split, as 47.2% of them think that consumers might decrease or stop consumption of agricultural products from the area, and 52.8% of them think that use of recycled water in the area would either have no affect on consumer behavior or increase consumption of agricultural products from the area. Finally, farmers' attitudes and perceptions with regards to the most important agricultural problems in Cyprus were investigated. The results reveal that 36.3% of the farmers think that low land fertility is a very important or an extremely important agricultural problem in Cyprus, whereas 41.7% think that low prices for agricultural goods; 44% consider that high water salinity; 45% think that low water quality due to pesticide and fertilizer use, and 53.9% think that insufficiency of water resources is a very important or extremely important agricultural problem in Cyprus.

The mean and median WTP results are reported in Table 1. Six respondents stated zero WTP for each one of the recycled water use programs. In order to differentiate true zero WTP values from protest responses, five follow-up questions in close-ended response format were asked [6]: (1) I shouldn't be asked to pay for the water under my land; (2) I do not believe that the system will succeed in improving conditions; (3) I have no interest for water quantity and quality in the aquifer; (4) I don't believe that recycled water is safe and appropriate for farming, and (5) It is not profitable for me to participate. Those respondents that have agreed with the statements (1), (2), or (4) were classified as protesters of the recycled water program and were removed from the sample. Consequently, all of the six respondents, i.e., 6.2% of the sample were classified as protestors, and the remainder of the sample, i.e., 93.8% believe that the recycled water use program would succeed.

The average and median WTP values for the remaining 91 respondents, who are in favor of the recycled water use programs, are reported in Table 2. On average farmers are WTP CYP 0.37 per m^3 of water to have the recycled water use program A implemented. This program would ensure that the farmers have as much water as they do now within the next 10 years. Although under this program the quality of treated water is low, this significantly high WTP reveals farmers' concerns with regards to decreasing quantity of water in the aquifer. Farmers are WTP an additional CYP 0.03 per m^3 of water for program B, in order to secure medium quality treated water alongside a medium quantity of water in the aquifer within the next 10 years. Farmers' WTP for higher quantity of water in the aquifer alongside medium quality treated water is CYP 0.03 more, i.e., CYP 0.43 per m^3 of water. Finally, they are WTP an additional CYP 0.07 per m^3 of water for recycled water use program D, which ensures high quantity of water in the aquifer, as well as the use of high quality treated water for replenishment of the aquifer.

Table 1. Mean and median WTP values for recycled water use programs (in CYP/m^3 of water)

Recycled water use program	Mean (standard deviation)
Recycled water use program A	0.37 (0.46)
Recycled water use program B	0.40 (0.46)
Recycled water use program C	0.43 (0.47)
Recycled water use program D	0.50 (0.47)
	Median
Recycled water use program A	0.1
Recycled water use program B	0.2
Recycled water use program C	0.2
Recycled water use program D	0.4

Source: Akrotiri Recycled Water Use Programme Survey, 2006.

The relationship between farmers' WTP for recycled water use programs and program attributes, controlling for farmer characteristics, attitudes, and perceptions was further examined. An ordinary least squares (OLS) stacked regression model was estimated, where respondents' WTP was specified to be a function of medium and high water quantity in the aquifer, and medium and high-quality treated water used to replenish the aquifer, taking low quality of treated water and low quantity of water in the aquifer as base levels. The results reveal that farmers' WTP significantly increases with the use of high quality treated water used to replenish the aquifer, as well as with medium and high levels of water quantity in the aquifer. The only significant determinants of WTP are total irrigated land area, percentage of water obtained from the well on the land, and farmers' perceptions on the importance of water quality and salinity in Cyprus. Farmers who have higher total areas of land irrigated, and those who obtain higher percentages of their water from wells on their lands, are WTP more for higher levels of water quality and quantity. Further, farmers who think that low water quality is a very or extremely important agricultural problem in Cyprus are WTP more for the recycled water use programs, inline with the findings for higher WTP values for recycled water use program D, which supports high-quality treated water. Those who consider high salinity to be a very or extremely important agricultural problem in Cyprus are WTP less for the recycled water use programs, reflecting that for this new technology to succeed in Cyprus, special water treatment technology, which keeps salinity of recycled water at low levels, should be employed, so as to not to affect the salinity of groundwater.

For the purposes of the cost-benefit analysis of the recycled water use programs, WTP values for programs B and D were calculated from regression parameters.

For recycled water use program B, which supports medium quality of recycled water and medium quantity of water in the aquifer, an average farmer is WTP CYP 0.369 per m^3 of water. In order to investigate the variation of the WTP across the sample, four farmer types were selected from the sample to represent (1) a large-scale farmer, with 873 stremmas of irrigated land, who obtains 90% of his irrigation water from the well on the land; (2) a large-scale farmer with 1120 stremmas of irrigated land, who obtains all of his irrigation water from dams and reservoirs; (3) a small-scale farmer, with 11.9 stremmas of irrigated land, who obtains all of his irrigation water from the well on its land; (4) a small-scale farmer, with 70.5 stremmas of irrigated land, who obtains all of his irrigation water from dams and reservoirs. Accordingly, the WTP of farmer 1 for program B is calculated to be CYP 0.485 per m^3 of water; for farmer 2 this figure is CYP 0.490 per m^3 of water; for farmer 3, it is CYP 0.317 per m^3 of water, and for farmer 4 it is as low as CYP 0.279 per m^3 of water. Finally, program D, which supports high-quality treated water and high quantity of water, the average farmers' WTP is CYP 0.509 per m^3 of water, whereas this figure is CYP 0.611 per m^3 of water for farmer 1; as high as CYP 0.615 per m^3 of water for farmer 2; CYP 0.44 per m^3 of water for farmer 3, and CYP 0.405 per m^3 of water for farmer 4.

4. CONCLUSIONS

In this paper a contingent valuation (CV) exercise is undertaken to examine (a) farmers' attitudes toward adoption of a new technology, namely use of recycled water to replenish an aquifer used for irrigation, and (b) their willingness to pay (WTP) for different levels of water quantity in the aquifer, and for different levels of treated water quality used to replenish the aquifer. The case study is the Akrotiri aquifer, a common-pool resource located in Cyprus, an arid country with chronic water shortages. The aquifer, similarly to several open access water resources, is facing rapid deterioration of its water quality and quantity, and is in need of drastic economic and other measures to ensure its efficient and sustainable management.

The results of this study reveal that 93.8% of the randomly selected farmers located in the Akrotiri area are willing to participate in and also WTP significant amount for recycled water use programs. Farmers' are WTP higher amounts for those programs, which generate higher water quantity in the aquifer, and use higher-quality recycled water for the replenishment of the aquifer. Farmers, however, are WTP even for those recycled water use programs, which use low-quality treated water, revealing the gravity of the water quantity scarcity problem faced by farmers in this area. Finally, those farmers who have larger areas of irrigated land, and those who obtain most of their irrigation water from wells on their lands are WTP more for water quality and quantity of water.

Table 2. Ordinary least squares regression on determinants of WTP for recycled water use programs

Variable	Coefficient (Standard Error)	
Constant	–0.178**	(0.11)
Medium quality water	0.025	(0.047)
High quality water	0.096*	(0.067)
Medium quantity water	0.370***	(0.037)
High quantity water	0.401***	(0.060)
Total area of land irrigated	0.0002***	(0.1)
% of irrigation water from well on land	0.0005***	(0.0001)
% of irrigation water from dams and reservoirs	–0.0002	(0.0004)
Farming experience of the main decision maker	0.001	(0.001)
Education level of the main decision maker	0.016	(0.019)
Total expenditure of the household	–0.00001	(0.00003)
Consumers' perceptions favourable	–0.016	(0.025)
Low water quantity most important problem	0.014	(0.027)
High salinity of water most important problem	–0.103***	(0.042)
Low water quality most important problem	0.134***	(0.041)
Adjusted R2	0.34	
Sample size	91	

Source: Akrotiri Recycled Water Use Programme Survey, 2006.
***1% significance level, **5% significance level, *10% significance level with two-tailed tests.

These results could have important implications for efficient and equitable pricing of aquifer water in the area, as well as for adoption of the appropriate recycled water use program, which maximizes the social welfare. In order to be able to draw sound policy recommendations, however, revenues that would be generated under each recycled water use program should be compared to the costs of the programs [7].

Acknowledgments: Please note that throughout the paper, "Cyprus" refers to Greek Cypriot area of the island, controlled by Cyprus Government. We gratefully acknowledge the European Union's, DG Research for financial support through the AQUASTRESS Integrated Project, under the 6th Framework Program. We would like to thank Alessandro Battaglia, Costantino Masciopinto, and Dimitris Glekas, for valuable comments, suggestions, and fruitful discussions. We would also like to thank Nikolaos Syrigos and Charialos Giannakidis for their assistance in data collection. Finally, we would like to thank NATO for their financial support for our participation at the NATO Advanced Research Workshop on Wastewater Reuse: Risk Assessment, Decision-Making and Environmental Security, held in Istanbul, Turkey, on October 12–16, 2006.

5. REFERENCES

1. Young RA. (2005). Determining the Economic Value of Water: Concepts and Methods, RFF Press, Washington, DC, USA.
2. Cornes R and Sandler T. (1996). The Theory of Externalities, Public Goods and Club Goods, Second Edition, Cambridge University Press, Cambridge.
3. Mazi K, Koussis AD, Restrepo PJ, and Koutsoyiannis D. (2004). A groundwater-based, objective-heuristic parameter optimisation method for a precipitation-runoff model and its application to a semiarid basin. J. Hydrology, 294/3-4, 243–258.
4. Republic of Cyprus, Ministry of Agriculture, Natural Resources and the Environment (2004). Report for the Implementation of the Directive Concerning the Protection of Waters Against Pollution Caused by Nitrates from Agricultural Sources.
5. Kontoleon A and Swanson T. (2003). The Willingness to Pay for Property Rights for the Giant Panda: Can a Charismatic Species Be and Instrument for Nature Conservation? Land Economics, 79(4): 483–499.
6. A Special Analysis of the **Eurobarameter** 2000 Survey. Vienna: SORA. Fogh Rasmussen, A. 1999. Dansk Folkeblad 3 (2): 10–11. Gaasholt, O. and Togeby, L. 1995.
7. Haab T. (1999). Nonparticipation or Misspecification? The Impacts of Nonparticipation on Dichotomous Choice Contingent Valuation, Environmental and Resource Economics, 14: 443–461.

EVALUATING THE WORLD NEW HEALTH ORGANIZATION'S 2006 HEALTH GUIDELINES FOR WASTEWATER

Hillel Shuval
Department of Environmental Health Sciences
Hadassah Academic College, Jerusalem
and The Hebrew University of Jerusalem, Israel

Corresponding author: hshuval@vms.huji.ac.il

Abstract:

After several years of intensive research, study, and consultations with world experts in public health, epidemiology, agronomics, environmental sciences, and engineering, as well as other UN agencies, the WHO, in 2006, has published a revised updated volume of the guidelines for the safe use of wastewater, excreta, and gray water. These new guidelines drafted by a panel of 35 experts are recognized as representing the position of the United Nations system on issues of wastewater, excreta, and gray water use and health by "UN-Water."

Keywords: WHO, guidelines, wastewater, agriculture, health risks.

1. INTRODUCTION

This paper will relate only to those wastewater treatment guidelines and standards concerning control of health risks and will not relate to other factors such as effects of wastewater quality on soil structure and possible pollution of groundwater and aquifers. In some areas these factors may be critical and require additional chemical guidelines and requirements needed to protect the environment from degradation. This is particularly true for Israel which faces problems of highly permeable aquifers particularly susceptible to pollution [1].

There are a number of different views as to what is an appropriate degree of wastewater treatment. As early as 1933, The California State Board of Health established strict microbial standards for wastewater irrigation of vegetables eaten uncooked at 2 coliform bacteria/100ml of wastewater effluent [2]. There was no particular scientific or epidemiological foundation for these standards, which were designed to be equivalent to those required for drinking water and thus were considered to be fail safe. Many countries around the world including developing countries, copied these standards, including Israel. However, they are rarely

M.K. Zaidi (ed.), Wastewater Reuse–Risk Assessment, Decision-Making and Environmental Security, 279–287.

enforced in developing countries and adjacent to most urban areas irrigation with raw, totally untreated wastewater is practiced, resulting in major health risks.

The World Bank and the World Health Organization (WHO) joined forces in 1982 to sponsor a series of major independent and parallel research studies to provide a scientific basis for the establishment of new more rational guidelines [3–5]. These studies provided the health risk – epidemiological basis for the expert committee of leading epidemiologists and water scientists who drafted the WHO 1989 Recommended Guidelines for Wastewater Use in Agriculture. These recommended guidelines established a much more liberal and feasible goal of 1,000 faecal coliforms/100ml for the irrigation of vegetables eaten uncooked.

However, the USEPA together with the USAID established their own more rigorous guidelines in 1992 at zero fecal coliforms/100 ml, a BOD of 10 ppm, a turbidity of 2 NTU, and a chlorine residual of 1 mg/l. These guidelines where prepared by one of the leading American consulting engineering firm, under contract to USEPA/USAID and represented to a great extent the technical and economic approaches of the consulting engineering profession that favors the construction of advanced, high tech, wastewater treatment facilities. Again the new American guidelines were not based on any new health risk information or specific epidemiological studies but were essentially based on those required for drinking water as a "no-risk/fail safe" approach. The fact that little if any natural river water or water at approved bathing beaches in the United States or Europe could meet these recommended irrigation guidelines did not seem to bother those who drafted and approved the new American guidelines. The UNEP/WHO Global Environmental Monitoring System (GEMS) report points out, that the faecal coliform level of European rivers ranged from 1,000 to 10,000/100 ml with occasional peaks exceeding 100,000/100 ml. No one has suggested that such river water should not be allowed for irrigation purposes, neither has any health risk from such irrigation been reported [6].

For most countries of the world the effect of such very strict microbial guidelines, which are very costly to meet and require large investments in high tech/equipment-intensive treatment systems which require well-developed technological services and infrastructure to operate and maintain, is to limit initiatives to recycle and reuse wastewater. Another result is that in those very numerous cases where the urban wastewater flows untreated, to adjacent arid agricultural areas where, squatters and poor farmer illegally irrigate vegetables and salad crops, eaten raw, nothing is done to improve the situation since the official standards, which often follow the original California standards or the new USAID guidelines are too costly to achieve. This is a tragic case of where insisting on the best prevents achieving the good.

2. THE FIRST QUANTITATIVE MICROBIAL RISK ANALYSIS (QMRA) STUDY

The debate over the appropriateness of the various guidelines has in the past been on a qualitative/polemic level with some author's holding that the WHO/World Bank guidelines are too liberal. Our study was the first attempt at developing a scientific quantitative microbial risk analysis (QMRA) and cost-effectiveness approach based on a mathematical model and experimental data, to arrive at a comparative risk analysis of the various recommended wastewater irrigation microbial health guidelines for unrestricted irrigation of vegetables normally eaten uncooked [7]. The guidelines we compared were those of the World Health Organization [8], and those recommended by the USEPA/USAID [9]. Israel, has guidelines more or less similar to those of the USEPA.

First, through laboratory experiments, we determined estimates of the risk of ingesting enteric pathogens from the consumption of wastewater-irrigated vegetables. With this data it was possible to estimate the risk of infection and disease based on the risk of infection and disease model developed for drinking water by Haas and his colleagues in 1993 as modified for the special case of wastewater reuse in irrigation of crops eaten uncooked.

The basic risk model [10] for the probability of infection/disease from ingesting pathogenic microorganisms in water is:

$$PI = 1 - [1 + N/N_{50}(21/\alpha - 1)] - \alpha. \qquad (1)$$

PI = The risk of infection by ingesting pathogens in drinking water/or on vegetables
N = Number of pathogens ingested
N_{50} = Number of pathogens that will infect 50% of the exposed population
α = The ratio N/N_{50} and PI

Based on this model, we could calculate the annual risk of virus disease—infectious hepatitis for example—from regularly eating vegetables irrigated with raw wastewater which was shown to be as high as 10^{-2} to 10^{-3}(one person per 100 or per 1,000 might become ill). This disease rate as estimated by the simulation model approximated the actual disease rates we found in the investigation of typhoid fever and cholera transmission by raw wastewater irrigation in Santiago, Chile [11].

Our study further indicated that the annual risk of succumbing to an infectious enteric disease from regularly eating vegetables irrigated with treated wastewater effluent meeting the World Health Organization (WHO) Guidelines of 1,000

FC/100 ml [8] is negligible and of the order of 10^{-6} to 10^{-7} (one person per million or 10 million/year). The USEPA considers an annual risk of 10^{-4} (one person per 10,000/year) to be acceptable for microbial contamination of drinking water [12]. Thus according to our initial QMRA study, the 1989 recommended WHO Guidelines for Wastewater Reuse in Agriculture were some 100 to 1,000 times safer than what the USEPA itself recommends as the degree of safety required for drinking water.

According to our cost-effectiveness estimate, treating wastewater to meet the USEPA/USAID guidelines would result in an additional cost, over and above the cost of treatment to the WHO guideline levels, of some 1 million dollars per case of disease prevented. This analysis applies, more or less, to the Israeli recommended standards (Ministry of Environment-Inbar Report) as well which call for 10 FC/100 ml and additional strict requirement for biological oxygen demand (BOD), suspended solids (SS), turbidity, and chlorine residual. It was concluded that it is questionable if such a high level of wastewater treatment is justified for the irrigation of crops eaten uncooked, from a public policy, economic and/or public health point of view.

3. THE 2006 WHO GUIDELINES FOR THE SAFE USE OF WASTEWATER IN AGRICULTURE (WHO, 2006)

The WHO recommended guidelines of 1989 have been very influential, and many UN agencies such as the Food and Agriculture Organization (FAO), United Nations Development Programme (UNDP), the World Bank as well as number of countries including France and Mexico have adopted or adapted them for their wastewater use practices. However, some countries such as the USA, Australia, and Israel have recommended much stricter health guidelines for unrestricted agricultural reuse of wastewater in the range of 1–10 faecal coliforms/100 ml resulting in continued divergence of views as to the appropriate guidelines.

After some 15 years of intensive research study [13] and consultations with world experts in public health, epidemiology, agronomics, environmental sciences, and engineering, as well as other UN agencies, the WHO, in 2006, has published a revised updated volume of the Guidelines for the safe use of wastewater, excreta, and gray water. These new Guidelines drafted by a panel of 35 experts are recognized as representing the position of the United Nations system on issues of wastewater, excreta, and gray water use and health by “UN-Water,” the coordinating body of the 24 United Nations agencies and programs concerned with water issues.

Since the publication of the second edition of the WHO Guidelines in 1989, the development of QMRA methodology has enabled increasingly sophisticated analysis of health risks associated with wastewater use in agriculture. QMRA can estimate risks from a variety of different exposures and/or pathogens that would be difficult to measure through conventional epidemiological investigations due to the high cost and necessity of studying large populations.

The 2006 WHO report is based on new, detailed, advanced QMRA studies by Professor Duncan Mara and colleagues which provide further detailed scientific information as a basis to evaluate the infection risks associated with the exposure to crops irrigated by wastewater as well as exposure of irrigation workers [14]. The WHO report (2006) states that these studies are based to great extent on the pioneering studies of Shuval and colleagues done in 1997, which have served as the corner stone of the methodology in the development of the new WHO recommended Guidelines [15]. A combination of standard QMRA techniques and 10,000-trial Monte Carlo simulations was used [10]. The risk estimates were determined by using the β-Poisson dose–response model for bacterial and viral infections and the exponential dose–response model for protozoan infections.

The new and innovative WHO approach assumes that the most appropriate metric for expressing the burden of a disease is disability adjusted life years (DALYs) [16].

WHO has adopted, in the third edition of the Guidelines for drinking water quality, a tolerable burden of waterborne disease from consuming drinking water of $\leq 10^{-6}$ DALY per person per year. This value corresponds to a tolerable excess lifetime risk of fatal cancer of 10^{-5} per person (i.e., an individual has a 1 in 100,000 lifetime chance of developing fatal cancer) from consuming drinking water containing a carcinogen at its guideline value concentration in drinking water. This level of disease burden can be compared with a mild but more frequent illness such as self-limiting diarrhea caused by a microbial pathogen from wastewater irrigation. The estimated disease burden associated with mild diarrhea at an annual disease risk of 1 in 1,000 (10^{-3}) (approximately 1 in 10 lifetime risk) is also about 1×10^{-6} DALY (1 μDALY) per person per year. Such a high level of health protection is required for drinking water, since it is expected to be "safe" by those who drink it.

Since food crops irrigated with treated wastewater, especially those eaten uncooked, are also expected to be as safe as drinking water by those who eat them, the new WHO Reuse Guidelines require the same high health protection level of $\leq 10^{-6}$ DALY per person per year for wastewater reuse in agriculture.

3.1. Box-Disability Adjusted Life Years (DALYs)

DALYs are a measure of the health of a population or burden of disease due to a specific disease or risk factor. DALYs attempt to measure the time lost because of disability or death from a disease compared with a long life free of disability in the absence of the disease. DALYs are calculated by adding the years of life lost to premature death to the years lived with a disability. Years of life lost are calculated from age-specific mortality rates and the standard life expectancies of a given population. Years lived with a disability are calculated from the number of cases multiplied by the average duration of the disease and a severity factor ranging from 1 (death) to 0 (perfect health) based on the disease (e.g., watery diarrhea has a severity factor ranging from 0.09–0.12, depending on the age group). DALYs are an important tool for comparing health outcomes, because they account not only for acute health effects but also for delayed and chronic effects—including morbidity and mortality. When risk is described in DALYs, different health outcomes (e.g., cancer vs giardiasis) can be compared and risk-management decisions can be prioritized.

The new WHO guidelines for monitoring and validation purposes to assure safe wastewater reuse are based on analysis by QMRA methods and the criterion that a very high level of health protection of $\leq 10^{-6}$ DALY per person per year for wastewater use in agriculture be assured. They are based as well, on authoritative research studies on indicator organism and pathogen removal by various treatment processes, irrigation practices, and environmental factors. They assume that the level of F. coli in raw wastewater is in the range of 107/100 ml. The WHO report concludes: "The Monte Carlo–QMRA" results for unrestricted irrigation, based on the exposure scenario of lettuce consumption together with the relevant epidemiological evidence show that, in order to achieve $\leq 10^{-6}$ DALY per person per year a total pathogen reduction of 6 log units for the consumption of leaf crops (lettuce) and 7 log units for the consumption of root crops (onions) is required.

In the new WHO Guidelines, a pathogen reduction of 6–7 log units is used as the performance target for unrestricted irrigation to achieve the tolerable additional disease burden of $\leq 10^{-6}$ DALY per person per year. However, the WHO report points out that " ...a 6–7 log unit pathogen reduction may be achieved by the application of appropriate health protection measures, each of which has its own associated log unit reduction or range of reductions. A combination of these measures is used, such that, for all combinations, the sum of the individual log unit reductions for each health protection measure adopted is equal to the required overall reduction of 6–7 log units."

To be on the conservative side the WHO assumes that the 7 log reduction of pathogens (or 99.99999%) or an equivalent reduction of F. coliform bacteria is required to assure the safe consumption of wastewater irrigated lettuce and root crops. However, they do not assume that all of the pathogen/F. coli removal must be achieved solely through wastewater treatment processes. They assume that the degree of post wastewater treatment pathogen reduction and/or removal resulting from irrigation and exposure to sun, soil and hostile environmental factors is some 0.5–2 logs/day or at least a reasonable minimum estimate for mean wastewater irrigation conditions of 2 logs (99% reduction). They also assume a minimum pathogen/F. coli removal of 1 log (90%) by simple home rinsing and washing of wastewater irrigated salad crops and vegetables. If detergents and mild dis-infectants are use the pathogen removal might be as high as 99%. They also assume a post treatment 99% bacterial reduction in the case that drip irrigation is used.

Thus, the new WHO Guidelines point out that if these additional environmental factors, resulting in significant levels of pathogen and indicator organism inactivation and/or removal are taken in account the degree of F. coli removal required to monitor and validate the efficacy of the wastewater treatment process need not be more than 4 logs or 99.99%. As an example of one possible scenario the WHO report assumes an initial F. coliform concentration in the raw wastewater of 107/100 ml then a 7 log unit pathogen reduction may be achieved by the application of appropriate health protection measures, each of which has its own associated log unit reduction or range of reductions. One of the examples given indicates that when a minimum 2 log reduction is assumed for pathogen die-away under field conditions (higher in warm, sunny climates) and a 1 log reduction minimum is assumed for pathogen removal by home rinsing/washing then the wastewater treatment would require only a 4 log removal (99.99%) by the wastewater treatment process or in other words an effluent quality of about 1,000 F. coliforms/100 ml. The report states: This is similar to the recommended required effluent quality of 1,000 thermotolerant coli/100 ml in the second edition of these guidelines [8]. Other scenarios assuming higher microbial removal by irrigation techniques and/or environmental factors resulting in lower levels of effluent requirements, such as 10,000 F. Coli/100 ml are also presented, such as with the use of drip irrigation.

"This option does not take into account pathogen reduction due to a) natural die-off between final irrigation and consumption and b) specific food and/or cooking and overall health protection is therefore greater than even 10^{-6} DALY per person per year. The very high costs and operational complexities of the wastewater treatment processes required by this option will generally preclude its applicant in many countries. Even in countries where this option is affordable it should be subject to robust cost-effectiveness analysis."

The meeting of experts held in Geneva that drafted these new guidelines concluded in a resolution approved by consensus that: While each country can and should select the combination of risk reduction elements that suit its epidemiological, social, and economic needs, however, the in-depth risk analysis studies carried out by the group, provide a sound epidemiological basis for concluding that the options presented (with the lower levels of wastewater treatment and microbial effluent quality) provide a high degree of health risk reduction and health protection which should meet the needs of most countries in a reasonable cost effective manner. The group has concluded that these new risk assessment studies validated the 1989 WHO Guideline recommendation of 1,000 *E. coli*/100 ml for unrestricted irrigation of most vegetable and salad crops eaten uncooked.

4. CONCLUSION

In general, WHO Reuse Guidelines provide a sound scientific basis for establishing appropriate guidelines which provide a high degree of public health protection. Additional guidelines to protect soil quality or prevent groundwater and aquifer pollution should be applied only as needed, depending on conditions.

5. REFERENCES

1. Elchanati S. (2006). Upgrading Effluent Quality – Hydrological Aspects, Engineering and Water – The Israeli Water J. No. 46, 26–27 (in Hebrew).
2. Ongerth HJ and Jopling WF. (1977). Water reuse in California. In Water Renovation and Reuse H. I. Shuval (ed), Academic Press, New York, 219–256.
3. Feachem RG, Bradley DH, Garelick H and Mara DD. (1983). Sanitation and Disease: Health Aspects of Excreta and WW Management. Wiley, New York.
4. Shuval HI, Adin A, Fattal B, Rawitz E and Yekutiel P. (1986). Wastewater Irrigation in Developing Countries: Health Effects and Technical Solutions. World Bank Tech. Paper No. 51. World Bank, USA.
5. Strauss M and Blumenthal UJ. (1989). Health aspects of human waste use in agriculture and aquaculture utilization practices and health perspectives. Intn. Reference Center for Waste Disposal, Dubendorf (Report No.08/88).
6. UNEP/WHO (1996). Global Environmental Monitoring System-GEMS Report for 1996, WHO, Geneva
7. Shuval HI, Lampert Y and Fattal B. (1997). Development of a risk assessment approach for evaluating wastewater reuse standards for agriculture. Water Science and Technology 25:15–20.

8. WHO-World Health Organization (1989). Health guidelines for the use of wastewater in agriculture and aquaculture - Group Chairman H. Shuval, Tech. Report Series 778, WHO Geneva, 74 pp.
9. USEPA/USAID (1992). Guidelines for Water Reuse US Environmental Protection Agency, Wash. Technical Report *no* 81 September 1992), 252.
10. Haas CN, Rose JB, Gerba C and Regli S. (1993). Risk assessment of virus in drinking water. Risk Analysis 13:545–552.
11. Shuval HI. (1993). Investigation of typhoid fever & cholera transmission by raw wastewater irrigation, Water Science Tech. 27(3/4):167–174.
12. Regli S, Rose JB, Hass CN and Gerba CP. (1995). Modelling risk for pathogens in drinking water. J. Am. Water Works Association 83:76–84.
13. Blumenthal UJ, Duncan M, Peasey A, Ruiz-Palacios G and Stott R. (2000). Guidelines for the microbiological quality of treated WW used in agriculture: recommendations for revising WHO Guidelines 78 (9): 1104–1116.
14. Sleigh PA and Mara DD (2003). Monte Carlo Program for estimating disease risks in wastewater reuse. University of Leeds, Water and Envn. Engn. Res. Group, Trop. Pub. Health Engn. (http//www.efm.leeds.ac.uk).
15. WHO-World Health Organization (2006). Guidelines for the safe use of WW, excreta and greywater - 2: WW use in agriculture, WHO, 221.
16. Murray CJL and Lopez AD. (1996). The global burden of disease. Boston, WHO, World Bank, Global Burden of Disease and Injury Series, I. 1–98.

SESSION 8. CURRENT RISK MANAGEMENT PRACTICES

NEW STANDARDS FOR TREATED WASTEWATER REUSE IN ISRAEL

Yossi Inbar
Senior Deputy Director General, Industries
Ministry of Environmental Protection, Jerusalem, 95464 Israel

Corresponding author: yossii@sviva.gov.il

Abstract:

The combination of severe water shortage, densely populated urban areas, and highly intensive irrigated agriculture, makes it essential that Israel put wastewater treatment and reuse high on its list of national priorities. Sewage treatment effluent is the most readily available water source and provides a partial solution to the water scarcity problem. National policy calls for the gradual replacement of freshwater because of the decision to increase the use of effluent and set up a committee to review existing regulations and to recommend new regulations for effluent use for irrigation or disposal to stream and receiving water. The recommended values were designed to minimize potential damage to water sources.

Keywords: Wastewater treatment, standards, risks, regulations, decision making.

1. INTRODUCTION

The total area of arable land in Israel has increased from 1,600 km^2 in 1948 to approximately 4,200 km^2 in 2001. Irrigated land has increased from 300 km^2 in 1948 to 1,866 km^2 in 2001. Water scarcity is the main limiting factor in Israeli agriculture. Annual rainfall ranges from about 800 mm in the north to 25 mm in the south of the country. Agriculture is the number one factor in the protection of open space and prevention of desertification. It also serves as a sink for waste produced in the urban sector, including effluents, sewage sludge, or compost.

The combination of severe water shortage, densely populated urban areas, and highly intensive irrigated agriculture, makes it essential that Israel put wastewater treatment and reuse high on its list of national priorities. In fact, national policy calls for the gradual replacement of freshwater allocation to agriculture by reclaimed effluents [1]. Currently about 70% (>300 million cubic meters [MCM]) of the wastewater produced in Israel is reclaimed for agricultural reuse. A new standard for unlimited use of effluents has been formulated. The standard, encompassing 36 parameters, takes account of public health, soil, hydrological

M.K. Zaidi (ed.), Wastewater Reuse–Risk Assessment, Decision-Making and Environmental Security, 291–296.

and flora considerations. This standard will enable the reallocation of nearly 50% of the freshwater (about 500 MCM), from agriculture to the municipal and industrial sectors.

2. WASTEWATER

Out of a total of 500 MCM of sewage produced in Israel, about 96% is collected in central sewage systems and 67% of the effluent is reclaimed (300 MCM). By law, local authorities are obligated to treat municipal sewage. In recent years, new or upgraded intensive treatment plants were set up in municipalities throughout the country. The ultimate objective is to treat 100% of Israel's wastewater to a level enabling unrestricted irrigation in accordance with soil sensitivity and without risk to soil and water sources [2].

2.1. Some Facts and Figures

- Some 500 MCM of wastewater are produced in Israel every year, of which 450 MCM/year is treated.
- Some 300 MCM per year of the effluent is reclaimed (about 64%)
- Some 4% of the wastewater is discharged to cesspools (20 MCM)
- Some 96% of the waste is collected in central sewage systems
- Some 33% of the wastewater/effluents is discharged to the environment (approx. 160 MCM)

3. WASTEWATER TREATMENT PLANTS

There are upwards of 500 sewage treatment facilities in Israel today, of which some 35 are advanced wastewater treatment plants (purifying over 360 MCM/year) with minimum annual capacity of more than 0.5 MCM each. Regulations promulgated by the Ministry of Health in 1992 require secondary treatment to a minimum baseline level of 20 mg/liter biochemical oxygen demand (BOD) and 30 mg/l total suspended solids (TSS) in urban and rural centers with populations exceeding 10,000 people. Local authorities are responsible for the construction and operation of wastewater treatment plants. Israel's wastewater treatment plants use intensive (mechanical/biological) and extensive treatment processes. Intensive treatment plants use the activated sludge method while extensive processes are based on anaerobic stabilization ponds, which are integrated with shallow aero bic ponds and/or deep facultative polishing reservoirs. Treatment facilities may include nitrogen and phosphorous removal. After treatment, the effluent is placed in seasonal reservoirs, which also serve to regulate the constant flow of treated wastewater and the seasonal demand for irrigation.

4. WASTEWATER TREATMENT

Because of the combination of severe water shortage, contamination of water resources, densely populated urban areas, and intensive irrigation in agriculture, wastewater treatment and reuse are high on Israel's list of national priorities. In 2001, some 46% of the effluents produced in the country (200 MCM) complied with the standards set in regulations (20/30 BOD/TSS). This number reached 60% (256 MCM) in 2002 and 72% (300 MCM) in 2005.

The organic load in Israel's municipal wastewater is much higher than in the Western world. Furthermore, due to the high rate of effluent reuse for irrigation purposes, environmental sensitivity to the salt content of sewage is especially high.

The adverse environmental impacts of domestic sewage may be reduced through the following activities:

1. Reduction of salt emissions to the sewage system through discharge of industrial brines to sea, as well as reduction in the use of salt in dishwasher and laundry detergent.
2. Changes in the chemical composition (especially reduction of boron) of detergents to environment-friendly materials.
3. Legislation to limit the use of domestic garbage grinders (in Israel, each person generates some 0.5 kg of organic waste per day); use of garbage grinders and disposers would increase the organic load in wastewater treatment plants tenfold.
4. Steps to assure that industrial sewage discharged to municipal treatment systems will undergo pretreatment to remove toxic or harmful materials.

5. EFFLUENT DISPOSAL AND REUSE

Sewage treatment effluent is the most readily available water source and provides a partial solution to the water scarcity problem. National policy calls for the gradual replacement of freshwater allocations to agriculture by reclaimed effluent. In 1999, treated wastewater constituted about 22% of the consumption by the agricultural sector. It is estimated that effluent will constitute 40% of the water supplied to agriculture in 2005, 45% in 2010, and 50% in 2020. The Ministry of Health maintains a permit system designed to ensure that irrigation with effluent is limited to nonedible crops such as cotton, fodder, etc. Only highly treated effluent, after disinfection, is used for irrigation of orchards, such as citrus groves avocado and others. Effluent is not used for irrigating crops in which there is direct contact between the water and the edible part of the plant (e.g., lettuce).

6. UPGRADED EFFLUENT QUALITY STANDARDS

Because of the decision to increase the use of effluent to a total of 500 MCM, the Ministerial Economics Committee (Decision 46, July 2000) decided to appoint an Inter-Ministerial Committee (Inbar Committee) for the purpose of reviewing existing regulations and recommending new regulations for effluent use for irrigation or for disposal to streams and receiving waters [3].

The recommended values, designed to minimize potential damage to water sources, flora, and soil, call for much higher treatment levels in existing and future wastewater treatment plants. An agreement in principle has been reached on the new effluent quality standards, and a technoeconomic review of the standard has been conducted. The objective is to treat 100% of the country's wastewater to a level enabling unrestricted irrigation in accordance with soil sensitivity and without risk to soil and water sources. The proposed regulation includes 36 biological and chemical parameters classified in three groups (Table 1):

- Organics, nutrients, and pathogens: BOD, TSS, chemical oxygen demand (COD), Fecal coliforms, dissolved oxygen, residual chlorine, mineral oil, pH, total nitrogen, ammonia, and total phosphorus.
- Salts: electrical conductivity (TDS), specific absorption rate (SAR), chloride, sodium, boron, and fluoride.
- Heavy metals: arsenic, barium, mercury, chromium, nickel, selenium, lead, cadmium, zinc, iron, copper, manganese, aluminum, molybdenum, vanadium, beryllium, cobalt, lithium, and cyanide.

To achieve the threshold values recommended for the parameters in the regulation, the quality of the effluent must be upgraded. The way to reach this objective will be different for any group of parameters. The group of organics, nutrients, and pathogens can be treated at the wastewater treatment plants, under present conditions or with some technical upgrading. Salts and heavy metals, at the present level of wastewater treatment, have to be treated at the source. Therefore, recent years have seen a flurry of new regulations (by the Ministry of the Environment in collaboration with other ministries) designed to improve

Table 1. Proposed new Israeli standards for effluent (average levels) *

Parameter	Units	Unrestricted Irrigation*	Rivers
Electric conductivity	dS/m	1.4	n/a
BOD	mg/l	10	10
TSS	mg/l	10	10
COD	mg/l	100	70
N-NH4	mg/l	20	1.5
Total nitrogen	mg/l	25	10
Total phosphorus	mg/l	5	1.0
Chloride	mg/l	250	400
Fluoride	mg/l	2	n/a
Sodium	mg/l	150	200
Faecal coliforms	Unit per 100 ml	10	200
Dissolved oxygen	mg/l	>0.5	>3
pH	mg/l	6.5–8.5	7.0–8.5
Residual chlorine	mg/l	1	0.05
Anionic detergent	mg/l	2	0.5
Mineral oil	mg/l	n/a	1
SAR	(mmol/L)0.5	5	n/a
Boron	mg/l	0.4	n/a
Arsenic	mg/l	0.1	0.1
Mercury	mg/l	0.002	0.0005
Chromium	mg/l	0.1	0.05
Nickel	mg/l	0.2	0.05
Selenium	mg/l	0.02	n/a
Lead	mg/l	0.1	0.008
Cadmium	mg/l	0.01	0.005
Zinc	mg/l	2	0.2
Iron	mg/l	2	n/a
Copper	mg/l	0.2	0.02
Manganese	mg/l	0.2	n/a
Aluminum	mg/l	5	n/a
Molybdenum	mg/l	0.01	n/a
Vanadium	mg/l	0.1	n/a
Beryllium	mg/l	0.1	n/a
Cobalt	mg/l	0.05	n/a
Lithium	mg/l	2.5	n/a
Cyanide	mg/l	0.1	0.005

* From soil, flora, hydrological and public health considerations.

wastewater quality. In some instances, regulations are based on European standards (e.g., regulations limiting the discharge of heavy metals); in others, they are specifically developed to address conditions that are unique to Israel (e.g., regulations prohibiting the discharge of brines into municipal sewage systems and detergent standards setting limits on concentration of chlorides, boron, and sodium). Special attention is currently being given to problems relating to high salinity of municipal sewage. This is an issue of particular importance in Israel, where wastewater recovery for agricultural purposes is imperative.

7. CONCLUSIONS

Israel's experience with wastewater reuse suggests that it can be an invaluable component in water management strategy for dry lands. However, there are strong public health and environmental implications, which must be considered prior to adopting a final policy. A water management system, which is not based on extremely high treatment levels, will not be sustainable or beneficial in the long run. Inadequate sewage treatment limits the range of crops that can be safely grown with wastewater irrigation. Consequently, the Government of Israel decided in April 2005 to upgrade its treatment of effluents to the above proposed advanced standards enabling unrestricted irrigation in accordance with soil sensitivity and without risk to human health, flora, soil, and water sources.

Acknowledgment: I am thankful to the NATO ARW committee to award the financial support to attend the meeting and present this paper.

8. REFERENCES

1. Aharoni A, Cikurel H (2006). Mekorot's research activity in technological improvements for the production of unrestricted irrigation quality effluents, Desalination 187, 347–360.
2. Paths to Sustainability, Presented to UN, April 2005.
3. Research News. (2002). Besa Bulletin, May 2002, 14, 1–4.

WASTEWATER REUSE IN ISRAEL—RISK ASSESSMENT

Yosef Dreizin
Water Desalination Administration
Water Commission, Tel Aviv, Israel

Corresponding author: jossefdr10@water.gov.il

Abstract:

Wastewater treatment plants in Israel treat approximately 500 million cubic meters a year. The high quality treated wastewater is used mainly for large-scale agricultural irrigation. Israel has acquired much experience in adjusting the treatment level of the wastewater treatment plants and the qualities and characteristics of the treated wastewater to land and crops. The economic advantage of utilizing the treated wastewater in agriculture substantiates the policy of assisting recycling plants and farmers, with the reuse of treated wastewater enabling the conversion of expensive freshwater.

Keywords: Wastewater, treatment plant, Israel, agriculture, irrigation.

1. INTRODUCTION

Reuse of treated wastewater involves several types of risks, such as health, environmental, economic, and strategic. It is difficult to measure these risks. However, when making such an analysis, one must consider the cost of avoiding the development of a treated wastewater supply project. In Israel, such a risk is being taken because as we are convinced that this cost is much higher [1–3].

1.1. The Need

Over the last decade there has been a rapid growth in environmental public awareness. A lot of investment is being made to solve environmental issues such as sea and rivers contamination by sewage flow. Israel is one of the driest countries in the world, along with other Middle East and North African countries (MENA—Middle East North Africa), it is facing an environmental issue—a continual lack of fresh water. The drought is a meteorological and hydrologic natural phenomena. Treating wastewater is meant to deal with these two issues:

M.K. Zaidi (ed.), Wastewater Reuse–Risk Assessment, Decision-Making and Environmental Security, 297–303.

a. It significantly reduces sanitary–ecological nuisances, thereby allowing a sustainable policy.
b. Reclaimed wastewater can be used for agriculture and city irrigation, if treated properly.

The process, according to which, more agricultural area will be transferred from being irrigated with treated wastewater, is powered using a combination of economic means, legal tools, and advantages of reliability. Apparently, the way has been cleared for the process and the treated wastewater will serve as an appropriate resource for agricultural development in Israel [4–6].

1.2. The Quantities of Treated Wastewater

The use of treated wastewater has been an integral part of Israel's water resources. The total amount of wastewater produced in Israel is approximately 500 million cubic meters a year (mcmy) including agriculture, industry, and other wastewater consumers. Almost all wastewater produced in Israel flows into the main sewage collection systems, while only 2.5% of the wastewater still flows into cesspits. Approximately 450 mcmy is being treated at 465 mechanical facilities and stabilization basins, using a variety of technologies.

Approximately 350 mcmy of the effluent, which is 80% of the total wastewater, is at least at 20 (BOT)/30 total suspended solids (TSS) quality, and approximately 150 mcmy of the effluent, which is 20% of the total quantity, is at improved, third degree quality. 300 mcmy of wastewater is reclaimed for irrigation (Table 1).

Over the last ten years the following significant changes have occurred:

- A 20.6% growth in the amount of the wastewater.
- A 24% increase in the capacity of the sewage systems.
- A 67% decrease in the disposal of wastewater by cesspits.
- A 37% increase in the amount of wastewater treated by the facilities.
- A 70% decrease in the untreated wastewater flowing to the rivers.
- A 55% increase in the reclamation and reuse of effluent.
- A 76% growth in the areas irrigated by effluent.

Effluent reservoirs make it possible to supply all year long, and to adjust the supply according to the irrigation season. Using effluent gives the consumer supply stability, with no concern about shortage due to drought. In addition, the limited amount of available fresh water for irrigation, the growth of the urban and industrial sectors, which require freshwater, and the high price of desalinated seawater, cause the effluent to become a growing part of the available water sources for the agricultural sector (Table 2).

Table 1. Quantities of wastewater, effluent, and reuse over the years 1963–2004

Year	Million cubic meter a year			% of Total	Irrigated area (dunams)
	Total wastewater	Treated wastewater	Reclaim & reuse		
1963	137.3	41.1	10.2	7%	15,000
1967	139.4	53.5	16.4	12%	20,700
1971	183.2	68.7	22.2	12%	30,400
1975	209.3	108.3	34.2	16%	51,000
1978	213.9	121.8	43.4	20%	68,600
1980	225.8	133.4	52.3	23%	88,200
1982	251.2	137.3	57.1	23%	100,000
1985	259.7	151.7	91.8	35%	163,000
1987	270.1	212.3	180.4	67%	257,700
1989	293.1	232.4	194.9	66%	278,400
1994	389.0	309.4	254.2	65%	363,100
2000	422.4	360.0	285.2	68%	440,900
2004	469.3	427.7	395.0	84%	705,252

By next decade the wastewater treatment facilities in Israel will be dealing with approximately 509 million cubic meters of treated wastewater a year, from which approximately 496 million cubic meters will be reclaimed for agricultural irrigation use. The development of the effluent quantities is more moderate at the end of the decade, because it evolves from natural growth of effluent in existing projects, rather than from initiating new projects, at the beginning and middle of the present decade.

1.3. Qualities of Treated Wastewater

A special committee has submitted its recommendations for the use of treated wastewater for unlimited irrigation and improvement of the effluent flow to the rivers and sea. These recommendations were widely adopted. Although the quality standards recommended will be required in future to achieve these quality levels (Table 3).

2. RISK ASSESSMENTS

The experience acquired in the water sector concerns the broadest entirety of the work comprising the initiative, planning, financing, construction, and operation of the treatment plants and the use of the recycled water in agriculture. This experience presents certain risks and plan is to map out these risks, and to examine consequences and evaluations in order to learn more about this subject.

Table 2. Summary of the national water balance for four years

Year	Water resources (mcmy)		Effluent consumption (mcmy)			
	Reclaimed wastewater	Total demand	Industry		Agriculture	
			Effluent	Total	Effluent	Total
1999	278	1,915	0	126	278	
2002	298	1,966	0	129	298	1,010
2005	403	2,417	0	140	403	1,062
2010	509	2,541	13	167	496	1,122

2.1. Human Health Risks

Wastewater irrigation poses a number of potential risks to human health via consumption or exposure to pathogenic microorganisms, heavy metals, harmful organic chemicals such as endocrine disrupting compounds and pharmaceutically active compounds. Of these, pathogenic microorganisms are generally considered to pose the greatest threat to human health. Household sewage contains a high percentage of organic materials and pathogenic microorganisms, including bacteria, viruses, protozoans, and parasitic worms.

Table 3. Maximum level of the main parameters for unlimited irrigation/rivers flow, according to the committee

Parameter	Units	Unlimited irrigation	Rivers flow
Electrical conductivity	dS/m	1.4	
BOD	mg/l	10	10
TSS	mg/l	10	10
COD	mg/l	100	70
Nitrogen (ammonia)	mg/l	20	1.5
Nitrogen (general)	mg/l	25	10
Phosphorous (general)	mg/l	5	1.0
Chloride	mg/l	250	400
Koli	Units/100 ml	**10**	**200**
Boron	mg/l	0.4	

The symptoms and diseases associated with such infections are also diverse and include typhoid, dysentery, gastroenteritis, diarrhea, vomiting and malabsorption. Any human contact with the treated wastewater might be hazardous. In addition, sprinkler irrigation could cause spray drops that contain pathogenic materials, which might cause a health risk to the population, if irrigated next to roads or inhabited areas. The process at the treatment facilities reduces the pathogenic microorganism contents, but it does not eliminate it completely. This problem can be solved by desalination of the treated wastewater, but this is an expensive process and is usually not required from the aspect of agricultural use. Different crops that are irrigated by recycled wastewater pose various threats to human health. There are crops with pathogenic microorganisms contamination that do not seem to pose any health risk:

- Industrial crops such as cotton or fodder.
- Fruits that are dried in the sun for at least 60 days after last irrigation.
- Watermelons grown for edible grains or for seeds, that are irrigated only before blooming.
- Groves or flora, without human access.

The agricultural system, which includes land and crops, is defined in the planning stages of the recycling supply system. Each land and crop has its own irrigation quality permit. The law requires the receipt of a permit from the Ministry of Health, for every treated wastewater irrigated land or crops. As mentioned above, the established committee regulations pose a set of high quality standards, for unlimited irrigation, which are supposed to prevent any potential health risk from eating treated wastewater crops or exposure to the effluent.

An additional potential health threat might happen if a cross connection between the effluent and the fresh water piping systems occurs. This situation poses a risk of a massive disease outburst, because it could insert the microbiological contamination directly into the fresh water piping system. Regulations were made in order to prevent this hazardous situation.

2.2. Environmental Risks

Irrigation with treated wastewater might add certain contaminents, such as chlorides, to the groundwater. This risk has an accumulative nature as the contaminates appear in the water supply systems, flow to the treatment plants and back to the aquifer. The risk in this respect have a long-term influence and are difficult to evaluate [7].

According to the health regulation, treated wastewater irrigation is forbidden in the vicinity of drinking water wells (except for effluent that does not pose any risk). In addition, treated wastewater irrigation along the national water carrier route (from the Sea of Galilee) is completely forbidden. Irrigation above freshwater pipes can be approved, only if the treated wastewater is at the needed quality level and the water pipeline is in good condition and there is no risk of under-pressure in the pipe. Untreated wastewater system must have a minimal distance of 3 m from the irrigated area end point to the freshwater pipeline.

In addition to microorganisms, household sewage also contains substantial salts additions. Besides the risk to the quality of the groundwater due to seepage, irrigation with treated wastewater causes land salinity. Treated wastewater irrigation also causes land sealing and sodium accumulation, which could cause increased runoffs and lands erosion. The above-mentioned regulations for the treated wastewater quality, deal with these issues as well. Other potential environmental effects:

- Poor quality treated wastewater, or treated wastewater that is being sucked from unaerobic layers of the treatment plant's reservoir could cause a strong odor nuisances.
- Irrigation with treated wastewater, if not properly controlled, could cause a decrease in yield, as well as in the quality of the crops.

2.3. Economic Risks

Another condition that may cause a significant loss of money to the entrepreneur is a change in the wastewater quality. The quality of the source water is determined according to the quality of the source sewage and the treatment technology. The control over this quality is not absolute because the sewage characteristics are not homogenous, due to a possible inappropriate treatment plant management. The recycling plant's obligation toward its consumers usually include quality specifications. A sizable deviation from these specifications could cause great damage, and even prevent use of the water. In addition to all health, environmental, and economical risks, using treated wastewater in such extended volumes, and investing resources in developing and establishing treated wastewater facilities, pose one more great risk to the State of Israel—a strategic risk.

The water sector reduces the exposure to these risks using the various methods in its possession. "Large projects" are established by Mekorot. The State insures the project, and eventually the risks are its responsibility.

The State's participation in the investments, while the project's entrepreneur is only partially involved in the financing, enables the entrepreneur to create a premium as part of its income, to cover those risks that are not under its control.

However, when making such an analysis, one must consider the cost of avoiding the development of a treated wastewater supply project. In Israel, we are taking the risk because we are convinced that this cost is much higher. One should keep in mind that the alternatives are either desalinated water supply, or decreased irrigated agricultural area. Both are unfavorable.

3. CONCLUSIONS

The dependency of Israel's agriculture on the wastewater treatment system exposes it to a risk at national level. Israel is a pioneer at a global level in recycled treated wastewater irrigation. If an essential problem will appear, environmental, public health, or regulative, that will prevent the use of the recycled wastewater, the potential damage could be enormous.

Acknowledgment: The paper was presented with the financial support of NATO Advanced Research Workshop grant.

4. REFERENCES

1. Goldfarb Or, Kislev Y (2002). A sustainable salt regime in the Coastal aquifer, The Water Commission and the Center for Agricultural Economic Research (Hebrew).
2. Kislev Y (2006). The water economy of Israel, in: K. David Hambright, F. Jamil Ragep, and Joseph Ginat (eds.). Water in the Middle East, Oklahoma U. Press, Norman, pp. 127–150.
3. Simon CP, Blume L (1994). Mathematics for Economists, w. w. Norton, New York and London.
4. Amiran DHK (1978). Geographical aspects of national planning in israel: The management of limited resources. Transactions of the Institute of British Geographers, 3(1), 115–128.
5. Friedler E, Lahav O, Jizhaki H, Lahav T (2006). Study of urban population attitudes towards various wastewater reuse options: Israel as a case study. J. environmental management, 81(4), 360–370.
6. Rattan RK, Datta SP, Chhonkar PK, Suribabu K, Singh AK (2005). Long-term impact of irrigation with sewage effluents on heavy metal content in soils, crops and groundwater—a case study. Agriculture, Ecosystems & Environment, 109(3/4), 310–322.
7. Brill E, Hochman E, Zilberman D (1997). Allocation and pricing at the water district level. American J. Agricultural Economics, 79(3), 952–963.

LIFE CYCLE ASSESSMENT, A DECISION-MAKING TOOL IN WASTEWATER TREATMENT FACILITIES

Mohamed Tawfic Ahmed
Suez Canal University, Ismailia, Egypt

Corresponding author: motawfic@tedata.net.eg

Abstract:

Wastewater reuse, along with wastewater treatment facilities are endowed with a number of risks that may inflect some serious damage to man and his environment. Decision construct and decision making of wastewater issues are of prime importance for sound use of wastewater and treatment processes. Equally important is the sound economical attributes wastewater treatment and wastewater alternative methods and construction should manifest. Measures of common use in this respect are mostly dominated by environmental impact assessment, risk assessment, and cost benefit analysis. Life cycle assessment (LCA) is one of the newly emerging techniques with wide application in the field of wastewater reuse and wastewater treatment facilities.

Keywords: Life cycle assessment, wastewater, decision making, developing countries.

1. INTRODUCTION

The provision of an infrastructure, which supplies drinking water and removes/ treats wastewater and storm water, is essential for an urban society. During this century global water use has increased by more than double the population increase. Today, about one-third of the world's population live in countries experiencing moderate water stress, i.e., where the use of freshwater is greater than 10% of renewable freshwater resources [1]. Water is becoming one of the most strategic issues in a number of Middle Eastern and North African countries. Countries of this region have 5% of the world's population but have less than 1% of the world's renewable freshwater. The region is one of the driest in the world and poorly endowed with natural freshwater supplies. The main source of water for Egypt is the Nile river, which represents 97% of the country's freshwater. The annual per-capita water availability in 1960 was about 1550 m^3 and has fallen by 40% to about 995 m^3 today and it is expected to be about 600 m^3 in 2025. The inclusion of treated wastewater to national water budget in these countries is becoming a vital necessity. Wastewater could be a potential source of risks to

M.K. Zaidi (ed.), Wastewater Reuse–Risk Assessment, Decision-Making and Environmental Security, 305–314.

humans and their environment. Stringent measures should be observed to avoid such risks. Construction and running wastewater treatment facilities tend to pose different impact and would require high energy level. Wastewater treatment is the single biggest electricity use in most governments. A sound decision for the selection wastewater treatment technology is entangled with a number of interconnected factors and consideration. With the spatial and temporal dimension LCA would provide, comparison of different methods and or scenarios based on chemical and energy consumption, quantity of sludge generation, emission of green house gases, capital cost (civil construction and mechanical installation), maintenance cost, and land requirement would be transparent and based on factual figures. A number of methods are used to assess the impact of wastewater usage and wastewater facilities. Many of these methods use economic and environmental criteria which only take into account the direct effect of effluent on receiving waters, disregarding indirect and cumulative economic and environmental effects, with the possibility of displacing impacts from one segment to the other. As a result, the true environmental and social costs of wastewater treatment are often not included in decision making.

1.1. Life Cycle Assessment (LCA)

LCA is a technique for assessing the potential environmental impacts associated with a product or a service, by compiling an inventory of relevant environmental exchanges of the product throughout its life cycle (LC) and evaluating the potential environmental impacts associated with those exchanges. LCA attempts to predict the overall environmental impact associated with a product, function, or service. It also allows you to compare the environmental performance of two products. One of the main features of LCA is to ensure that environmental effects are not displaced from one stage of a process to another. To achieve this, it requires the appreciation of: all inputs of materials and energy; all emissions, including solid and liquid wastes; and all inputs and waste associated with use, reuse, recycling and disposal of the product. It also provides a framework, and methods for identifying and evaluating environmental burdens associated with the LC of materials and services, from cradle-to-grave, including the recyclability of the materials. LCA is one of the newly emerging tools used to help decision makers coming up with a sound decision about wastewater treatment facilities. The use of LCA in the area of wastewater is gaining more momentum and is being used in a number of developed and developing countries. The LC approach' main feature is to ensure that improvement in one part of the product's LC is not compensated by another part of the cycle. Some of the benefits of LCA are that it is an operational, standardized method and that software exists. The LCA community has progressed in the discussion of evaluating results i.e., in aggregating data and weighting different environmental impacts. Although the focus of LCA has been mostly on products it is obvious that LCA has potential for comparing different processes and services [2]. There are an increasing number of LCA studies for water and wastewater systems, parts of systems, or components such as pipes or chemicals. The magnitude of studies will increase the knowledge of the environmental impact from the systems but also on how to use LCA. Methodological experience has been gained on how to choose system boundaries and what parameters to focus on. Data is now available for different unit processes, from construction of a toilet to a whole wastewater system, from the use of a single chemical to supplying drinking water to a whole city. There are an increasing number of LCA studies for water and wastewater systems, parts of systems or components such as pipes or chemicals. The magnitude of studies will increase the knowledge of the environmental impact from the systems but also on how to use LCA. Methodological experience has been gained on how to choose system boundaries and what parameters to focus on. Data is now available for different unit processes, from construction of a toilet to a whole wastewater system, from the use of a single chemical to supplying drinking water to a whole city. Various reports have illustrated the role LCA can play in wastewater decision making process [3–9].

1.2. Definition

The International Standards Organization (ISO) has defined LCA as: "A technique for assessing the environmental aspects and potential impacts associated with a product by:

- Compiling an inventory of relevant inputs and outputs of a product system,
- Evaluating the potential environmental impacts associated with those inputs and outputs,
- Interpreting the results of the inventory analysis and impact assessment phases in relation to the objectives of the study" (ISO 14040).

The technique examines every stage of the LC, from the winning of the raw materials, through manufacture, distribution, use, possible reuse/recycling and then final disposal. For each stage, the inputs (in terms of raw materials and energy) and outputs (in terms of emissions to air, water, soil, and solid waste) are calculated, and these are aggregated over the LC. These inputs and outputs are then converted into their effects on the environment, i.e., their environmental impacts. The sum of these environmental impacts then represents the overall environmental effect of the LC of the product or service.

1.3. LCA Background

The concept of LCA first emerged in the late 1960s, but did not receive much attention until the mid 1980s. In 1989, the Society of Environmental Toxicology and Chemistry (SETAC) became the first international organization to begin oversight of the advancement of LCA. In 1994, the International Standards Organization (ISO) began developing standards for the LCA as part of its 14,000 series standards on environmental management. The standards address both the technical details and conceptual organization of LCA.

- ISO 14040—a standard on principles and framework
- ISO 14041—a standard on goal and scope definition and inventory analysis
- ISO 14042—a standard on life-cycle impact assessment
- ISO 14043—a standard on life-cycle interpretation

Several of the methods described as LCA methods follow the LCA framework defined in ISO 14040, involving an inventory similar to that described in ISO 14041, and assessment of impacts to some degree as described in ISO 14042, while a smaller number take on the normalization and weighting also discussed in ISO 14042. Still, methods based on the ISO standards may differ greatly, given that the ISO standards allow flexibility to customize characterization and normalization factors and weighting methods to suit the values and conditions of a particular location or sector. LCA is a method that aims to analyze and evaluate the environmental impacts of products or services [2]. The whole chain of activities, required for the production of a certain product or service is taken into consideration. Both emissions of potentially harmful substances from these activities and their consumption of natural resources are analyzed. In this way, different technical systems, producing the same utility (product or service) can be followed meaning that all activities are included; covering the mining of minerals, transports, production and disposal, reuse or recycling) and compared with regard to their impacts on the environment.

1.4. Goals and Scope

This is the first stage of the study and probably the most important, since the elements defined here, such as purpose, scope, and main hypothesis considered are the key of the study. The scope of the study usually implies defining the system, its boundaries (conceptual, geographical and temporal), the quality of the data used, the main hypothesis, and a priori limitations. A key issue in the scope is the definition of the functional unit. This is the unit of the product or service whose environmental impacts will be assessed or compared. It is often expressed in terms of amount of product, but should really be related to the amount of product needed to perform a given function. During the goal definition process, the following issues should be considered:

1. Why is the study being conducted (i.e., what decision, action, or activity will it contribute to or affect)?
2. Why is LCA needed for this decision, action, or activity? What, specifically, is it expected to contribute?
3. What additional analytical tools are needed and what will they be expected to contribute?
4. Who is the primary target audience for the study (i.e., who will be making the decision, taking or directing the action, or organizing or participating in the activity)?
5. What other audiences will have access to the study results? What uses might these audiences make of the study findings?
6. What are the overall environmental goals, values, and principles of the sponsoring organization and intended audience?
7. How does the intended application of the study relate to these goals, values, and principles?

In most wastewater treatment facilities studies, the primary goal would be to evaluate the environmental impact of the treatment technologies and possibly their combinations that are capable of producing water quality required for particular end uses, including discharge into nearby water body. Such end use could also vary to a great extent from the cultivation of particular crop, or the use of treated water as cooling water or boiler feed water [9].

1.5. System Boundaries

System boundaries are generally drawn so that environmental impacts elsewhere, for example, to the waters where a key input is manufactured, or to the air or soil anywhere, are not counted. System boundaries in time are generally drawn to consider effects of a wastewater treatment alternative immediately after it opens. The onsite wastewater treatment system is installed according to specifications, and probably functions well the day it starts up, or after a short time, when the microbial communities become established. The many jurisdictions that do not have any sort of management program for onsite systems implicitly either assume that the system will continue to perform like new without regular maintenance or do not count environmental effects that occur after a number of years. Similarly, with centralized wastewater treatment plants, the effects of short-term overflows, infiltration and inflow, and pipe breakages are sometimes not considered. Kirk et al. had pointed out the influence of system boundaries on LCA results, since setting system boundaries in different ways can tip the scales in favor of one technology over another [5]. They showed how the concepts of system boundaries and parameters help illuminate why wastewater decisions may only move problems in time and space, rather than solve them.

1.6. System Function and Functional Unit

The functional unit is a measure of the performance of the product system. The primary purpose of the functional unit is to provide a reference to which the inputs and outputs are related and is necessary to ensure comparability of results. The function is related directly to the questions that the study is designed to answer, and the functional unit must be selected as the basis for the study. One of the primary purposes for a functional unit is to provide a reference for the system inputs and outputs. A well-defined functional unit that assures equivalence also allows for more meaningful comparisons between alternative systems. In their study of LCA of wastewater treatment technologies treating petroleum process waters, Vlaopolous et al. had considered a process water flow of 10,000 m^3/day for a time period of 15 years (system design life) as the function unit used in order to compare the different wastewater treatment processes [10].

1.7. Inventory Analysis

The inventory analysis is a technical process of collecting data, in order to quantify the inputs and outputs of the system, as defined in the scope. Energy and raw materials consumed, emissions to air, water, soil, and solid waste produced by the system are calculated for the entire LC of the product or service. In order to make this analysis easier, the system under study is split up in several subsystems or processes and the data obtained is grouped in different categories in a LCI table.

1.8. Impact Assessment

Life cycle impact assessment (LCIA) is a process to identify and characterize the potential effects produced in the environment by the system under study. The starting point for LCIA is the information obtained in the inventory stage, so the quality of the data obtained is a key issue for this assessment. LCIA is considered to consist of four steps that are briefly described below. The first step is classification, in which the data originated in the inventory analysis are grouped in different categories, according to the environmental impacts they are expected to contribute. Indicators of impact categories include:

1. Climate change
2. Acidification
3. Eutrophication
4. Photochemical smog
5. Fossil fuel depletion
6. Ecotoxicity
7. Ozone depletion
8. Human toxicity

The second step, called characterization, consists of weighting the different substances contributing to the same environmental impact. Thus, for every impact category included in LCIA, an aggregated result is obtained, in a given unit of measure. The third step is normalization, which involves relating the characterized data to a broader data set or situation, for example, relating SO_x emissions to a country's total SO_x emissions.

The last step is weighting, where the results for the different impact categories are converted into scores, by using numerical factors based on values. This is the most subjective stage of an LCA and is based on value judgments and is not scientific. For instance, a panel of experts or public could be formed to weight the impact categories. The advantage of this stage is that different criteria (impact categories) are converted to a numerical score of environmental impact, thus making it easier to make decisions.

2. INTERPRETATION

This is the last stage of the LCA, where the results obtained are presented in a synthetic way, presenting the critical sources of impact and the options to reduce these impacts. Interpretation involves a review of all the stages in the LCA process, in order to check the consistency of the assumptions and the data quality, in relation to the goal and scope of the study. All necessary inputs and emissions in many stages and operations of the LC are considered to be within the system boundaries. This includes not only inputs and emissions for production, distribution, use and disposal, but also indirect inputs and emissions—such as from the initial production of the energy used—regardless of when or where they occur. If real environmental improvements are to be made by changes in the product or service, it is important not to cause greater environmental deteriorations at another time or place in the LC. LCA offers the prospect of mapping the energy and material flows as well as the resources, solid wastes, and emissions of the total system, i.e., it provides a "system map" that sets the stage for a holistic approach. The power of LCA is that it expands the debate on environmental concerns beyond a single issue, and attempts to address a broad range of environmental issues, by using a quantitative methodology, thus providing an objective basis for decision making. Unfortunately, LCA is not able to assess the actual environmental effects of the system. The ISO 14042 standard, dealing with LCIA, specially cautions that LCA does not predict actual impacts or assess safety, risks, or whether thresholds are exceeded. The actual environmental effects of emissions will depend on when, where, and how they are released into the environment, and other assessment tools must be utilized. For example, an aggregated emission released in one event from one source, will have a very different effect than releasing it continuously over years from many diffuse sources. Clearly, no single tool can do everything, so a combination of complementary tools is needed for overall environmental management.

2.1. LCA in Developing Countries—an African Perspective

In developing countries, the use of LCA in wastewater sector along with other sectors is still in infancy. However, in these countries, there is a growing need to construct wastewater treatment facilities to cope with population increase and the regular development of urban districts, the use of LCA in these countries can provide a viable tool for sound decision making, based on both environment and economics. One of the most important obstacles to implementing LCA in developing countries is the complexity and difficulty of the many of the methods in the LCA toolbox, which tend require special training and experience. The attainment of sustainable development goals of initiatives such as the Plan of Implementation of World Summit on Sustainable Development (WSSD). The application of LCA is most likely to make a positive impact if they are incorporated in selected development programs targeting key areas such as energy development, wastewater treatment, and others. Major difficulties in applying the tools in developing countries context are as follows:

1. Low level of awareness of the usefulness of the tool among policy makers
2. No internal capacity in industry and government
3. Academics tend to be the LCA competence in most of African countries
4. Lack of accessible background data
5. Quality assurance
6. Need for adequate impact categories
7. Lack of collaboration among LCA experts in the region

Nevertheless, most agree that LC thinking, a major component of LCA, should be considered in the development of environmental policies in developing countries, even if the full spectrum of LCA and its related tools are not implemented. The LC perspective is an elemental factor to achieve sustainable development but there is a need for simplified methods. Several proposals exist for the simplification of LCAs [10–12]. One approach is a screening process to find which parameters should be included or omitted in the study.

3. CONCLUSION

If a number of LCAs have been carried out on a sector or product group it should be possible to select the most environmentally important parameters or categories. Another way to simplify LCA is at the system level where up-stream or down-stream processes are excluded, e.g., by focusing on the steps in the LC that could specifically be controlled by the organization conducting the LCA.

Acknowledgment: I am grateful to the kind support of the NATO Advanced Research Workshop for the financial support provided.

4. REFERENCES

1. UNCSD (1997), Comprehensive Assessment of the Freshwater Resources of the World, Fifth session, 7–25 April, 1997. United Nations, E/CN.17/1997/9.
2. Romero-Hernandez O (2005), Applying Life Cycle Tools and Process Engineering to Determine the Most Adequate Treatment Process Conditions. a tool in Environmental Policy. Int. J. LCA. 10 (5), 355–363.
3. Lundin M (1999), Assessment of Environmental Sustainability of Urban Water System, a PhD thesis, Department of Technical Environmental Planning, University of Goteborg, Sweden.
4. Hospido A, Moreira1 T, Martín MM, Rigola M, and Feijoo G (2005), Environmental Evaluation of Different Treatment Processes for Sludge from Urban Wastewater Treatments: Anaerobic Digestion versus Thermal Processes, Int. J. LCA. 10 (5), 336–345.
5. Kirk B, Etnier C, Kärrman E, and Johnstone S (2005), Methods for Comparison of Wastewater Treatment Options. Project No. WU-HT-03-33. Prepared for the National Decentralized Water Resources Capacity Development Project. Washington University, St. Louis, MO, by Ocean Arks International, Burlington, VT.
6. Tarantini M, Ferri F (2001), LCA of Drinking and Wastewater Treatment Systems of Bologna, Final Results. 4th IRCEW Conference, Fortaleza, Brazil.
7. Bridle T, Skrypski-Mantele S (2000), Assessment of Sludge Reuse Options: a Life-cycle Approach. Wat. Sci. Tech. 41 (8) 131–135.
8. Houillon G, Jolliet O (2004), Life Cycle Assessment for the Treatment of Wastewater Urban Sludge: J. Cleaner Prod. 13 (3) 287–299.

9. Hospido A, Moreira MT, Fernández-Couto MP, Feijoo G (2004), Environmental Performance of a Municipal WWTP Int. J. LCA. 9 (4) 261–271.
10. Vlasopoulos N, Memon F, Butler MR (2006), Life Cycle Assessment of Wastewater Treatment Technologies, Treating. Sci. Total Environ. (367), 58–70.
11. Weitz KA, Todd JA, Curran MA, and Malkin MJ (1996), Streamlining Life Cycle Assessment, Considerations, Int. J. LCA 1 (2) 79–85.
12. Schmidt WP, Beyer HB (1999), Environmental Considerations on Battery-Housing Recovery, Int. J. LCA 4 (2) 107–112.

RADIATION-THERMAL PURIFICATION OF WASTEWATER FROM OIL POLLUTION

I. Mustafayev, R. Rzayev, and N. Guliyeva
Institute of Radiation Problems, Azerbaijan National Academy of Sciences, 31-a H. Javid ave Baku-AZ1143, Azerbaijan

Corresponding author: imustafaev@iatp.az

Abstract:

The results of study oxygen's influence on the radiation-thermal decomposition of n-heptane admixtures in water medium are adduced. The main parameters of radiolysis were changed within the limits: temperature 20–400°C, absorbed dose– 0÷16.0 kGy at dose rate 3.2 kGy/h. As a product of decomposition are observed H_2, CO, CH_4, C_2H_4, C_2H_6, C_3H_8, C_3H_6, C_4H_8, hydrocarbons C_5–C_6. The changes of n-heptane concentration in the reactor also are established. The increasing of n-heptane decomposition rate and decreasing of gas formation rate in presence of O_2 are observed. As a result of summarizing of experimental data, the kinetic model of processes has been proposed.

Keywords: Water, heptane, oxygen, radiolysis, purification.

1. INTRODUCTION

The problem of purification of water resources from oil and oil products has great importance as for extraction of additional resources of oil from oil-polluted waters, as well as for environmental protection. In Apsheron peninsula of Azerbaijan as a result of exploitation of oil-fields within 150 years it was formed more than 250 artificial oil-polluted lakes in which concentration of hydrocarbons sometimes exceeds 25 mg/l. Annually from oil industry are discharged more than 45 million tons (by production 10 million tons of oil) sewage. Now it is applied various physical and chemical methods to purification of sewage from oil pollution [1], including radiation–chemical methods [2–6]. In [2] the γ-radiation of the isotope source of ^{60}Co and accelerated electrons for decomposition of various fractions of oil in the water medium at the temperature 30–40°C are used. At the absorbed dose 25–32 kGy is achieved 85–96 % of cleaning.

M.K. Zaidi (ed.), Wastewater Reuse–Risk Assessment, Decision-Making and Environmental Security, 315–322.

Before [7] we are established the chain decomposition of hydrocarbons at the radiation-thermal processes. In this work with the purpose of reduction of the absorbed dose and increase of purification degree we used such approach, i.e. the joint action of radiation and temperature on oil polluted waters. In [8], we have established the dependence of the radiation-chemical yield of the decomposition of n.heptane from its initial concentration. It is shown, that with increase of initial concentration of *n*-heptane grows radiation-chemical yield of decomposition. At low concentration of heptane ($<10^{-3}$%) change of its concentration on the 10 times does not influence on rate of decomposition. Similar result is observed also in the formation of products.

At the radiolysis of binary mixtures of water–heptane the process of decomposition happens due to transfer of energy from molecules of water to heptane as a result of reaction of recharge. The difference in the potential of ionization—2.7 eV energy is distributed between components. As a neutral active product of recombination of ions are alkyl radicals [9]. These radicals play the basic role in formations of the stabile products of radiolysis. The role of medium in course of processes should be essential. In present study we investigated the influence of oxygen on course of process of decomposition of hydrocarbon in the water medium and formation of products. Kinetic calculations on radiation-thermal transformation of heptane in the water medium are lead also.

2. EXPERIMENTAL

The experiments on study of the kinetics of radiation-thermal process were conducted on mixtures of water with *n*-heptane as a oil hydrocarbon model. The simulated mixtures have prepared in vacuum equipment. The water vapor, *n*-heptane and oxygen sequentially were feed in the glass ampoule of 30 ml volume and were sealed. The total concentration of the mixture in the ampoule was 10^{20} mol/ml, and the components ratio changed in the limits $[C_7H_{16}]/[H_2O] = (1–100) \times 10^{-5}$. Irradiation of the ampoule has been carried out in γ-radiation source of ^{60}Co, by dose rate 3.2 kGy/h and adsorbed dose D = 3.2–16.0. In the all studied interval of temperature the n-heptane was in the vapor phase. Determination of the rate of the dose of γ-radiation was conducted by ethylene and ferrosulphate dosimeters, and results of the measurements were agreed within the limits of 12–15%. The concentration of gas products of the radiation-thermal decomposition of mixtures of water with *n*-heptane and oxygen was determined chromatographically. In all cases kinetics of the process has been investigated, the generation rate and the radiation-chemical yields of products were determined on initial segments of kinetic curves.

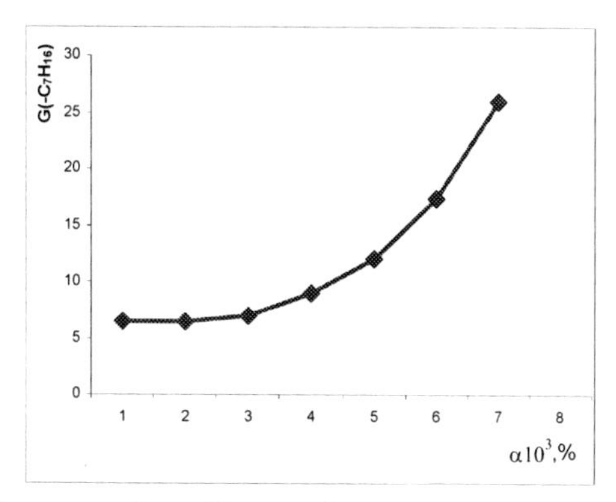

Figure 1. Influence of concentration of heptane in water medium on radiation-chemical yield its decomposition. $T = 400°C$, $I = 3.2–16$ kGy

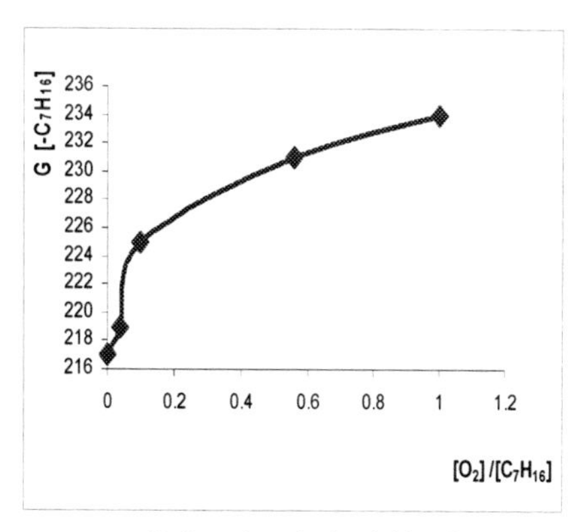

Figure 2. Influence of oxygen on radiation-chemical yield of heptane decomposition in the water medium

3. RESULTS AND DISCUSSION

The dependence of the radiation-chemical yield of decomposition of heptane in the water medium on its initial concentration is presented in the Figure 1. It is visible, that at concentration below 3×10^{-3}% there is no concentration dependence. As shown in [9] at concentration above 10^{-2} % there is sharp increase G ($-C_7H_{16}$) up to 216 mol/100 eV. Under such circumstances the yield of gases also grows. The mechanism of decomposition of initial hydrocarbon and formation of products include the reactions of recombination, separation, and dissociation of radicals. Depending on the temperature, concentration and, dose rate prevail one of the reactions and changes in rate of gas formation are observed. If these reasoning lawful at presence of oxygen, we should observe reduction of products due to fast reactions of capture of these radicals by oxygen

$$R + O_2 \rightarrow RO_2, ROOH, etc.$$

Probably also direct radiation-chemical oxidation of initial hydrocarbon

$$RH + O_2 \rightarrow ROOH$$

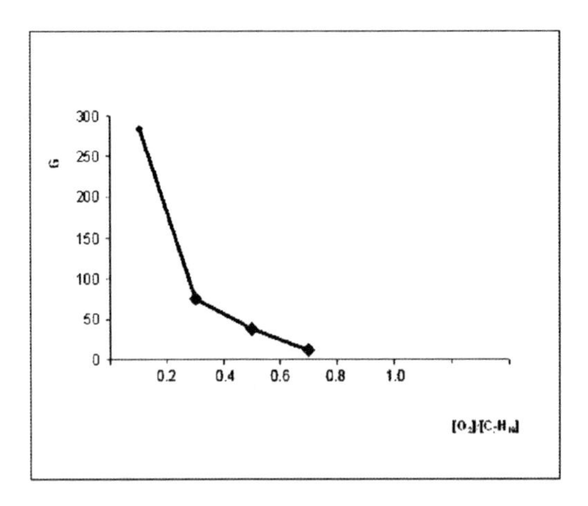

Figure 3. Influence of oxygen on radiation-chemical yield of methane at the decomposition of heptane in water medium

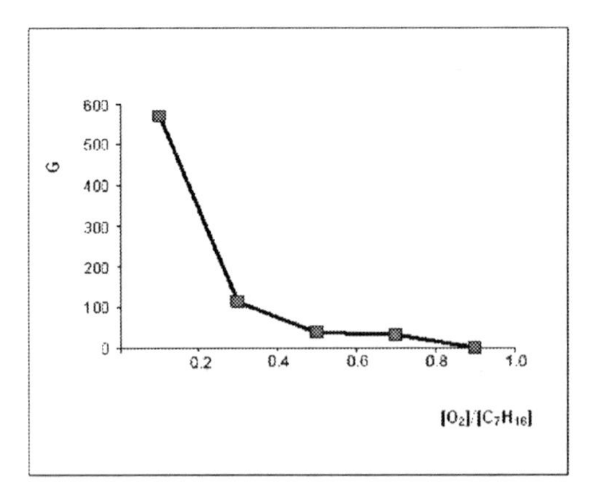

Figure 4. Influence of oxygen on radiation-chemical yield of etylene at the decomposition of heptane in water medium

As well as it was supposed at statement of this study due to capture of radicals by oxygen occurs reduction of hydrocarbonic gases yields. The organic oxygen-containing substances are probably formed in this case. In spite of the fact that rate of decay of initial hydrocarbon increases, from the point of view of cleaning this process is nonrational. As a result of such process, liquid organic compounds such as ketones, acids, aldehydes, alcohol, etc. can be formed. Certainly the microadmoxtures of oxygen can give the positive effect as a result of participation of oxygen in destructive process. The dependence of the ratio of the yield of products in the liquid and gas phase (G_{gas}/G_{liq}) from temperature of the process is presented at absence of oxygen (1) and its presence (2) adduced in Table 1. The yield of gas products with rise in temperature grows in both cases, in case of participation of oxygen this growth is slower.

In the investigated system except of hydrocarbon gases, it is observed also carbon monoxide (Figure 5), as a product of incomplete oxidation of hydrocarbons.

Table 1. Dependence of G_{gas}/G_{liq} on temperature

	T°C	100	200	300	400
G_{gas}/G_{liq}	H_2O-C_7H_{16}	0.07	52	5.2	15.8
	H_2O-C_7H_{16}-O_2	0.05	0.1	0.8	2.5

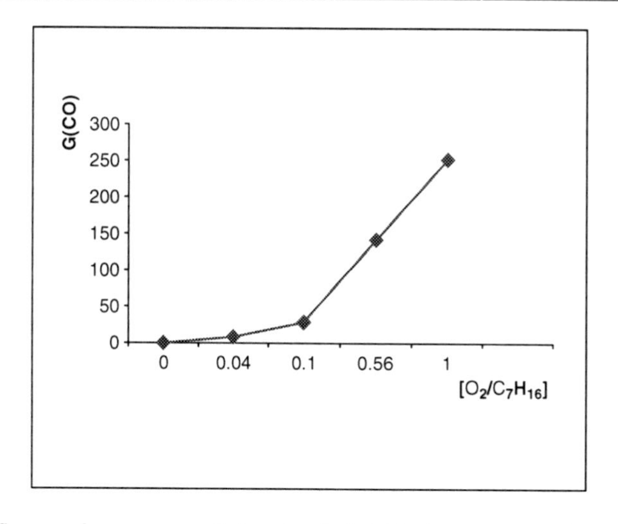

Figure 5. Influence of oxygen on radiation-chemical yield of carbonmonoxide at the decomposition of heptane in water medium

At creation of the condition for selective transformation of oxygen in carbon monoxide can lead to increase of efficiency of process of cleaning. In this case rate of formation of liquid oxygen-containing products decreases. As a result of the analysis of experimental data, we offer the simplified kinetic scheme of proceeding processes and presented (Table 2).

It is seen that in 19 reactions, about 30 particles participate. Reactions can be divided into the following groups:

- Processes of generation of active particles (radicals and ions) at influence of radiation on system water–*n*-heptane (1–3)
- Recombination reactions of radicals with formation of the saturated hydrocarbons and olefins (18, 19)
- Reactions of disintegration of larger alkyl radicals to olefins and small alkyl radicals (6–11)
- Separation reactions of alkyl radicals with initial heptane (4, 5, 12–17).

Table 2. Reaction scheme of radiation-thermal transformations of heptane admixtures in water medium. $[C_7H_{16}]/[H_2O] = 10^{-5}–10^{-1}$, I = 3, 2 kGy/h, D = 3.2–16.0 kGy, P = 0, 1 MPa, T = 375–675K

No.	Reactions	Type of reactions
1	$H_2O \rightarrow H_2O^+ + e$, $H_2O \rightarrow H + OH$, etc.	Radiolysis
2	$C_7H_{16} \rightarrow C_7H_{16}^+ + e$, $C_7H_{16} \rightarrow C_7H_{15} + H$, etc.	Radiolysis
3	$H_2O^+ + C_7H_{16} \rightarrow C_7H_{16}^+$, $C_7H_{16} \rightarrow C_7H_{15} + H$	Radiolysis
4	$H + C_7H_{16} \rightarrow H_2 + C_7H_{15}$	Separation
5	$OH + C_7H_{16} \rightarrow H_2O + C_7H_{15}$	Separation
6	$C_7H_{15} \rightarrow H + C_7H_{14}$	Dissociation
7	$C_7H_{15} \rightarrow CH_3 + C_6H_{12}$	Dissociation
8	$C_7H_{15} \rightarrow C_2H_5 + C_5H_{10}$	Dissociation
9	$C_7H_{15} \rightarrow C_3H_7 + C_4H_8$	Dissociation
10	$C_7H_{15} \rightarrow C_4H_9 + C_3H_6$	Dissociation
11	$C_7H_{15} \rightarrow C_5H_{11} + C_2H_4$	Dissociation
12	$CH_3 + C_7H_{16} \rightarrow CH_4 + C_7H_{15}$	Separation
13	$C_2H_5 + C_7H_{16} \rightarrow C_2H_6 + C_7H_{15}$	Separation
14	$C_3H_7 + C_7H_{16} \rightarrow C_3H_8 + C_7H_{15}$	Separation
15	$C_4H_9 + C_7H_{16} \rightarrow C_4H_{10} + C_7H_{15}$	Separation
16	$C_5H_{11} + C_7H_{16} \rightarrow C_5H_{12} + C_7H_{15}$	Separation
17	$C_6H_{13} + C_7H_{16} \rightarrow C_6H_{14} + C_7H_{15}$	Separation
18	$C_7H_{15} + C_7H_{15} \rightarrow C_{14}H_{30}$	Recombination
19	$C_7H_{15} + C_7H_{15} \rightarrow C_7H_{16} + C_7H_{14}$	Disproportion

The kinetic regularities of the formation of gases and disintegration of an initial component have been calculated according to this mechanism. Numerical integration of system of the rigid ordinary differential equations was made with use of a package of applied programs "Kinetika-90" by Gere's method with a choice of a step and the order of the scheme. Absorption of radiation by products of radiation-thermal decomposition of heptane at this stage neglected. The rate of constants of monomolecular disintegration stole up so that the calculated yield of products in the best degree corresponded to the experimental data. On the example of formation of methane in system H_2O-C_7H_{16} are lead comparison of experimental data.

In Figure 6 the line shows calculation G (CH_4) by means of analytical expression, points—experimental data. It is visible that the experimental and numerical calculation give about one or two. It means that for the description of yields ethane, propane, etc. it is possible to use the resulted approach also. The results obtained are approved on process of radiation-thermal purification of wastewaters from oil pollution. It is shown, that use of electron accelerator with power of beam 100 kW can clean up about 70 cubic m/h the sewage, containing 0.1% of oil.

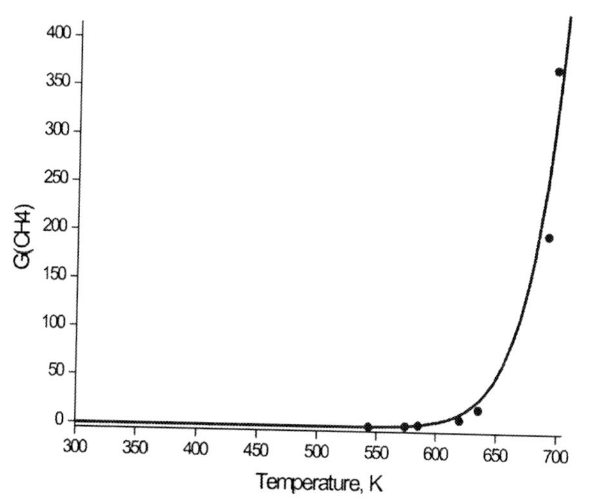

Figure 6. Comparison of experimental and calculation data on formation of CH_4 at the radiation-thermal conversion in the system of C_7H_{16}-H_2O

4. CONCLUSION

In framework of the offered kinetic scheme of radical reactions the experimental results on radiation-thermal cleaning of oil polluted waters is explained. At the radiation-thermal influence at absorbed doses of radiation 5–6 kGy concentration of hydrocarbon in the water medium decreases from 0.1% to 0.001%. It shows the perspectives of the processes of radiation-thermal cleaning of water from oil hydrocarbons. At the application of this approach for cleaning of water from natural oil we have a deviation from the results, received on model hydrocarbon up to 30%. Presence of radiation-resistance polycyclic aromatic hydrocarbons and other compounds in natural oil decreases of process rate.

Acknowledgment: We express our acknowledgements to NATO-Science Program for support of participation of the author in the Workshop.

5. REFERENCES

1. Minhalma Muguel, De Pinho Maria Norberta. (2001). Flocculation/flotation/ultra-filtration integrated processing wastewater. Environ. Sci. and Technol., 35 (24), 4916–4921.
2. Pikaev AK. (2001). New Environmental Application of radiation technology. J. Khimiya vysokikh Energiya, 35 (3), 175–187.
3. Huaying Bao, Yuanxia Liu, Haishun Jia (2002). A study of irradiation in the treatment of wastewater. Radiation Physics and Chemistry 63, 633–636.
4. Fang XW, Wu JL. (1999). Some remarks on applying radiation technique combined with other method to the treatment of industrial wastes. Radiat. Phys. Chem. 55, 465–468.
5. Getoff N. (1999). Radiation chemistry and the environment. Radiat. Phys. Chem. 54, 377–384.
6. Pikaev AK. (2002). The contribution of Radiation technology to Environmental Protection. J. Khimiya vysokikh Energiya, 36 (3), 163–175.
7. Mustafaev I, Gulieva N. (1999). Radiation-thermal transformation of pentadecane. J. Khimiya vysokikh Energy, 33 (5), 354–359
8. Dzantiev BG, Ermakov AN, Jitomirskiy BM, Popov VN. (1979). Termoradiation decomposition water vapour. Atom. Energy, 46 (5), 359–361.
9. Mustafayev I, Guliyeva N, Aliyev S, Mamedyarova I. (2004). Radiation-thermal Purification of Wastewater from Oil Pollution. NATO ARW.

ANALYSIS OF ANTIBIOTIC RESISTANCE PATTERNS OF FECAL *ENTEROCOCCUS* IN COASTAL WATERS OF PUERTO RICO

Nydia J. Rodríguez and Baqar R. Zaidi
Department of Marine Sciences
University of Puerto Rico, Mayaguez, PR 00681–9013, USA

Corresponding author: bzaidi@uprm.edu

Abstract:

Multiple antibiotics resistance of *Enterococcus* species *(ES)* was compared in Guánica, Guayanilla, and Mayagüez coastal water. Only 10, 12, and 12% of *ES* respectively were resistant to chlortetracycline at 10 μg/ml and less than 5% were resistant at 50 μg/ml. However, in Barceloneta coastal water 48% and 28% of *ES* were resistant to chlortetracycline at 10 and 50 μg/ml respectively. Similar results were obtained using oxytetracycline and salinomycin. Numbers of *Enterococcus* species resistance to multiple antibiotics were highest in Barceloneta and lowest in Guánica coastal water. Results of multiple antibiotic resistances in Barceloneta River, estuary, and coastal water are presented.

Keywords: Dairy industry, multiple antibiotic resistances (MAR), *ES*, coastal waters, and Puerto Rico.

1. INTRODUCTION

For the last 50 years, high levels of antibiotics are commonly used for treatment and prevention of diseases in humans, animals, and fish. This has led to increase in the occurrence of antibiotic resistance among bacteria from areas where antibiotics are heavily used. However, occurrence of antibiotic resistance in bacteria is also increasing in aquatic environments e.g., pond, stream, and rivers [1–3]. There is potential risk that antibiotic resistance genes may be transferred into a wide range of aquatic environmental bacteria [4].

Antibiotics are also used in cattle, swine, and poultry farms for treating and preventing infection and to promote growth and reduce the cost of livestock [2]. United States Environmental Protection Agency (USEPA) establishes that in some cases as much as 80% of the antibiotics administered orally to animal's passes through their body unchanged into the environment. Investigations suggest that the greatest potential risk is presented by the transfer of plasmid-encoded resistance

M.K. Zaidi (ed.), Wastewater Reuse–Risk Assessment, Decision-Making and Environmental Security, 323–330.

genes between the aquatic bacteria to bacteria causing infections in humans [4]. Antibiotics resistances in bacteria mediated by plasmid were also found in marine coastal environment of Puerto Rico [5]. A number of plasmid carrying strains and bacteria with multiple resistances have been isolated from marine air–water interfaces. It has been shown that plasmid transfer between bacteria occurs in a variety of natural habitats that include wastewater, sewage, seawater, river water, lake water, sediments, soil, and gastrointestinal tracts [6]. The conjugation and transfer of plasmids can occur between bacterial strains of human, animal and fish origin [6].

Enterococci are regarded as one of the leading causes of nosocomial infections, and cases of endocarditis, bacteremia, urinary tract infection, and neonatal sepsis have often been reported [7]. *Enterococcus* species has been isolated from different aquatic habitats and used as an indicator of the occurrence and transfer of antimicrobial resistance [3]. This microorganism may survive longer in a marine environment by its capacity to tolerate high salts concentrations [8]. Increased number of multiple antibiotic resistant of Enterococci in fresh and marine waters increases the health risk to human because the treatment of infection caused by these bacteria will be difficult if they are already resistant to multiple antibiotics.

Enterococci are known to acquire antibiotic resistance with relative ease and are able to spread these resistant genes to other species. Tetracycline resistant enterococci are commonly isolated from humans, sewage, aquatic habitats, agricultural runoff, and animal sources. Widespread distribution of tetracycline resistant genes supports the hypothesis that the tetracycline resistant genes are exchanged by bacteria from many different ecosystems and between humans and animals. Thus, bacteria exposed to antibiotics in the environment or in animals can ultimately influence antibiotic resistance in bacteria of human origin [9]. Therefore, it is important to understand the bacterial antibiotic resistant pattern and survivability of these bacteria as they move from terrestrial to coastal marine environment. This information may be important in assessing potential health risks associated with the consumption of raw or partially cooked seafood.

2. MATERIALS AND METHODS

Samples were collected from Barceloneta, Guánica, Guayanilla, and Mayagüez coastal waters and from river flowing through dairy industry. Sample sites are shown in the map.

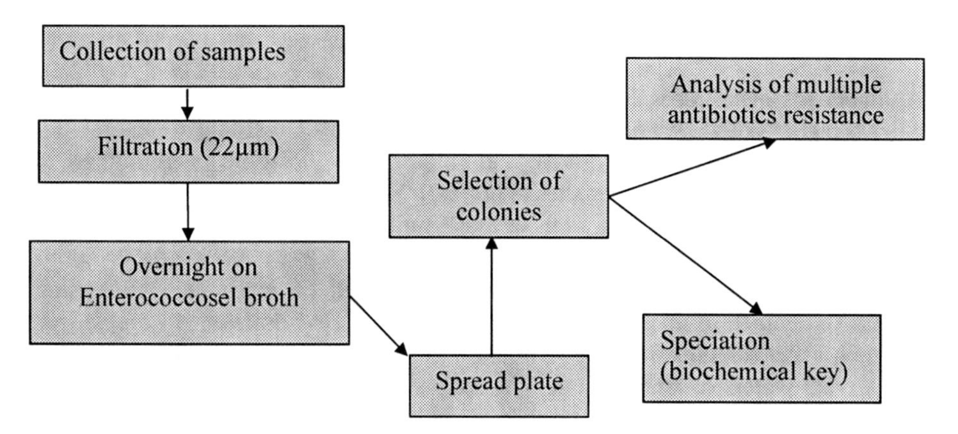

Figure 1. Steps in isolation and identification of antibiotic resistant *ES*

2.1. Selection of Antibiotics

Streptomycin, chlorotetracycline, oxytetracycline, and salinomycin were initially used to compare antibiotic resistance in coastal environment in Puerto Rico. In Barceloneta penicillin, tetracycline, and, vancomycin were used. These antibiotics are selected because of their widespread use in dairy industry.

2.2. Isolation of Bacteria Resistant to Antibiotics

One liter of the water sample was filtered through 0.22 μm filter. The filters were transferred to a flask with enterococcosel media. Filters were incubated for 48 h at 25°C and plated on agar plates with appropriate dilutions. Colonies were picked at random and transferred to microplate containing 0.2 ml of Enterococcosel media. After 48 h the bacteria showing positive growth were transferred to brain heart medium and the catalase test was performed. *ES* was identified by biochemical tests [10]. The characterized *ES* was transferred to microplates containing Enterococcosel broth with antibiotics. Flow chart indicates the different steps in isolation and identification of *ES* (Figure 1).

2.3. Minimum Inhibition Concentration (MIC)

MIC of each antibiotic was determined by using parameters developed by National Council of Clinical Laboratory Standards (NCCLS). The antibiotics and concentration used were tetracycline (0.01 μg, 4 μg, 10 μg, 30 μg, 40 μg,), penicillin (0.01 μg, 0.05 μg, 1 μg, 10 μg, 15 μg,) and vancomycin (0.05 μg, 1 μg, 10 μg, 20 μg, 30 μg). Antibiotics were tested in combination at their MIC.

3. RESULTS

Antibiotic resistant *ES* bacteria were isolated from Barceloneta, Guánica, Guayanilla, and Mayagüez coastal water samples. The percentage of *ES* resistant, to chlorotetracycline, oxytetracycline (40μg/ml), and salinomycin (15μg/ml) was highest in Barceloneta compared to other sites (Figure 2). Similar results were obtained with 10, 20, and 50 μg/ml concentrations of these antibiotics (data not shown).

After finding of higher number of antibiotic resistant *ES* in Barceloneta coastal water, we decided to investigate in detail the coastal environment of Barceloneta. We took water samples from the river close to the dairy industry as well as estuary and coastal waters of Barceloneta. We isolated 1.0×10^9 cells/l of *ES* from river running through dairy industry, 1.4×10^8 cells/l from estuarine water, and 1.9×10^3 cells/li from coastal marine water in Barceloneta, Puerto Rico (Figure 3).

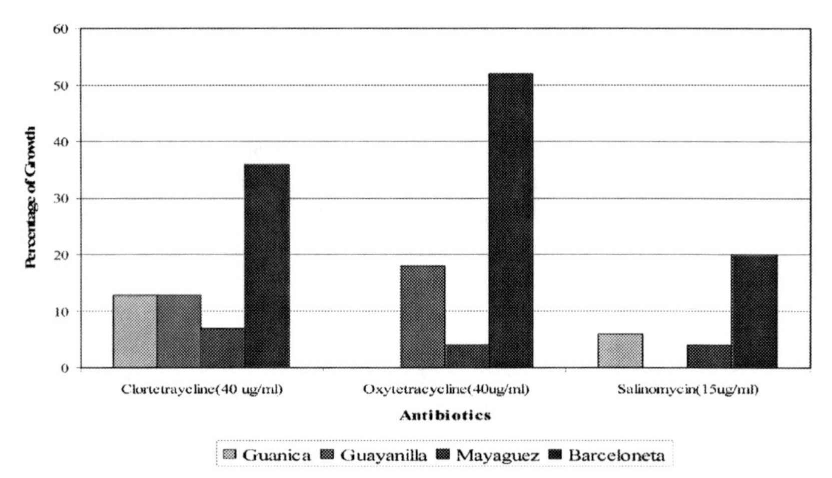

Figure 2. *ES* resistance to different antibiotics in coastal waters of Puerto Rico

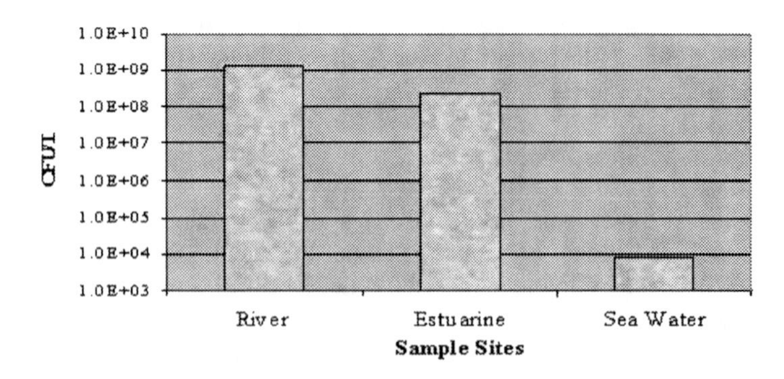

Figure 3. Number of *ES* isolated from Barceloneta waters

Analysis of percentage of different species of *Enterococcus* identified by standard microbiological and biochemical techniques indicated that 71% of *Enterococcus faecalis* was found in river water samples (Figure 4). Only data on the distribution of *ES* found in all three environments i.e. river, estuarine, and seawater is included in Figure 4.

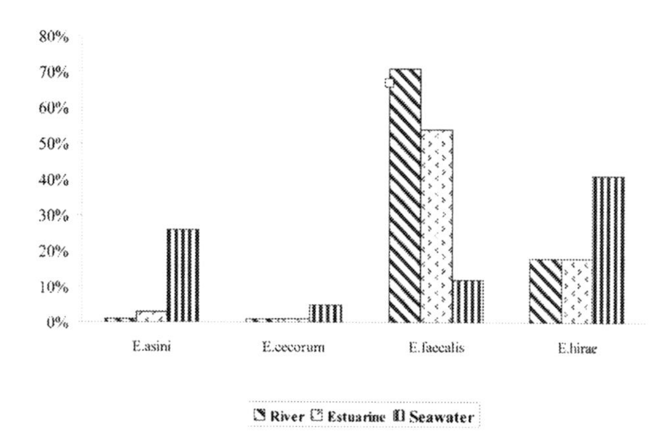

Figure 4. Different *ES* in Barceloneta environmental sample

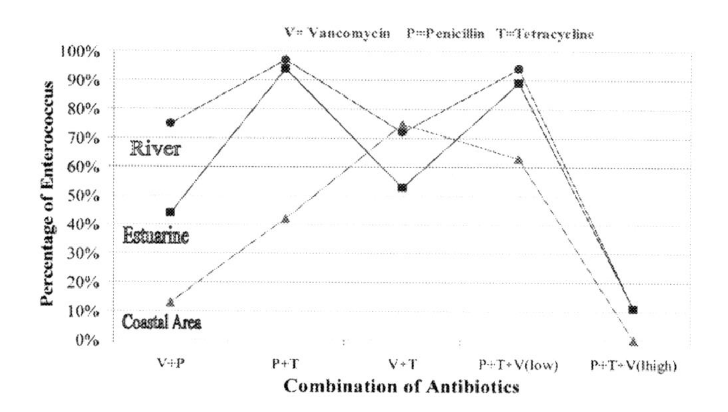

Figure 5. Percentage of *ES* resistant to multiple antibiotics

E. faecalis number decreased to about 54% in estuary while only 12% were found in seawater. Surprisingly, highest percentage 41% of *E. hirae* was found in seawater. The percentage of *E. hirae* in river and estuarine environment were only

18%. The percentage of *E. asini* in seawater was also high (26%), but only 3% were found in estuarine and even less (1%) in river water samples. In comparison even though *E. cecorum* was isolated from all the three environments but the percentage was small. However, higher percentage of this bacterium was found in seawater (Figure 4).

Multiple antibiotic resistance of *ES* isolated from the three environments was analyzed. The antibiotics penicillin, tetracycline, and vancomycin were used. MIC for each of these antibiotics was determined. MIC for both tetracycline and vancomycin was 10 μg/ml and for penicillin 5 μg/ml. Percentage of *ES* resistant to combination of these antibiotics is shown in Figure 5. Resistance to vancomycin + penicillin at MIC was highest (75%) in river samples, 44% in estuarine and only 13% in seawater. For penicillin + tetracycline it was 97% in river, 94% in estuarine and 42% in coastal seawater and for vancomycin + tetracycline it was 75% for both river and coastal seawater and 53% in estuarine samples. Resistance of *ES* to three antibiotics (penicillin + tetracycline + vancomycin) at MIC was highest at 94% in estuarine water, 89% for river, and 63% for seawater. Resistance dropped to 11% in both river and estuarine water samples when these antibiotics were used at high concentrations of 15, 40, and 30 μg/ml. In coastal seawater however resistance was 0% (Figure 5).

4. DISCUSSION

Most large industrial cities in Puerto Rico are located on the coast. Therefore, the multiple antibiotic resistances of *ES* in coastal waters of industrial cities of Barceloneta, Guayanilla, and Mayagüez were compared with Guánica coastal water, where no industry is located. As expected lowest multiple antibiotic resistances were found in Guánica. However, after observing highest multiple antibiotics resistances in *ES* in Barceloneta coastal water, we decided to study it in detail (Figure 1). Barceloneta is the site of one of the biggest pharmaceuticals industries in the world. It also has a large dairy industry.

Knowledge of predominant species present in specific site could be useful in developing methods using enterococci as microbial source tracking. A comparison of enterococci distribution in river, estuarine, and coastal seawater showed that *E. faecalis* dominates in river and estuarine water samples but not in coastal seawater (Figure 4). Predominant species in tropical Barceloneta coastal water, in order of occurrence were *E. hirae*, *E. asini,* and *E. faecalis*. *E. hirae* is a member of animal microflora, but is known to cause infections in humans [11]. Species distribution in tropical Barceloneta coastal water is different than found in California coastal water, where distribution of *ES* in order of occurrence was *E. faecalis*, *E. faecium*, and *E. hirae* [12].

Large quantities of antibiotics are used in dairy industry in Barceloneta, Puerto Rico as growth promoters and prevention and control of infections. Mostly tetracycline (T) and penicillin (P) are used in large quantities. This is reflected in higher resistance of *ES* to these two antibiotics at minimum inhibitory concentrations, while resistance to vancomycin is relatively low in all three environments. The distribution of tetracycline + penicillin-resistance enterococci (TPE) in order of occurrence is river, estuarine and coastal seawater i.e. highest near the dairy industry. It is interesting to note that resistance to high concentrations of three antibiotics could be due to disturbance caused by heavy rain. It is known that rain causes mixing of bacteria from the sediments and more bacteria are found in the water column (12). However, further studies are needed to establish transfer of resistance to antibiotics from *ES* to marine bacteria.

5. CONCLUSIONS

The study clearly indicates more multiple antibiotic resistant *ES* in Barceloneta coastal environment. Comparison of dominance of *ES* between samples of river water flowing through dairy industry and coastal marine water of Barceloneta indicated different species of *Enterococcus* dominate in coastal water.

Acknowledgment: Authors wish to acknowledge partial travel support from Office of Dean Arts and Sciences, University of Puerto Rico and NATO to attend the symposium and to AGEP and NASA Space Graduate fellowship to Nydia J. Rodríguez.

6. REFERENCES

1. Claudio DM, C. Kehrenberg Ulep C, Schwarz S MC. Roberts 2003. Diversity of Tetracycline Resistance Genes in Bacteria from Chilean Salmon Farms. Antimicrob. Agents and Chemother. 47: 883–888.
2. Gebreyes WA, C Altier, 2002. Molecular characterization of Multidrug Resistance *Salmonella enteric* subsp. *enterica* Serovar Typhimurium Isolates from Swine. J. Clin. Microbil. 40: 2813–2822.
3. Peterson A, Andersen J.S, Kaewmak T, Somsiri T, Dalsgaard. 2002. Impact of Integrated Fish Farming on Antimicrobial Resistance in a Pond Environment. Appl. Environ. Microbiol. 68: 6036–6042.
4. Rice EW, JW Messer, CH Johnson, DJ Reasoner. 1995. Occurrence of high-level aminogycoside resistance in environmental isolates of enterococci. Appl. Environ. Microbiol. 61: 374–376.

5. Baya AM, Brayton PR, Brown VL, Grimes DJ, Russek-Cohen E, Colwell RR. 1986. Coincident plasmids and antimicrobial resistance in marine bacteria isolated from polluted and unpolluted Atlantic Ocean samples. Appl. Environ. Microbiol. 51: 1285–1292.
6. Kruse H, Sorum H. 1994. Transfer of Multiple Drug Resistance Plasmids between bacteria of Diverse Origins in Natural Microenvironments. Appl. Environ. Microbiol. 60: 4015–4021.
7. Huycke MM, Sahm DF, Gilmore MS. 1998. Multiple-drug resistant enterococci: the nature of the problem and an agenda for the future. Emerg. Infect. Diseases. 4: 239–249.
8. Hardwood VJ, Whitlock J, Withighton V. 2000. Classification of Antibiotic Resistance Patterns of Indicator Bacteria by Discriminant Analysis: Use in Predicting the Source of Fecal Contamination in Subtropical Waters. Appl. Environ. Microbiol. 66: 3698–3704.
9. Tannock GW, Cook G. 2002. Enterococci as members of the intestinal microflora of humans. In The Enterococci: Pathogenesis, Molecular Biology, and Antibiotic Resistance ed. Gilmore, M.S., Clewell, D.B, Courvalin, P., Murray, B.E. and Rice L.B. p. 105. Washington, DC.
10. Manero A, Blanch A. 1999. Identification of ES with biochemical key. Appl. Environ. Microbiol. 65(10): 4425–4430.
11. Schmidt AS, Bruun MS, Dalsgaard I, Petersen K. 2001. Incidence, Distribution and Spread of Tetracycline Resistance Determinants and Integrons- Associated Antibiotic Resistance Genes among Motile Aeromonads from a Fish Farming Environment. Appl. Environ. Microbiol. 67: 5675–5682.
12. Ferguson DF, Moore Getrich, MA, Zhowandai MH. 2005. Enumeration and speciation of Enterococci found in marine and intertidal sediments and coastal water in southern California. J. Appl. Microbiol. 99: 598–608.

EFFECTS OF WASTEWATER IRRIGATION ON CHEMICAL, MICROBIOLOGICAL AND VIROLOGICAL STATUS OF SOIL

Filip Zdenek and Katherina Demnerova
Institute of Chemical Technology, Department of Biochemistry and Microbiology, Technicka 3-5, 166 28, Prague 6, Czech Republic

Corresponding author: zdenek.filip@t-online.de

Abstract:

A long-term wastewater irrigation on soil composed of glacial sand resulted in the transformation of the soil profile, which included chemical, microbiological, biochemical, and virological soil characteristics. The individual soil samples contained organic particles by weight, but these particles represented an important storage of nutrient and energy for microorganisms. The wastewater irrigation on a sandy soil resulted in positive effects on soil activities, and did not create any risk through pathogenic bacteria or viruses to a local deep groundwater aquifer.

Keywords: Wastewater, aquifer, irrigation, soil profile, biochemical activity.

1. INTRODUCTION

Indeed, wastewater irrigation on land belongs to the ancient anthropogenic activities that affect soil quality. Mainly, soil has been used as a black box in which a waste was incorporated on the surface and there was a hope that what moved through the soil profile was not detrimental for the terrestrial ecosystem, the groundwater or human health. As a low-cost technology irrigation still remains one of the frequently used wastewater treatments in many countries [1].

In the outskirts of Berlin (Germany), wastewater irrigation on land was started in early 1878, and in the following decades up to 700,000/m^3/d wastewater was applied on almost 14,000 ha of land [2]. Individual plots received about 1,500 mm/a wastewater until 1992 [3]. In the last 20 years, the total area became an object of an extensive scientific research.

1.2. Chemical Characteristic of Wastewater and Soil

Blume gave the following mean chemical characteristics of the municipal waste-water used in the territory as described above (mg/l): C_{tot} 115, N_{org} 80, P 14.2, NH_4 71, NO_3 0.8, Cu 0.06, Fe 1.1, Mn 0.06, Zn 0.34, pH 8.0 [4]. Both the

M.K. Zaidi (ed.), Wastewater Reuse–Risk Assessment, Decision-Making and Environmental Security, 331–336.

Cambisol and Haplic Luvisol which never received wastewater (controls) were slightly acidic and weakly furnished with carbon and nitrogen. The basic chemical properties of the soil plots are given in Table 1 on the example of a sandy Haplic Luvisol (Ah Horizon).

Table 1. Chemical characteristics of Haplic Luvisol irrigated or not with wastewater

Treatment	pH	Total-C	Total-N	C:N
	H_2O	%	%	
No irrigation	5.55	0.555	0.0294	18.9
No irrigation since 20 years	6.05	0.751	0.0590	12.7
Long-term irrigation	5.80	2.147	0.204	10.5

The data in Table 1 show that the wastewater irrigation affected the pH values only moderately. Carbon and nitrogen content, however, was increased by factor 3.9 (C) and 6.9 (N), respectively. Nevertheless, when considering large volumes of municipal wastewater long-term applied, carbon and nitrogen accumulated also rather moderately in the top soil horizon. In fact, the C:N ratio decreased from 18.9 to 10.5 after a long-term irrigation, indicating an improved utility of the organic matter for soil microorganisms.

2. MICROBIOLOGICAL, BIOCHEMICAL, AND VIROLOGICAL CHARACTERISTICS OF THE SOIL

The biochemical transformation of organic matter and the elemental output from the wastewater irrigated soil through mineralization pathways are operated by both the soil autochthonous and secondary introduced microorganisms. Thus, the development of different groups of microorganisms and their biochemical, i.e., enzymatical activities undoubtedly can be used as indicators of the soil purification activity. Goyal confirmed an enhancement of microbial biomass, dehydrogenase activity, and a denitrification rate in wastewater-treated soils was confirmed by Goyal [5] and it was also postulated that soil irrigation can impose, to a certain degree, an osmotic stress to microorganisms, which might trigger osmoadaptive mechanisms that lead to a higher immobilization of C and N in microbial biomass. A land application of untreated sludge introduced large numbers of heterotrophic bacteria into leachates at 60 cm soil depth [6].

In our study, we determined the content of adenosine triphosphate (ATP) in the individual soil fractions which is well known to correlate both with the total living microbial biomass and the microbial activity in soil [7]. In fact, bacterial numbers increased considerably in the wastewater treated soil. The copiotrophic bacteria which require a high concentration of nutrients for their growth demonstrated a ratio of

1.5:1 when compared their inhabitance in the OP 1–7 fractions with that of the silt–clay fraction. An opposite ratio (0.4:1) has been calculated for less nutrient dependent oligotrophic bacteria. The numbers of both groups of bacteria and also those of fungi clearly decreased with the decreasing size of soil organic particles. The differences between the coarsest and the finest fractions (OP1 vs. OP 7) were up to 2 orders of magnitude. Mineral soil particles (MP 5–7) were only slightly inhabited by microorganisms. In comparison to OP 1, e.g., their microbial counts were up to 4 orders of magnitude lower. A strong inhabitance of microorganisms in relation to the dry weight of soil particles could be observed in the silt-clay fraction of the individual soil samples. In comparison with never irrigated soil, the long-term wastewater irrigation raised here the counts of copiotrophic bacteria by 234%, oligotrophic bacteria by 217%, actinomycetes by 234%, and fungi by 206%. With the only exception for actinomycetes, the silt-clay fraction of the soil sample from the site which remained 20 years without irrigation was absolutely higher in microbial counts than that from never irrigated soil. A depression of microorganism which could be expected because of the accumulation of heavy metals seemed effectively suppressed in a wastewater treated soil by the high content of organic matter [8]. A microbial biomass-N in wastewater irrigated soil to be affected rather by moisture that by a content of heavy metals was reported [9].

Microbial transformation of organic matter is catalyzed by various enzymes. In our study we estimated the activity of (i) ß-glucosidase, which acts in the hydrolysis of various ß-glucosides such as cellobiose and cellulose, (ii) ß-acetylglucosaminidase, which affects hydrolysis of compounds such as chitobiose, chitin, and glycoproteins, and (iii) proteinase which hydrolyses various peptides and proteins. For activity of these enzymes, a significant correlation with concentration of specific substrates was found in soil [10]. Thus, the enhanced enzyme activities, as measured in our soil samples, imply an increased substrate concentration in the wastewater treated soil, and also a high biochemical capacity in these soil fractions to hydrolyze different organic compounds. The fact that the enzyme activities were higher in soil not treated since 20 years in comparison to controls may be due to the presence of stable enzyme-organic-mineral complexes in which enzymes themselves are protected against degradation. An increase in different groups of microorganisms, and an enhancement in amylase, cellulase, dehydrogenase, and phosphatase enzymes in soil after a long-term (15 years) paper-mill effluent irrigation was reported [11]. On the other hand reports exist on enzymes to become inhibited in soil in the dependence on the quantity and frequency of wastewater application on soil.

A long-term wastewater irrigation on soil has also a health associated dimension. This is because a close proximity of water-draw points to wastewater disposal sites in an environment of sandy soils and shallow groundwater aquifers might create a potential risk of contamination of the water supply by pathogens. Bacterial migration up to 830 m and viral migration up to 408 m have been reported by Keswick and Gerba [12,]. More recently, however, McCarthy [13] reported that sandy soil exhibit significant filtration effects on bacteria. In a field assay, the authors were not able to detect coliform bacteria and *Escherichia coli* in groundwater within a distance of 20 m from the disposal point. In model experiments and using soil columns, Kocaer [14] observed an apparent increase in heterotrophic bacteria and total coliform numbers only in a depth of 15 cm of a soil amenden with with a fly-ash-sludge mixture. Down to a 30–45-cm soil depth, gradually decreased.

We investigated the distribution of coliform and potentially pathogenic bacteria in a long-term wastewater irrigated soil already earlier. As documented by data in Table 2, there was a range in distribution of these microorganisms in nonirrigated and irrigated soil, with maximum counts obtained immediately after wastewater irrigation. Various potentially pathogenic bacteria could be detected on a selective McConkey nutrient growth medium in both irrigated and control soils down to 60 cm soil depth. In a 90–12-cm soil layer, however, neither coliform nor potentially pathogenic bacteria have been detected.

Table 2. Distribution of coliform and other facultative pathogenic bacteria in a Haplic Luvisol long-term irrigated with municipal wastewater

Microorganisms	Soil depth (cm)	Treatment – soil irrigated			
		before irr.	after irr.	Not Irr. Since 1962	Never irr (Control)
	0 – 5	9.7×10^3	3.0×10^4	1.5×10^3	2.9×10^2
Coliform	5 – 30	1.7×10^3	5.3×10^3	1.5×10^2	None
Bacteria	30 – 60	6.2×10^1	6.1×10^2	None	None
	60 – 120	None	None	None	None
	0 – 5	1.7×10^4	7.2×10^4	6.6×10^3	6.5×10^3
Facultative	5 – 30	1.3×10^4	2.2×10^4	5.0×10^2	5.3×10^2
Pathogenic	30 – 60	3.4×10^2	6.0×10^2	5.0×10^1	1.1×10^2
Bacteria	60 – 120	None	None	None	None

The contents of enteric viruses in the soil that was under a long-term wastewater irrigation are presented in Table 3. From the total 87 soil samples tested viruses were detected in only seven samples. They were distributed almost uniformly from the surface soil layer (0–5 cm) to 210 cm depth. Samples taken immediately after wastewater was irrigated on soil did not show any substantial difference. All viruses were of the Coxsackie and poliovirus type as determined serologically. No viruses were detected in the control nonirrigated soil, as well as in soil had no received wastewater since 20 years. Samples of groundwater from aquifer located in a depth of 10–12 m were found free of both enteric viruses and potentially pathogenic bacteria. Similar results could be found in an older literature: The effluent renovation did not detect viruses or *Salmonella* species, in well water samples, and the counts of other bacteria were decreased by 99.9% after wastewater passage through a 9-m soil layer [13].

Generation of waste, including wastewater, is increasing worldwide and strategies for its environmentally safe use should be developed and/or the old ones further optimized. This is because in general waste undoubtedly can contribute to chemical and biological pollution of soil. Though our experiences obtained from investigations as reported above belongs to rather positive ones, we understand the necessity of continuing studies in this field. To cope effectively with possible risks as connected with the utilization of organic wastes, suggested to (i) further develop standards for different pollutants, (ii) improve treatment of sewage and sewage sludge to reduce the contents of pollutants, (iii) adapt waste application rates to soil properties, and (iv) harmonize methods and protocols for the monitoring and control the fate of waste which are to be applied onto soils [14]. In Europe, respective standards should be primarily develop for the European Union in order to make land application of wastewater and other municipal wastes a practical way of recycling without any secondary risks to be created.

Table 3. Distribution of enteric viruses in a Haplic Luvisol irrigated with municipal wastewater Conclusion C

Soil depth (cm)	Irrigation		Virus type
	Before	After	
0 – 5	Yes	Yes	Coxsackie B3
5 – 30	Yes	Yes	Coxsackie B3 and B5
30 – 60	Yes	Yes	Poliovirus I and III
90 – 120	No	No	
180 – 210	Yes	No	Coxsackie B1

3. CONCLUSION

The wastewater irrigation on a sandy soil resulted in positive effects on soil activities, and did not create any risk through pathogenic bacteria or viruses to a local deep groundwater aquifer. The highest numbers of bacteria and fungi were estimated in organic particles of the long-term irrigated soil. Similar was true for the ATP content, as a measure of metabolic active microbial biomass, and the enzyme activities of ß-glucosidase, ß-acetylglucosaminidase, and proteinase.

Acknowledgments: The research results reported in this chapter were obtained during the service of the senior author (Z.F.) to the German Federal Environmental Agency. Both the authors wish to express their gratitude to the European Commission, Brussels, for the granting of a Marie Curie Chair in Environmental Microbiology and Biotechnology, tenable at the Institute of Chemical Technology, Prague, Czech Republic. The financial support by the NATO Advanced Research Workshop organizing committee on Water Reuse, Decision making and Environmental Security is acknowledged.

4. REFERENCES

1. McCarthy J, Duse AG. 2004, Wastewater disposal at safari lodges in the Okavango Delta, Botswana, Water SA 30: 121–128.
2. Nestler W, Sowa E, Luckner L, Sarfert F. 1990, Umweltgerechte Nutzung ehemaliger Rieselfelder Berlins, Wasserwirtsch.-Wassertech. 5: 106–108.
3. Blumenstein O. 1995, Ausgewählte theoretische Grundlagen, in: Wenn Abwasser die Landschaft verändert, Arbeitsgruppe Stoffdynamik Geosystemen der Univ. Potsdam 1: 4–35,
4. Blume HP, Horn R, Alaily F, Jaykody AN, Mershef H. 1980, Sand cambisol functioning as a filter. Soil Sci. 130: 186–192.
5. Goyal S, Chander K, Kapoor KK. 1995, Effect of distillery wastewater application on soil, Microbiological properties. Environ. Ecol. 13: 89–93.
6. Kocaer FO, Alkan U, Baskaya HS. 2004, The effect of alkaline-stabilized-sludge application on the Microb. quality. J. Plant Nutr. Soil Sci, 167: 704–712.
7. Nannipieri P, Ceccanti B, Grego S. 1990, Ecological significance of the biological activity in soil, in: Soil Biochemistry, vol. VI, Bollag, J.-M., and Stotzky, G., eds., Dekker, New York, 239–355.
8. Schneider J. 1995, Mikroorganismen in Rieselfeldböden, in: Wenn Abwasser die Landschaft verändert. Stoffdynamik in Geosystemen, Arbeitsgruppe Stoffdynamik der Univ. Potsdam, Bd. 1, 92–108.
9. Kanazawa S. 1979, Characteristic of soil organic matter and respiration in subalpine coniferous forest of Mt. Shigayama. VI. Aminosugars content and ß-acetylglucosaminidase activity. J. Sci. Soil Manure (Japan) 50: 297–303.
10. Burns RG. 1982, Enzyme activity in soil: Location and possible role in microbial ecology, Soil Biol. Biochem. 14: 423–427.
11. Kanan, K and Oblisami G, 1990, On Fluence of Papermill Effluent Irrigation, Soil. Biol. Biochem. 22: 923–926.
12. Keswick BH and Gerba CP. 1980, Viruses in groundwater, Environ. Sci Technol. 14: 1290–1297.
13. McCarthy TS, Gumbricht T, Steward RG, Brandt D, Hancox PJ, McCarthy J and Duse A.G. 2004, Wastewater disposal at safari lodges in the Okavango Delta, Botswana, Water SA 30: 121–128.
14. Kocaer FO, Alkan U and Baskaya HS. 2004. The effect of alkaline-stabilized-sludge application on the Microbiological quality of soil and leachate, J. Plant Nutr. Soil Sci., 167: 704–712.

SESSION 9. QUANTITATIVE RISK ASSESSMENT AND POLICY MAKING–IMPACT ON SOILS AND HEALTH

RISK ASSESSMENT ON POPULATION HEALTH

A. Bayramov
Institute of Physics, National Academy of Science of Azerbaijan
G.Javide av.33, Baku Az1143, Azerbaijan

Corresponding author: bayramov_azad@mail.ru

Abstract:

The toxicology of a human environment is based on the assumption that toxicity of pollutants is directly proportional to their concentration (in wastewater, in aerosphere, and in soil). The calculation method of probability of effect (risk) of toxic pollutants action on the population health is described. The algorithm of risk assessment of pollution impact on a population health is considered. The risk factor (coefficient of a ratio of risk increase) is a function of concentration of toxic substances in an environment: atmosphere, wastewater, soil.

Keywords: Pollutants concentration, risk assessment, toxic effect, probability of effect, population health.

1. INTRODUCTION

In many situations in developing countries, especially in arid and semiarid areas, wastewater is simply too valuable to waste. It can be profitably reused to grow crops and/or fish. But first the wastewater should be treated to reduce pathogenic and faecal indicator microorganisms to acceptable levels, to ensure there is no threat to human health. This is using treated wastewater for crop irrigation. Crop yields are higher than with freshwater irrigation as treated wastewater not only supplies water, but also plant nutrients (especially nitrogen and phosphorus). Restricted irrigation is the irrigation of all crops except salad crops and vegetables that may be eaten uncooked. For restricted irrigation, World Health Organization (WHO) recommends the treated wastewater should contain no more than one human intestinal nematode egg per liter. These accrue mainly because, when wastewater is treated and reused, it is not discharged (often untreated) into streams, rivers, or lakes where it would cause aquatic pollution of varying severity. In extreme cases (which are not uncommon in developing countries) the receiving water can become seriously deoxygenated, even anaerobic, with a literally deadly effect on aquatic life, and on the lives of those who use the polluted water for domestic, including drinking, purposes [1].

M.K. Zaidi (ed.), Wastewater Reuse–Risk Assessment, Decision-Making and Environmental Security, 339–346.

The quality of water, whether it is used for drinking, irrigation, or recreational purposes, is significant for health in both developing and developed countries worldwide. In responding to the challenge of improving water quality, countries develop standards intended to protect public health. Recognizing this, the WHO has developed a series of normative "guidelines" that present an authoritative assessment of the health risks associated with exposure to health hazards through water and of the effectiveness of approaches to their control. To date, the various WHO guidelines (Guidelines for drinking-water quality, Guidelines for the safe use of wastewater and excreta in agriculture and aquaculture, and Guidelines for safe recreational water environments) have been developed in isolation from one another. However, their common primary quality concern is for health hazards derived from excreta. Addressing their specific areas of concern together will tend to support better health protection and highlight the value of interventions directed at sources of pollution, which may otherwise be undervalued.

The potential to increase consistency in approaches to assessment and management of water-related microbial hazards was discussed by an international group of experts between 1999 and 2001 [2]. These discussions led to the development of a harmonized framework, which was intended to inform the process of development of guidelines and standards. Subsequently, a series of reviews was progressively developed and refined, which addressed the principal issues of concern linking water and health to the establishment and implementation of effective, affordable, and efficient guidelines and standards.

Quantifiable risk assessment was initially developed, largely, to assess human health risks associated with exposure to chemicals (NAS 1983) and, in its simplest form, consists of four steps:

- Hazard assessment
- Exposure assessment
- Dose–response analysis
- Risk characterization

The output from these steps feeds into a risk management process. As will be seen in later sections this basic model (often referred to as the chemical risk paradigm) has been extended to account for the dynamic and epidemiologic characteristics of infectious disease processes [3–6].

The toxicology of a human environment is based on the assumption that toxicity of pollutants is directly proportional of their concentration in an environment (in wastewater, aerosphere, and in soil). Besides this, the more duration of contact of pollutants with biological object, the probability of toxic effect is more, and the risk (hazard) for human health is more. In this paper the algorithm of risk assessment of pollution impact on population health is considered. For assessment of risk it is offered to use the equation, where the risk factor (coefficient of a ratio of risk increase) is function of concentration of toxic substances in an environment: atmosphere, wastewater, and soil.

$$R = a - b \cdot \exp(-U_R \cdot C)$$

here: R is a risk of occurrence of unfavorable effect (it is probability of occurrence of this effect under the given conditions); C is a real concentration of the pollutant, having an effect on a human organism; U_R is an unity of risk (it is coefficient of a ratio of increase of risk depending on value of acting concentration); a and b are the coefficients determined by empirical [7]. The estimation of pollution is carried out according to multiplicity of overflow of maximum permissible concentration (MPC) which is determined from the formula [8]

$$K = \frac{C_M}{\text{MPC}},$$

here C_M is the maximum pollutants concentration in environment.

2. PROCEDURE AND METHODS

The behavior of contaminants in water is caused by the complex regularities. Therefore, precise prediction and estimation of risk here is inconvenient. In most cases to forecast behavior of hazardous impurities, it is necessary to accept simplifying assumptions. It's concern and account of a real dose of harmful substances which acts on the human should be made. For determination of toxicity of this or that contaminant it is necessary to know its dose which has got into an organism due to potable water. The kinetics of toxic substance in biological objects follows under the law:

$$A_t = A_{t0} \cdot \exp[\lambda(t - t_0)] \qquad (1)$$

here A_t is a concentration of toxic agent in organism at the t moment; A_{t0} is the concentration of toxic agent in organism at the t_0 moment; λ is the constant of excretion:

$$T = \lambda / 0.693 \qquad (2)$$

Where T is an excretion half-life. On the basis of this, in Europe and USA for calculation of risk the following equation is used [9]:

$$R = [1 - \exp(-U_R \cdot C)] \tag{3}$$

Where R is the risk of occurrence of unfavorable effect (it is probability of occurrence of this effect under the given conditions); C is a real concentration of the pollutant, having an effect on a human organism; U_R is an unity of risk (it is coefficient of a ratio of increase of risk depending on value of acting concentration).

For calculation of U_R, the method of data analysis of epidemiological character about a level of health breakdown under certain dose-time requirements is used. For example, let accept, that at a water pollution lead a dose of 0.03 mg/m^3 the risk of morbidity sicknesses of nervous system is 20% relatively normal conditions: $R = 0.2$; $C = 0.03$ мг/м^3. Substitute these values in Equation 3, we get:

$$0.2 = [1 - \exp(-U_R \cdot 0.03)]$$

$$U_R = -\ln(0.8)/0.03 \approx 7.4$$

Thus, the equation of calculation of nervous system pathology, depending on dose will become

$$R = [1 - \exp(-7.4 \cdot C)]$$

The modern toxicological investigations point, that U_R also is function of concentration and for calculation of risk by more exact there would be usage of the equation

$$R = a - b \cdot \exp(-U_R \cdot C) \tag{4}$$

The estimation of pollution is carried out according to multiplicity of overflow of (MPC) which is determined from the formula [8]

$$K = \frac{C_M}{MPC} \tag{5}$$

here C_M is the maximum pollutants concentration in the environment. The biological effect with less favorable class of risk is calculated:

$$C_{np} = C_1 + C_2 \frac{\mathrm{MPC}_1}{\mathrm{MPC}_2} + ... + C_n \frac{\mathrm{MPC}_1}{\mathrm{MPC}_n} \quad (6)$$

The estimation of a water contamination level for a combination of substances is carried out on C_{pr}. From experimental investigations we draw the graph "concentration-effect" which on a logarithmic probit grid is approximated by a straight line. Further the tangent of tilt angle P_a is determined. The equation of linear dependence $Y = a + b \cdot X$ can be present as:

$$P_{rob} = \lg (EC_0) + \mathrm{tg}\, \alpha \cdot \lg C \quad (7)$$

here P_{rob} is effect probability (R), expressed in probits, according to the equation of normal probability distribution:

$$R = \frac{1}{\sqrt{2\pi}} \int_{-\infty}^{P_{rob}} \exp\left(-\frac{t^2}{2}\right) dt \quad (8)$$

here C is a contaminant concentration, EC_0 is a contaminant concentration with effect accepted as 0. The information on their average annual concentrations is used for calculation of the effects coupled to a long-term exposure of contaminants. The contamination level is calculated in view of overflow order of average annual MPC_{aa} of substance. Average annual MPC_{aa} is calculated from average daily MPC_{ad} under formula [8]

$$\mathrm{MPC}_{aa} = a\ \mathrm{MPC}_{ad}$$

Values of coefficient a for various pollutants are shown in Table 1.

Table 1. Coefficient a for various pollutants

Concentrations	Coefficient a
Ammonia, nitrogen oxide and dioxide, benzol, manganese dioxide, sulfur dioxide, synthetic bituminous acids, phenol	1
Trichloroethylene	0.4
Amines, aniline, suspended matter, chlorine	0.34
Sulphuric acid, phosphoric anhydride, fluorides	0.3
Acetaldehyde, acetone, toluene, ethylbenzene	0.2
Acrolein	0.1

For estimate a probable time of toxic effects beginning from the stored total dose at an estimation of overflow order of MPC, we use the following equation:

$$\lg T = \lg T_0 - \lg\left(\frac{C^b}{\mathrm{MPC}_T}\right) \quad (9)$$

here T is a probable time of toxic effects beginning; T_0 is an estimated time of assured ($p<0.05$) absence of toxic effect (this period is equal 25 years in medical–ecological investigations). C is an averaged concentration of toxic contaminants; MPC_T is the hygienic regulation; b is the coefficient considering singularities of toxicological properties of substance. In Table 2 the values of recommended coefficients b depending on exposure of contamination are shown [9].

Table 2. The values of coefficients b depending on exposure of contamination

Period of averaging	Coefficients b at danger class			
	1	2	3	4
20–30 min.	1.36	1.08	1.00	0.95
24 h	1.37	1.11	1.00	0.93
1 month	1.57	1.16	1.00	0.91
1 year	2.35	1.28	1.00	0.87
At extrapolation for 25 years	2.40	1.31	1.00	0.86
At concentrations < MPC	1.00			

3. DISCUSSIONS

At investigations on a regulation of harmful substances in the populated places, at choice T_0 it is necessary to consider, that the duration of round-the-clock exposure of experimental animals normally is 10–15% of a lifetime of laboratory animals. The estimation of a toxicological kinetics of a greater part of the regulated chemical compounds and comparative physiology of laboratory animals and human shows that in medical–ecological investigations this period should be fixing as 25 years. Let consider of toxic effects hazard as a result of chronic impact of contaminations. Chronic impact of a contaminant during human life in concentration in the order MPC can stimulate toxic effect with probability not above 5%, while impact in the order threshold concentration—with probability not less than 95%. In this case the Equation 9 can be presented in two alternatives:

$$\lg T_5 = 4 - \lg\left(\frac{C^b}{\mathrm{MAC}_T}\right) \quad (10)$$

$$\lg T_{95} = 4 - \lg\left(\frac{C^b}{\mathrm{MAC}_T} K\right) \quad (11)$$

here T_5 – is a period for beginning evolution of chronic intoxication with probability 5%; T_{95} – is a period for beginning evolution of chronic intoxication with probability 95%. For practical application we offer next equation of risk calculation:

$$P_{rob} = \frac{3.3 \cdot (\lg T_{95} - \lg T_5)}{2 \cdot \left[(\lg T_{95})^2 + (\lg T_5)^2\right]^2 - (\lg T_{95} + \lg T_5)^2} \cdot \left(\lg T_l - \frac{\lg T_{95} - \lg T_5}{2}\right) \quad (12)$$

here T_l is lifetime of human at established conditions, and P_{rob} is coupled with risk in according with Equation 8. At choice T_l it is necessary to consider, that in medical–ecological investigations average time of human life is assumed as 70 years or 25,000 days. At usage of other time period this value should be improved according to concrete requirements.

4. CONCLUSION

The toxicology of a human environment is based on the assumption that toxicity of pollutants is directly proportional of their concentration in an environment (in wastewater, an aerosphere, in soil). Besides this, the more duration of contact of pollutants with biological object, the probability of toxic effect is more, and the risk for human health is more. Hence, in this paper the calculation method of probability of effect (risk) of toxic pollutants impact on the population health is described. The algorithm of risk assessment of pollution impact on a population health is considered. For assessment of risk it is offered to use the equation, where the risk factor (coefficient of a ratio of risk increase) is function of concentration of toxic substances in an environment: atmosphere, wastewater, soil.

Acknowledgment: We express our acknowledgments to NATO-Science Program for support of participation of the author in the Workshop.

5. REFERENCES

1. Duncan Mara Wastewater. Use in Agriculture: Microsoft Producer presentations, 2006, http://www.personal.leeds.ac.uk/~cen6ddm/Reuse.html University of Leeds.
2. Water quality - Guidelines, standards and health: Assessment of risk and risk management for water-related infectious disease. Edited by Lorna Fewtrell and Jamie Bartram. (2001). IWA Publishing, London, UK.
3. Eisenberg JN, Seto EYW, Olivieri AW, Spear RC. (1996) Quantifying water pathogen risk in an epidemiological framework. Risk Analysis 16, 549–563.

4. Haas CN, Rose JB, Gerba C, Regli S. (1993). Risk assessment of virus in drinking water. Risk Analysis 13(5), 545–552.
5. Haas CN, Rose JB, Gerba CP. (1999). Quantitative Microbial Risk Assessment. Wiley, New York.
6. Macler BA, Regli S. (1993). Use of microbial risk assessment in setting United States drinking water standards. International Journal of Food Microbiology 18(4), 245–256.
7. Belyaev LS. (1978). Solving the Complex Optimization Tasks in Indeterminacy options. Nauka, Novosibirsk.
8. Engineering ecology. Edited by V.T. Medvedev. (2002).
9. Wark K, Warner CP. (1981). Air Pollution: Its Origin and Control. 2nd Ed. Harper & Row, New York.

THERMAL TREATMENT OF WASTEWATER FROM CHEESE PRODUCTION IN TURKEY

Lutz B. Giese[1], Alper Baba[2], and Asaf Pekdeğer[3]
[1] Geothermia Geochimica GGG, Waltersdorfer Str. 54, D-12526 Berlin, Germany
[2] ÇOMÜ, Department of Geological Engineering, Terzioğlu Campus, 17020 Çanakkale, Turkey
[3] FU Berlin, Ins. f. Geologische Wiss., Malteserstr. 74-100, D-12249 Berlin, Germany

Corresponding author: geotherm_lg@yahoo.de

Abstract:

During the last years, several cheese factories had been established in western Anatolia. Due to this positive industrial development, pollution from wastewater discharge has threatened the local environment. On the other hand, this area especially owns a significant renewable energy potential and thereby geothermal energy. These environmentally sound sources are to be applied in thermal wastewater treatment process.

Keywords: Wet oxidation, deep-shaft reactor, thermal treatment, cheese wastewater, geothermal energy.

1. INTRODUCTION

Recently some cheese factories were established in the Ezine district. Most of these factories discharge untreated wastewater and thus affect groundwater and soil negatively. So, an advanced treatment system is planned to be installed and could be driven by the environmental-friendly and renewable geothermal energy. This study is to prove how to apply the geothermal heat in such process to treat cheese wastewater in order to reuse it. It is expected to contribute to improve the local economy and to initiate environmental progress to the region.

1.1. Subjects of Investigation

The Tuzla field in Biga Peninsula is promising due to the direct use of geothermal heat in various industrial applications. The Tuzla geothermal basin is approximately 10 km to 5 km large and is located 80 km from Çanakkale. The geothermal

M.K. Zaidi (ed.), Wastewater Reuse–Risk Assessment, Decision-Making and Environmental Security, 347–355.

reservoir is liquid base and the temperature amounts to 173°C. These conditions tend to provide best conditions that those geothermal resources could be applied for thermal treatment of industrial wastewater.

There are several concepts for thermal treatment of wastewater and sewage sludge by wet oxidizing the organic matter at 150–300°C and under a pressure of 10–220 bar, some process super-criticallity up to 600°C and 250 bar. During such processes named as LOPROX, WETOX, or KENOX the proteins, lipids, and carbohydrates are eliminated or transformed by thermal decay, hydrolysis, and oxidizing. The residence time in the reactor amounts to 1–100 min under addition of air or pure oxygen. However, formerly Mannesmann offered the VerTech technology at 150–280°C and 100 bar. As subsurface reactor, a geothermal borehole based system is used called as deep shaft.

2. RESULTS

2.1. Geothermal Heat

2.1.1. Application of geothermal heat

Renewable energy source (RES) is sustainable and environmentally friendly. The economical feasibility of thermal applications decreases vs increasing distance. The applications of geothermal energy are

- District heating systems
- Industrial applications
  - Process heat, e.g., drying, treatment processes, cooling
- Agricultural utilization
  - e.g., Greenhouse heating
- Food production processes
  - e.g., Cooking, drying
- Fish farming

A good overview about geothermal technology and especially technical aspects of thermal applications [1, 2]. Furthermore, geothermal energy and fluids can be used to produce electricity and substances such as carbon dioxide and salts. Lindal defined for the geothermal energy the hierarchy of applications due to the process temperatures and also specified industrial applications [3–5].

- Saturated steam (140–180°C) for conventional electric generation or thermal processes
- Saturated steam (100–140°C) for binary fluid generation or thermal processes

- Hot water (20–120°C) for heating and thermal processes
- Hot water (20–60°C) for heating with heat pumps

Following the Lindal diagram, there is an utilization possible as (Figure 1):

Due to the reservoir temperature, geothermal fields can be distinguished as

- High enthalpy (>160°C),
- Low enthalpy (<160°C), or
- Shallow or very low-enthalpy (mostly utilized with heat pumps).

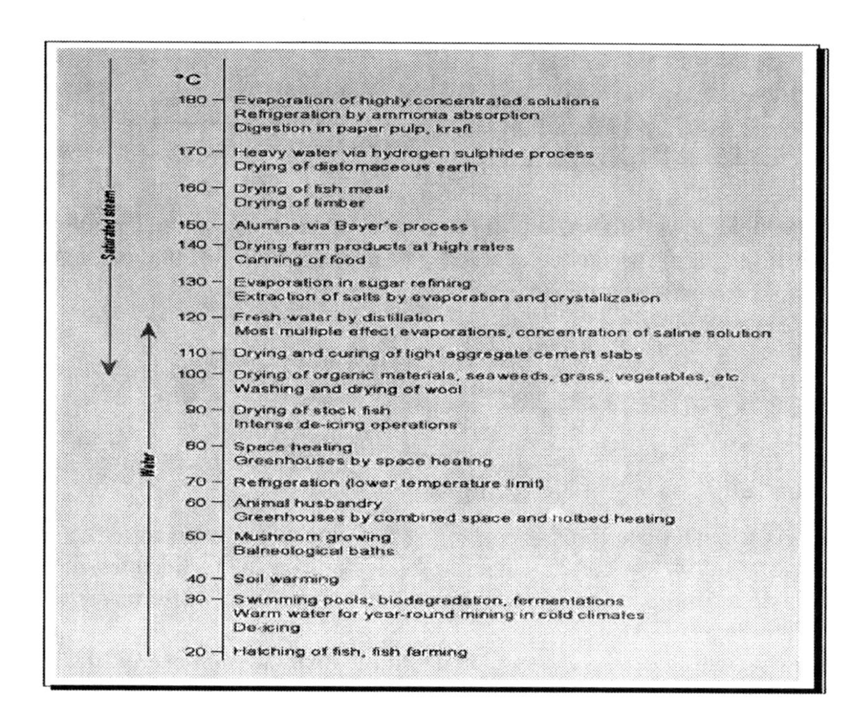

Figure 1. Lindal diagram for the application of geothermal energy

Usually, the geothermal gradient (grad *T*) and the terrestric heat-flow density (HFD) amount to 33 K/km and 65 mW/m^2, respectively.

$$\text{grad T} = \frac{dT}{dx} = \frac{T_2 - T_1}{x_2 - x_1} \qquad (1)$$

$$\mathrm{HFD} = \frac{\mathrm{d}Q}{\mathrm{d}t * \mathrm{A}} \tag{2}$$

Due to the geotectonic activity, the values are drastically increased, e.g., in Kizildere/Denizli up to 200 K/km and 120 mW/m^2. For hydrothermal systems, the extractable power P_{th} can be calculated using the steam table, furthermore the rate of adiabatic steam extraction can be received (Δh (H_2O,l): specific enthalpy of the liquid water; Δh_{b}: boiling enthalpy; data: Koglin [6]:

$$P_{\mathrm{th}} = \frac{\mathrm{d}m}{\mathrm{d}t} * \Delta h(\mathrm{H_2O,l},T_{\mathrm{upper}}) - \Delta h(\mathrm{H_2O,l},T_{\mathrm{lower}}) \tag{3}$$

$$X_{\mathrm{b}} = 100\% * \frac{\Delta h(\mathrm{H_2O,l},T_{\mathrm{upper}}) - \Delta h(\mathrm{H_2O,l},T_{\mathrm{lower}})}{\Delta h_{\mathrm{b}}(T_{\mathrm{lower}})} \tag{4}$$

The extractable power from geo heat probes can be estimated using the following formula (l: length of the probe; λ: specific heat conductivity of the rocks; ΔT/ ln (r_e/r_i): radial temperature gradient):

$$P_{\mathrm{th}} = \frac{2\pi * \mathrm{l} * \lambda * \Delta T}{\ln (r_e/r_i)} \tag{5}$$

2.1.2. Geothermal energy in Western Anatolia

Anatolia is a microplate moving toward W/SW and this movement is called as tectonic escape [7]. As consequence of the tectonic activity, magmatism, high heat-flow density, and active hydrothermal systems appear in Western Anatolia.

Turkey owns a high geothermal potential and by 1990, 30 high-energy fields had been discovered, up to 350 production wells had been drilled. In the same year, the installed electrical power amounted to 20 MW$_e$ in Turkey. In high-enthalpy geothermal fields as Germencik, Kizildere, and Salavatli with reservoir temperatures of 200–260°C, electric generation is likely, but technical problems such as scaling appears [8]. Geothermal fields with respect to their liquid reservoir temperatures around 160°C can be called as middle-enthalpy fields. Nevertheless, those reservoir temperatures easily enable thermal industrial process such as wet oxidizing process, e.g., the Tuzla geothermal field with middle-to-high-enthalpy conditions.

2.2. Thermal Treatment of Wastewater and Sewage Slugde

2.2.1. Thermal wet oxidation processes and reactors

Sewage sludge received from such process as well as wastewater directly can be treated thermally using the wet oxidation process. Table 1 shows an overview about some typical process and processing conditions including onsite and deep-shaft technologies. While the first can be suitable under conventional conditions, the latter offer specifically under limited space or in geothermal environment best results. Furthermore, deep shafts show a progressive *p*/*T* gradient downhole by nature.

Usually, the temperature and pressure range from under-critical state at 150°C and 10 bar up to the super-critical state at 600°C and more than 250 bar. Depending on the conditions, the residence time varies from 1 to 100 min. As oxidizer, air oxygen, pure oxygen, or liquid oxygen can be applied. Usually, the quality of the gas is 90–99% technical oxygen. The injection of oxygen gas into the reactor systems requires a specific injection probe and a compressor system such as cryo-pumps.

Table 1. Wet oxidation processes and their properties

Process	Temperature	Residence time	Pressure	Oxidant
	°C	min	MPa	–
LOPROX	190	90	1.7	Pure O_2
MODAR	600	1	25	Air O_2
Osaka Gas	250	8.4	9	Air O_2
OXIDYNE 1	288	55	13.8	Pure O_2
OXIDYNE 2	455	55	26	Pure O_2
VerTech	280	50	10.4	Pure O_2
WETOX	237	45	4.2	No data
ZIMPRO	260	96	9 to 12	Air O_2
KENOX	250	30	5	O_2
Stigsnäs	300	60	11	Liquid O_2

The thermochemical reaction consists of three steps, which are exothermic:

- Thermolysis
- Hydrolysis
- Oxidation

The specific reaction steps will be explained for the deep-shaft reactor process VerTech in Section 2.2.2. While the reaction is exothermic, the specific reaction enthalpy can be estimated using the dry residue of the sludge or wastewater, or better their chemical oxygen demand (COD):

$$\Delta h_{\text{reaction}} = \frac{13\ \text{MJ}}{\text{kg COD}} = \frac{3.6\ \text{kWh}}{\text{kg COD}} \tag{6}$$

The exceed enthalpy from the reactor can be used for

- Generation of electricity
- Process heat
  - Drying e.g., of residues
  - Preheating e.g., of biogas fermentation
  - Stripping of ammoniac
- District heating
- Greenhouse heating

2.2.2. Deep-shaft reactors

The deep-shaft process mainly is dedicated to treat sludge in a thermal wet oxidizing process. According to Mazumder and Dikshit [9], while the deep-shaft configuration increases the partial pressure of oxygen, thereby causing a high saturation concentration in the reactor, higher DOC rates can be treated. Furthermore, low energy and area requirements are resulting in lowered overall cost of treatment. The deep-shaft technology had been applied to treat polluted wastewater, as well as on wastewater containing toxic or slowly biodegradable pollutants. Figure 2 shows a deep-shaft system. Mannesmann formally offered the VerTech process [10].

During the main reaction (wet oxidation process), the organics decay by reaction with oxygen and form carbon dioxide and water. Nitrogen is converted into ammoniac and has to be removed later.

Wet Oxidation

$$C_5H_7O_2N + 5\ O_2 \rightarrow 5\ CO_2 + 2\ H_2O + NH_3 + \text{Energy} \tag{7}$$

Due to thermal conditions, up to 3% CO can be formed. By catalytic gas treatment, the carbon monoxide can be removed.

Removal of CO

$$CO + \tfrac{1}{2}\ O_2 \rightarrow CO_2 + \text{Energy} \tag{8}$$

Remained or by decay formed hydrocarbons are removed by catalytic reaction, as well.

Removal of Hydrocarbons

$$C_xH_y + (x+y/4)\ O_2 \rightarrow x\ CO_2 + y/2\ H_2O + \text{Energy} \tag{9}$$

Within the last part of the treatment process, nitrogen derived from the wet processing part which had been converted into ammoniac or ammonium, respectively, has to be converted into nitrate by oxidation. Within the following procedure, nitrate is turned into nitrogen gas by reduction with carbon.

Nitrification

$$2\ NH_4^+ + 3\ O_2 \rightarrow 2\ NO_2^- + 4\ H^+ + 2\ H_2O \tag{10}$$

$$H^+ + HCO_3^- \leftrightarrow H_2CO_3 \leftrightarrow H_2O + CO_2 \tag{11}$$

$$2\ NO_2^- + O_2 \rightarrow 2\ NO_3^- \tag{12}$$

Denitrification

$$2\ NO_3^- + 2\ H^+ + 5/2\ C \rightarrow 2\ N_2\uparrow + H_2O + 5/2\ CO_2 \tag{13}$$

According to Mannesmann Anlagenbau AG [10], the deep-shaft process VerTech has the following requirements:

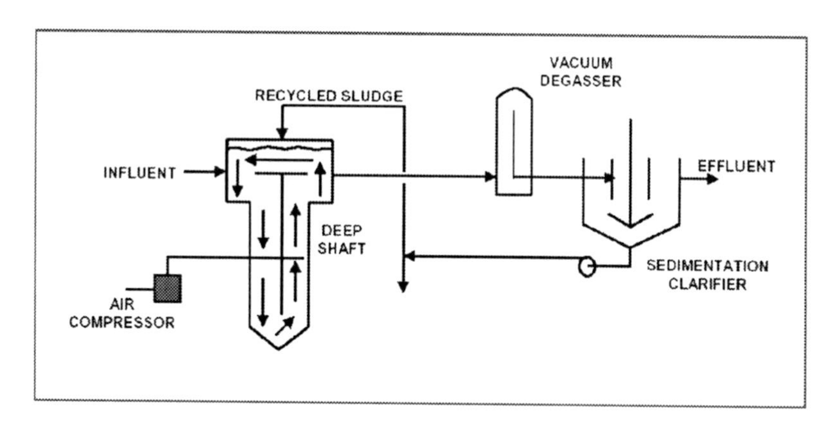

Figure 2. Deep-shaft system [11]

- Sludge dilution or wastewater concentration with preferably 5 wt% content of dry substance
- Grain diameter of solids of ≤5 mm, by conditioning usually 3–5 mm
- Adjustment of a sufficient content of dry substance of 4–6 wt%, or COD of 30 to 45 g/l, respectively
- Required reactor temperature (geothermal heat plus reaction enthalpy) of up 280°C (internal cooling)
- Reactor pressure 8–10 MPa, hydrostatically same as 1,000 m depth

3. CONCLUSIONS

The cheese factories discharge untreated wastewater and thus affect groundwater and soil negatively. An advanced treatment systems in planned to be installed and would be driven by the environmentally friendly and renewable geothermal energy. This process is to apply the geothermal heat to treat cheese wastewater in order to reuse it. It is expected to contribute to improve the local economy and to initiate environmental progress.

Acknowledgments: The authors are indebted to thank very much all colleagues from ÇOMÜ, Canakkale and FU, Berlin supporting this study. Furthermore, authors would like to thank very much the German Academic Exchange Service DAAD and NATO for providing financial support to present this paper.

4. REFERENCES

1. Dickson MH, Fanelli M. (Ed.) (1995). Geothermal Energy, 1. ed., Wiley, NY, USA.
2. Schneider D, Kraft I. (1991). Die obertägige Verfahrenstechnik geothermischer Heizzentralen - Die Einbindung Geothermischer Energie in Wärmeversorgungsaufgaben. In: Geothermie - Wärme aus der Erde - Technologie - Konzepte - Projekte, W. Bussmann, F. Kabus & P. Seibt (Ed.), 1. ed. Muller, Karlsruhe. pp. 138–154.
3. Lindal B. (1973). Industrial and other applications of geothermal energy, except power production and district heating. In: Geothermal energy, H.C.H. Armstead (Ed.). Earth Sci., 12, pp. 135–148.
4. Lindal B. (1992). Review of industrial applications of geothermal energy on future considerations. Geothermics, 21 (5/6), pp. 591–604.
5. Culver G. (1992). Industrial apps research and current industrial applications of geothermal energy in the USA. Geothermics, 21 (5/6), pp. 605–616.
6. Koglin W. (1954). Kurzes Handbuch der Chemie - Die Eigenschaften der Elemente und Verbindungen, V3/4. Vandenhoeck/Ruprecht,Gottingen. 1805.
7. Brinkmann R. (1976). Geology of Turkey, Elsevier, Amsterdam. pp. 157.

8. Giese LB, Pekdeger A, Dahms E. (1998). Thermal fluids and scaling in geothermal plants. In: Electricity production from geothermal energy. V.H. Forjaz, Proc. Int. Geothermal Days Azores 1998/Portugal, pp. 8.1–8.16.
9. Mazumder D, Dikshit AK. (2002). Applications of the deep-shaft activated sludge process in wastewater treatment. Int. J. Env. Poll., 17 (3), pp. 266–272.
10. Mannesmann Anlagenbau AG. (1995). VerTech – Ein neues, umweltverträgliches und wirtschaftliches. Presentation material, Düsseldorf. 17.
11. Aker Kvaerner Engineering Services Ltd. (2005). Deep Shaft Process Technology. Presentation Material, Stockton. pp. 2.

ENVIRONMENTAL ASPECTS OF WASTEWATER REUSE

Mohammed K. Zaidi
Idaho State University, College of Engineering
Pocatello, ID 83209-8060, USA

Corresponding author: zaidimk@gmail.com

Abstract:

Wastewater has a high potential for reuse in agriculture, especially in arid and semiarid areas. Wastewater reuse increases food production by providing a stable supply of water, and contributes to environmental security by reducing the pollution level of rivers and surface water. It also conserves a significant portion of river basin waters and allows for the disposal of municipal wastewater in a low-cost, sanitary way. A very small percentage of wastewater is currently being used for irrigation in the world. Wastewater collection, its treatment, disposal of waste, and use of treated wastewater is a big challenge to environmental security. This paper discusses the latest research developments in the field of wastewater reuse and the factors affecting environmental security.

Keywords: Wastewater, storage, process, environment and security.

1. INTRODUCTION

Humans have always realized the importance of water to life have given top priority to using it wisely. Benjamin Franklin said that "we know the worth of water when the well is dry." A twelfth century king of Sri Lanka, Parakrama Bahu, told his nation that "let not even a small quantity of water go to the sea, without benefiting the man" [1]. Almighty God revealed in Quran that "old civilizations vanished due to unwise use of water", "and we have distributed the (water) among them in order that they may celebrate our praises but most humans are averse but be ingratitude" [2]. The world's supply of freshwater is slowly running dry. Forty percent of the world's population is already reeling under the problem of water scarcity. According to United Nations, water is one of the most serious crises facing the world and things are only getting worse.

Uzbekistan, Kazakhstan, Chile, Mexico, Paraguay, Argentina, Peru, Brazil, parts of China, Middle East, and more than 25 countries of Africa are all suffering from varying degrees of desertification. In India, Pakistan, and some African and

M.K. Zaidi (ed.), Wastewater Reuse–Risk Assessment, Decision-Making and Environmental Security, 357–366.

Mediterranean countries, a major part of the country is desert, leading to a great shortage of water [3–6]. Many believe that the next wars will be fought on water issues. Even though the world is two-thirds water, most of it is not potable, and much of it is not usable for any other purpose as well. We are busy consuming and contaminating whatever is left of it, as if it were a nondepletable resource. A keen analysis of the increasing water shortage forces us to identify ways to make the best use of water, an increasingly scarce resource, by recovering it from wastewater, either for direct reuse or for aquifer recharge.

The use of urban wastewater or sewage in agriculture has a long history and is receiving renewed attention in the light of increased global water scarcity [7]. Wastewater is currently used to irrigate agricultural crops in Middle East, North and South Africa, Central and South America, Asia, Australia, and even in parts of Europe [8]. Countries and regions in which water reuse is on the rise include the US, Western Europe, Australia, and Israel [9]. South Australians use treated wastewater from various wastewater treatment plants for the irrigation of vineyards, trees to produce wood for houses and furniture, pasture, horticultural crops, parks, and golf courses [10]. In the USA, treated wastewater use is considered more suitable for irrigation compared to other uses [11].

Wastewater has always been used for irrigation, because its nutrient value promotes plant growth, and as a partial solution for water scarcity, especially in arid and semiarid regions. Comparative studies between farmers using wastewater and farmers using freshwater have shown that the former make a higher profit, because they spend less on fertilizers, and because the reliable wastewater supply allows them to grow short-cycle cash crops [12–14]. Farmers recognize the nutrient value but are also aware that the wastewater may be too ''concentrated'' at certain times for direct application. Dilution with groundwater or canal irrigation water is then carried out. Alternatively, when the original source groundwater or canal water becomes scarce or too expensive, farmers turn to wastewater, first using it conjunctively and then, in some cases, turning exclusively towards it.

In arid environments with scarce water resources, wastewater reuse represents a substantial percentage of the total resources of water allocated to irrigation [15]. In areas where all available water resources have been utilized, as for example in the Middle Eastern countries, wastewater can, and often is, recycled under planned conditions, and used for agriculture where it represent as much as 70% of irrigation water use [16]. Shortage of water in arid and semiarid areas of the world is so great that the governments in these areas have to think about developing methods to save and reuse the wastewater.

Farmers prefer to rely on freshwater as it is usually supplied at a low cost, but when they do not have an alternative they use untreated wastewater for irrigation. The microbial pathogens and chemicals in untreated wastewater are associated with severe risks to human health. The specific risks vary depending on the type of recycled water, the type of pathogen or chemical present and their numbers or concentration, the water treatment employed, and the resulting use of the recycled water [17]. For wastewater reuse to become a valuable part of the water cycle in both urban and rural communities, sensible regulations, improved health risk assessment and management, and adoption of economically viable treatment processes are essential.

Such systems have been installed in various parts of the world, especially in the developed world. Wastewater reuse can be planned through specifically designed projects to treat, store, convey, and distribute for irrigation. This planned reuse can be found in Israel (Middle East) and Tunisia (North Africa). However, in many countries wastewater is just released to water channels and reused for irrigating agricultural crops. The major unplanned reuse of wastewater is being practiced in Jordan, Gaza and West Bank, Syria, Yemen (Middle East), Algeria, and Morocco (North Africa) where most of the wastewater is untreated. In most countries of the region, wastewater treatment plants are not properly operated and maintained, making wastewater unsuitable for unrestricted irrigation even where it has passed through a treatment plant.

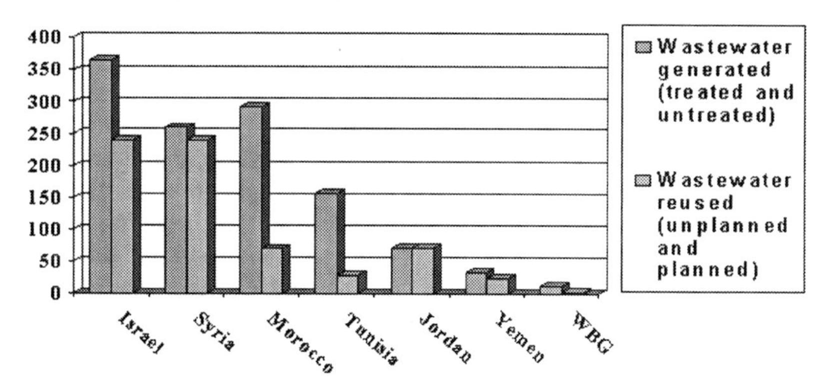

Figure 1. Wastewater generation and reuse in M.E countries (MCM) *Source*: [18]

The wastewater reuse table gives an overview of the quantities in million cubic meters per year of wastewater discharged through the sewerage network and the quantities that are being reused in seven countries [18]. Figure 1, below, illustrates the results. Israel, Morocco, and Syria generate the largest quantities of wastewater; Syria, Jordan, Yemen, and Israel reuse most of their wastewater. The quantities that are not being reused are directly or indirectly discharged into the sea or evaporate from streams and reservoirs. In rural and peri-urban areas of most developing countries, the use of sewage and wastewater for irrigation is a common practice and often is the only source for irrigation in these areas. Even in areas where other water sources exist, small farmers often prefer wastewater because its high nutrient content reduces or even eliminates the need for expensive chemical fertilizers.

2. ENVIRONMENTAL RISKS AND BENEFITS

Concerns for human health and the environment are the most important constraints in the wastewater reuse. While the risks do need to be carefully considered, the importance of this practice for the livelihoods of countless smallholders must also be taken into account. The aim of wastewater treatment and its reuse is to maximize the benefits to the poor who depend on this resource while minimizing the risks. Recently, the International Water Management Institute (IWMI) successfully completed a project on wastewater treatment and reuse in Guanajuato river basin, Mexico, whose mission was to reduce poverty and increase food security and environmental conservation [19].

In developing countries wastewater is discharged, often untreated, into streams, rivers, or lakes. The accumulation of nutrients and pollutants causes eutrophication, leading to the loss of aquatic life. This process results in an increased risk for those who use the polluted water for domestic use, including drinking. The untreated wastewater may also cause groundwater contamination.

Wastewater irrigation sites fall into three categories: (1) Unrestricted access sites or simply unrestricted irrigation is used to irrigate vineyards, trees used for wood production for houses and furniture, pasture, horticultural crops, parks, and golf courses; (2) Restricted access sites or simply the restricted irrigation is used for irrigation in fenced or isolated woodlands or meadows or to grow food products, except for salad crops and vegetable that may be eaten uncooked; and (3) Agricultural sites—areas where nonhuman food crops are grown.

When selecting a site for irrigation with wastewater, cost trade-off factors should be considered. These factors include the cost of the land and the cost of transporting the wastewater to the site. In addition, pumping treated wastewater to an upland site costs more than allowing gravity flow to a lowland site; and the cost of a stream crossing to reach a possible site may outweigh the higher land costs nearby. Restricted access and remote sites have lower treatment requirements, resulting in lower treatment costs [20].

The Tashkent Research Institute "VODGEO" presents a technology for treatment of wastewater from sugar refineries. Their offered purification technology includes accumulation of sewage in the sectional tank and biological purification in the air-tank-mixer; to improve characteristics of active silt, the regenerator is used in this technology. Recycling active silt gets to the regenerator and separation of active silt proceeds in the thin-layer precipitation tank. Deep after purification of biologically purified sewage proceeds in bioponds with superior water flora. Their technology makes it possible to get required norms during the chop of purified sewage into the superficial receiving waters [21].

Wastewater collection, its treatment, disposal of waste (biosolids), and use of treated water is a big challenge to environmental security due to the fact that there are so many variables and no two wastewater streams are identical. There will always be a need for continuous studies on wastewater availability, and treatment needs to be fine-tuned after determining the composition of the raw wastewater. For example, in Idaho Falls, a small city in Idaho with a population of 50,000 people, a wastewater collection plant of about 800,000 gallons per day is being designed and built. The total per capita water use in Idaho is 22,000 gallons daily; most of the water is used for agriculture and fish farming (97%), and the rest is used for industrial and mining (2%) and for domestic and commercial purposes (1%). For comparison, the average per capita water use in the United States is 1,408 gallons daily [22].

The high cost of plant construction and operation in developing countries receives less attention from the service providers. Landscaping is needed to improve the plant's appearance, and safe working conditions for the employees should be ensured. The consideration of noise, odor, and the location of the plant are serious issues to be resolved before such a plant is designed and installed. To avoid odor associated with sepsis, adequate volumes of air are required to the basin for mixing.

3. WASTEWATER TREATMENT PROCESSES

Typically, sewage treatment involves three stages, called primary, secondary, and tertiary treatment. First, the solids are separated from the wastewater stream. Then the dissolved biological matter is progressively converted into a solid mass by using indigenous, waterborne bacteria. Finally, the biological solids are neutralized, then disposed of or reused, and the treated water may be disinfected chemically or physically (for example by lagooning and microfiltration). The final effluent can be discharged into a stream, river, bay, lagoon, or wetland, or it can be used for the irrigation purposes. If it is sufficiently clean, it can also be used for groundwater recharge.

The primary treatment is the first stage and consists of coarse screening, fine screening, and grit removal. It undergoes oxygen treatment, which stabilize solids, reduce pathogens and biomass quantity by partial destruction and produces methane (CH_4) as a by-product. It undergoes hydrolysis formation of soluble organic compounds and lastly methane is formed as reusable gas. In some cases, phosphorous removal is needed. In order to prevent sludge from settling in the wastewater in the corners and bottom of the tank, a mixing method is used. The mechanical method needs too much maintenance. Therefore, the gas-circulation method is commonly used; this method is the most economical and free from routine maintenance cost.

Secondary treatment is designed to substantially degrade the biological content of the sewage, including human waste, food waste, soaps, and detergents. The majority of municipal and industrial plants use aerobic biological processes to treat the settled sewage. For this to be effective, the biota requires both oxygen and a substrate on which to live. This can be done in a number of ways, all based on the same principles: the bacteria and protozoa consume biodegradable soluble organic contaminants (sugars, fats, organic short-chain carbon molecules, etc.) and bind much of the less-soluble fractions into floc particles. Secondary treatment systems are classified as fixed film or suspended growth systems. In fixed film systems—such as rock filters—the biomass grows on media and the sewage passes over its surface. In suspended growth systems—such as activated sludge—the biomass is well mixed with the sewage.

Tertiary treatment is the final stage needed to raise the effluent quality to the standard required, before it is discharged to the receiving environment (sea, river, lake, ground, etc.). The major aim of this treatment in most of the treatment plants is to remove excessive nutrients, especially nitrogen and phosphorus. The removal of nitrogen is effected through the biological oxidation of nitrogen from ammonia (nitrification) to nitrate followed by denitrification, the reduction of nitrate to nitrogen gas. Nitrogen gas is released to the atmosphere and thus removed from the water. Phosphorus can be removed biologically in a process called enhanced biological phosphorus removal. In this process, specific bacteria, called polyphosphate accumulating organisms, are selectively enriched and accumulate large quantities of phosphorus within their cells (up to 20% of their mass). When the biomass enriched in these bacteria is separated from the treated water, these biosolids have a high fertilizer value. Phosphorus removal can also be achieved by chemical precipitation usually with salts of iron ($FeCl_3$) aluminum (alum).

More than one tertiary treatment process may be used at any treatment plant. If disinfection is practiced, it is always the final process. It is also called "effluent polishing."

4. AGRICULTURAL USE OF WASTEWATER

Vegetables and other crops where wastewater is used for irrigation are at risk of contamination from heavy metals and waste products. Studies have reported high heavy metal contamination in vegetables, such as arsenic, lead, and cadmium, in Pakistan [23], Bangladesh [24], and Egypt [25]. Cadmium (Cd) is considered a potential toxin that is principally dispersed in natural and agricultural environments through anthropogenic sources. Untreated municipal sewage, often a potential source of Cd, is generally used to irrigate urban agricultural soils in many developing countries. An alarmingly high concentration of heavy metals in vegetables grown under untreated municipal sewage-irrigated soils has been reported. The metal ion concentration in municipal sewage was found to be 3-fold (0.03 mg/l), while the permissible concentration in irrigation water is (<0.01 mg/l). The metal ion was found in leaf (0.17–0.24 mg/kg fresh weight) and fruit (0.07–0.18 mg/kg fresh weight) portions of all the sampled vegetables: bitter gourd, cauliflower, eggplant, fenugreek, okra onion, pumpkin, and spinach. Leafy tissue accumulated about twice as much Cd as that of the fruit portion. These results suggest that prolonged ingestion of sewage-irrigated leafy vegetables can produce Cd levels in the human body that may cause a number of illnesses.

Apart from heavy metals, pesticides residues were also found in fruits and vegetables cultivated with untreated municipal sewage-irrigation. Beside the effects on consumer health and the environment, the presence of pesticide residues also has significant trade implications. There is no regular testing of pesticide residues on food in Pakistan. Typically in the US and UK, where comprehensive monitoring schemes exist, maximum residue limits (MRLs) for fresh fruit and vegetables are exceeded by about 1–2% of the samples. A 4-year testing program on Pakistani fruits and vegetables has shown that the MRLs are regularly exceeded [26]. Of 550 samples analyzed during the late 1980s and early 1990s for organophosphate (OPs), organochlorine (OCs), and synthetic pyrethroid insecticide residues, 214 (39%) samples contained residues. Of these 79 (14%) samples were above MRLs set by Food and Agriculture Organization (FAO). This could, under some circumstances, pose a hazard to the consumers [27].

Exceeding an MRL indicates that good agricultural practices have not been carried out regarding the use of sewage wastewater. Some relatively high levels were also observed for food/commodities for which an MRL has not yet been set. The MRLs for organochlorine, dichlorodiphenyl trichloroethane (DDT) were exceeded by 10.3 times in cauliflower, 8.6 times in cabbage, and 8.1 times in okra. The MRL for the OP methyl parathion in coriander was exceeded by 7.5 times, for malathion in onion by 9.2 times, and in beet sugar by 8.6 times. The most significant overuse involved the synthetic pyrethroids. The MRL for fenvalerate was exceeded by 20.0 times and cypermethrin was exceeded by 30.0 times in turnip and 34.3 times in okra. In view of these findings, it has become essential to have a testing and monitoring of pesticide residues in food and regulations on pesticide use in Pakistan.

5. COMMERCIAL USE OF WASTEWATER

The discharge of wastewater containing heavy metals in the water bodies has been increasing as a result of major urbanization and heavy industrialization processes. This includes arsenic, cadmium, copper, chromium, lead, mercury, and zinc that have been classified as toxic heavy metals by the US Environmental Protection Agency (USEPA) [28]. When their concentration exceeds the permissible limits in soil, they accumulate in the food chain and disrupt the biological system. The sources of heavy metals include waste from the industries such as electroplating, metal fish, metallurgical, tannery operations, chemical, mine drainage, battery, leather tanning, fertilizer, pigmat, leachates from landfills, and contaminated groundwater from hazardous waste sites [29]. Heavy metals are also emitted from resource recovery plants in relatively high levels as fly ash particles [30].

6. CONCLUSIONS

Wastewater collection, its treatment, disposal of waste sludge, and use of treated water is a big challenge to environmental security due to the fact that there are so many variables and no two wastewater streams are identical. Use of untreated wastewater in agriculture is totally unacceptable as it may cause soil and ground water contamination. In new buildings, plumbing fixtures can be designed to reuse wastewater, as in the case of using gray water from washing machines and kitchen sinks to flush toilets and irrigate lawns. Improved public education to ensure awareness of the technology and its benefits, both environmental and economic, is recommended. Health problems, such as waterborne diseases and skin irritations, may occur in people coming into direct contact with reused wastewater. Some gases and sulfuric acid, produced during the treatment process can result in chronic health problems. In some cases, wastewater reuse is not economically feasible because of the requirement for an additional distribution system. Removal of heavy metals from the wastewater will contribute to environmental safety.

Acknowledgment: I am thankful to the NATO ARW organizing committee to provide financial support to make this presentation possible. I am thankful to Dr. Tahir Rashid and Professor Nava Haruvy to make it worth presenting.

7. REFERENCES

1. Worldwatch, (2000). Worldwatch Institute 1776 Massachusetts Ave., N.W. Washington, D.C. 20036-1904 USA; http://www.worldwatch.org.
2. Holly Quran, Al-Furqan Verse 50. Translated by Ali Yussuf.
3. Ghosh, D. (1999).Wastewater utilization in East Calcutta wetlands.
4. Rashid MT. (2002). Fighting Water Shortage. Pakistan Water Gateway, http://www.waterinfo.net.pk/artfws.htm.
5. Ensink JHJ, Mahmood T, Hoek WVD, Raschid-Sally L and Amerasinghe FP. (2004b). A nationwide assessment of wastewater use in Pakistan: an obscure activity or a vitally important one? Water Policy 6: 197–206.
6. Angelakis AN, Paranychianakis NV and Tsagarakis KP. (2003). Water Recycling in the Mediterranean Region.
7. Scott CA, Faruqui NI, Raschid-Sally L. (Eds.), 2004. Wastewater Use in Irrigated Agriculture: Confronting the Livelihood and Environmental Realities. CABI Publishing, Wallingford.
8. Bastian R. (2006). The future of water reuse. BioCycle, 47(5): 25–27.
9. Miller GW. (2006). Integrated concepts in water reuse: managing global water needs. Desalination 187: 65–75.
10. Radcliffe J. (2004). Water Recycling in Australia. Australian Academy of Technological Sciences and Engineering, http://www.atse.org.au/index.php?sectionid=600
11. US Environmental Protection Agency (2001). Water Recycling and Reuse: The Environmental Benefits.
12. Van der Hoek W, Hassan M, et al. (2002). Urban Wastewater: a Valuable Resource for Agriculture, a Case Study from Haroonabad, Pakistan. Research Report 63. International Water Management Institute: Colombo, Sri Lanka.
13. Ensink JW, Van der Hoek Y, et al. (2002). Use of Untreated Wastewater in Peri-urban Agriculture in Pakistan: Risks and Opportunities. Research Report 64. International Water Management Institute: Colombo.
14. Ensink J, Simmons RW, Van der Hoek W. (2004a). Wastewater use in Pakistan: the case studies of Haroonabad and Faisalabad. In Wastewater Use in Irrigated Agriculture: Confronting the Livelihood and Environmental Realities, Scott C, Faruqui NI, Raschid-Sally L (eds). Commonwealth Agricultural Bureau International, Int. Dev. Res. Centre: Canada; 91–99.
15. Raschid-Sally L, Carr R, and Buechler S, (2005). Managing wastewater agriculture to improve livelihoods and environmental quality in poor countries. Irrig. and Drain. 54: S11–S22.
16. Abu-Zeid KA, Abdel-Megeed O. (2004). Potential for water demand management in the Arab region. Paper presented at the International Water Demand Management Conference 30 May–3 June 2004, Dead Sea, Jordan. Organized by the Ministry of Water and Irrigation, Jordan, and USAID.

17. Toze S. (2006). Water reuse and health risks—real vs. perceived. Desalination 187: 41–51.
18. Saghir J, Schiffler M and Woldu M. (2000). Urban Water and Sanitation in the Middle East and North Africa: The Way Forward.
19. International Water Management Institute (IWMI), 127, Sunil Mawatha, Pelawatte, Battaramulla, Sri Lanka. www.iwmi.cgair.org.
20. Reuse of Reclaimed Wastewater Through Irrigation, Ohio State University, Bulletin # 860.
21. Avdeeva E, Katsenovich E, Shapovalova L, Markova N, Groisman E. (2004). Rational scheme of water use and sugar-refineries sewage purification. Tashkent Research Institute "VODGEO".
22. Boyd D. (2006). Water Wars, Idaho State Journal, July 2, (2006).
23. Qadir M, Ghafoor A, Murtaza G. (2000). Cadmium concentration in vegetables grown on urban soils irrigated with untreated municipal sewage. Environ. Develop. Sustain. 2: 13–21.
24. Alam MG, Snow MET and Tanaka A. 2003. Arsenic and heavy metal contamination of vegetables grown in Samta Village, Bangladesh. Sci. Total Environ. 306: 83–96.
25. Mansour SA. (2004). Pesticide exposure – Egyptian scene. Toxicology 198: 91–115.
26. Richardson M. (ed.), (1995). Pesticide residues in foodstuffs in Pakistan: organochlorine, organophosphorus and pyrethroid insecticides in fruit and vegetables, Environmental Toxicology Assessment, Taylor & Francis, UK.
27. FAO/WHO. (1986). Joint Food Standards Program, Codex Alimentarius Commission on Pesticide Residues. Vol. 13. Food and Agriculture Organization, Rome.
28. Keith LH, Telliard WA. (1979). Priority Pollutants. Env. Sci. Tech. 13, 416–424.
29. Reed BF. et al. (1994). Rem of lead and cadmium from aqueous waste streams using Grannular activated carbon column. Environ. Prog. 13, 60–65.
30. Igwe JC, Abia AA. (2006). A bioseperation process for removing heavy metals from wastewater using biosorbants. Afr. J. Biotech, 5 (12): 1167–1179.

WASTEWATER REUSE IN ISRAEL AND IN THE WEST BANK

Sarit Shalhevet
Economic Consultant, Brookline, MA, USA

Corresponding author: sarit.shalhevet@gmail.com

Abstract:

Improved wastewater treatment in the Israel and in the West Bank is an important step towards resolving the existing problems of water shortage and pollution in the region. This will enable to increase water availability and quality in the West Bank villages, thereby increasing the social, economic, and environmental sustain ability as well as decreasing the environmental threat of pollution of the groundwater resources. This paper examines different scenarios of water supply, demand, and allocation arrangements in order to reach the optimal decisions on wastewater treatment, based on principles of sustainable development.

Keywords: Water resources, West Bank, Israel, sustainable development, environmental risks.

1. INTRODUCTION

The Middle East is mostly classified as an arid or semiarid zone, with limited water resources. Demographic and economic trends, as well as the global climate change, are exacerbating the water shortage and water pollution in the region [1]. Israel's largest water source is Lake Kinneret (the Sea of Galilee) and the three transboundary water resources that are also shared by the Israelis and the Palestinians—they are the Mountain aquifer, the Coastal aquifer, and the Jordan River basin area [2, 3]. Maintaining and developing these water resources is essential to ensure future water supply for the region's population.

1.1. The Current Situation of Water Supply in the West Bank

The natural water resources are already overexploited, and many wells in the area are already polluted. Groundwater flows from the West Bank to the groundwater on the Israeli side of the Green Line on its way to the sea. The Mountain aquifer is already heavily overpumped, and additional pumping is expected to cause intrusion of seawater into the coastal aquifer [4]. The Palestinians and the settlers have very limited sewage treatment, and the nontreated sewerage in the West Bank is contaminating the Israeli groundwater [2].

M.K. Zaidi (ed.), Wastewater Reuse–Risk Assessment, Decision-Making and Environmental Security, 367–374.

20% of the West Bank is not connected to a public water network; Btselem reports that 197,000 residents (10% of the West Bank) are not connected to a water network [5]. These residents generally purchase water from tanks or collect rainwater. In the past several years, the water supply became increasingly erratic, and residents in the smaller municipalities are increasingly resorting to water sources from dirty cisterns used for watering livestock [6].

The situation is especially severe in the southern West Bank, where 37% of the population does not have running water and 20% of the population purchase water from tankers. The restrictions of mobility resulted in an increased cost of tanker transport in some villages from about $1 up to $5 per cubic meter [6] and have caused a significant decrease in available income. Some households now spend as much as a third of their total family income on water, causing them to sell the livestock they depend on for a living. [5, 7].

1.2. The Sustainability of the Existing Water Plans

The Israeli Water Commission controls all the water management decisions in Israel and the West Bank as well, and is responsible for planning the water supply according to projected demand. The Water Commission's master plan relies to a large degree on desalination projects to fill the gap between the forecasts of demand and supply [8, 9]. Fisher developed an economic model for optimal allocation of water between Israel, Jordan, and the Palestinian Authority, and concluded that desalination should not be used before a full exploitation of all the other, lower-cost options [10].

Most publications analyzing the Israeli water supply options do not address the social aspects. Gleick [11] stresses the importance of local group participation and equity in allocation as social aspects in sustainable development of water supply.

1.3. Implications of Wastewater Treatment

Wolf and Murakami [4] grade the overall feasibility of wastewater treatment for irrigation as slightly higher than seawater desalination, with a low environmental feasibility and a high economic and political feasibility. A cost-benefit analysis shows that when the social and environmental impacts (aquifer recharge, pollution, and health risks) are taken into account, wastewater irrigation in Israel saves US$0.5 1/m^3 as compared with the alternative of river disposal [12].

The implications of wastewater desalination depend on the treatment level chosen. Wastewater can be treated for industry, irrigation, or aquifer replenishment; the treatment level required varies by use. Urban sewage treatment is essential for safe environmental disposal, and the additional marginal cost of adapting it to agricultural use is low [13]. Wastewater treated to the minimal level required for crop irrigation provides a high reliability of quantity (not necessarily quality), a relatively low cost, and a solution for wastewater disposal. This level of treatment makes it possible to use the nitrogen and phosphorous in wastewater as plant nutrients, while high-level desalination would cause the double damage of desalination and transporting nutrients. However, it causes damage to crops, soil, and groundwater, including groundwater contamination through the creation of wastewater reservoir in the shore aquifer. It increases water consumption through evaporation and increased need for leaching, and increases the demand for freshwater for mixing with wastewater. Some researchers recommend desalinating wastewater to the level of drinking water to avoid environmental damages [14], but then most of the crops would become unprofitable.

2. METHODOLOGY

The paper examined two aspects of the current situation. First, the situation is socially and economically unsustainable from the demand side, as it does not fill the current needs in the West Bank and does not address the problems of water allocation of a common water resource between Israel and the West Bank. Second, the situation is environmentally unsustainable from the supply side – the current plan assumes large amounts of seawater desalination, which consumes scarce energy sources, causes air pollution, and has an effect on increasing the global warming trend. The paper examined whether implementing demand-based management and lower-cost, lower-impact options could delay the need to desalinate seawater.

2.1. Sustainability of Water Demand

Sustainable demand should take into account future possible agreements, based on existing international approaches for allocation of water between countries with shared water resources. There are many nonbinding approaches in international law and treaties to resolving water allocation between nations with shared water resources. Some of these approaches are absolute sovereignty (a state has absolute rights to water flowing through its borders); riparian rights (any territory along a riverway has rights to an unchanged river); prior appropriation (First in time, first in right); and development of the area as a single hydrologic unit [2]. According to the Helsinki rules (1966), each state is entitled to a "reasonable and equitable share, “taking into account the alternative resources available to each side and balancing the needs of each side [3]. Other approaches include maximizing economic efficiency (the ability to achieve the highest return per unit water); rights proportional to the amount of the water source within a nation’s territory; allocation on an equal per capita basis [2]; and meeting the minimum water requirements per capita as top priority [3]. Moore [15] calculated the “alternative equity standards” to arrive at an allocation which would best meet both sides’ needs, based on international law principles. He concluded that the Palestinian Authority’s allocation should be about double the amount it actually is. Elmusa’s estimate, based on similar factors, also concluded that the Palestinians’ allocation should be at least double the present allocation [16].

This research relied on the concept of water allocation on the basis of “reasonable and equitable share,” based on the factors mentioned in the Helsinki rules [3] and the Bellagio treaty [16]. This concept incorporates geographical and hydrological data as well as needs and alternative water sources. Needs may include forecasts of specific water uses, such as industrial and agricultural uses. According to this concept, a forecast for the future sustainable demand was made as a basis for planning the water supply needs.

The Water Commission’s master plan, published in 2002, is based on a population forecast of 6.79 million in 2005 and 7.3 million in 2010. According to Israel’s Census Bureau of Statistics, Israel’s population in September 2004 was 6.81 million and the forecast is for between 7–8 million in 2010. The Water Commission estimates the Palestinian Authority’s water use in 2005 as 100 MCM (70 from Israel and 30 by self-production), which is far from meeting the equity standards. The West Bank Palestinians’ water needs in 2005 were estimated to be 144.7 MCM for domestic use, 29 MCM for industry, and 244 MCM for agriculture [17]. This forecast was based on a population forecast of 2.24 million in 2005. In practice, by 2003 the West Bank Palestinian population was 2.37 million, and including the Gaza Strip the total population was 3.7 million [18].

To estimate the Palestinian Authority’s fair water allocation, Baskin’s estimates [17] were multiplied by the actual population/estimated population, assuming that the Gaza Strip needs are in the same proportion as the West Bank needs. Baskin’s estimates are relatively low, since they are based on the actual ability to consume water rather than per capita allocation. For example, this allocation assumes a Palestinian domestic per capita annual consumption of 64.6 MC, compared with an Israeli domestic per capita annual consumption of 115.5 MC. This allocation results in a per capita consumption of 340.6 MCM for Israel (total consumption for all local uses) compared to 118.2 MCM for the Palestinians.

To simplify the forecasts, it was assumed that Palestinian consumption between 2005 and 2010 would increase at the same rate as the Israeli consumption forecasts. This is again a low estimate, since the Palestinian population rate of growth is higher than the Israeli rate.

2.2. Sustainability of Water Supply

There are many ways to reduce the water shortage. The demand-side options include water-saving accessories, reducing conveyance losses, and market-side policies. The supply side options range from simple technologies such as rainwater harvesting to large desalination plants, and up to mega-projects involving cooperation with other countries in the region.

The available water supply options for Israel, Jordan, and the Palestinian Authority were ranked by an international committee [19] on the following criteria: impact on available water supply, technical feasibility, environmental impact (including impact on other water sources, impact on biodiversity, and other adverse or beneficial environmental impacts), economics (cost/benefit analysis), and the implications for intergenerational equity (effect on the water quantity and quality available for future generations). These rankings do not include the political feasibility, which is an important consideration in the Middle East. Not all the economically and environmentally preferred projects are politically feasible, and the political price for each option must be considered as well.

Wolf and Murakami [4] analyzed a variety of water management options for Israel and the Palestinian Authority, ranking each project on its technical (water quantity, quality, and reliability), environmental, economic (investment required and economic efficiency), and political viability. The total viability was estimated quantitatively using the Delphi process relying on members of the Middle East Water Commission, and weights were given to each factor to calculate a total value for each project's overall viability.

The contribution of available water supply options to sustainable development was evaluated based on its economic, environmental, and social impacts, as well as its political viability. According to the principles of sustainable development, a project is sustainable only if it satisfies the minimum constraints on each aspect. Therefore, rather than assigning a total viability grade as Wolf and Murakami [4] have done, projects that have a low ranking on any aspect are rejected.

3. RESULTS

3.1. Sustainable Demand Based on the Equitable Share Scenario

The results of the "reasonable and equitable share" allocation scenario for 2010 are shown in Table 1. The Palestinians' allocation should be at least double their current allocation, although this is still less than an equal per capita share basis. The Palestinian Authority's water consumption should be 401 MCM more than the Israeli Water Commission's forecasts [8], and the total water consumption of Israel and the Palestinian Authority should be 372 MCM more that the commission's forecasts. This allocation results in an annual per capita consumption of 340.6 CM for Israel (115.5 MC for domestic use) and 118.2 CM for the PA (64.6 MC for domestic use).

Table 1. Consumption under the equitable share scenario (2010, in MCM)

Area	Domestic	Industry	Agriculture	Total*
Israel	886	167	1,122	2,460
PA	173	37	273	482
Total	1,059	204	1,395	2,942

*The total consumption includes water allocation to Jordan, storage capacity restoration, and nature and landscape water use, which are not shown here.

3.2. Evaluation of Water Supply Projects

The alternative water supply options were ranked on an ordinal scale, on the following aspects: economic (investment, cost/benefit), environmental (greenhouse gases, groundwater pollution, using scarce water and energy sources), social (cost to domestic consumer, cost to farmers, effect on employment in the periphery, increasing water supply to Palestinian population), and political (likelihood to receive political support). The grades (low, medium, and high) are subjective, based on the general conclusions from the analysis in previous sections. Some options were not ranked because they were rejected in the preliminary analysis.

The first priority is to adopt options that involve a low cost and a high benefit/cost ratio, low environmental damage and positive or at least no negative social effects, and have no negative political percussions. These include installing water-saving accessories and a water-saving campaign in Israel, and rainwater harvesting in the West Bank. Brackish groundwater desalination is also considered desirable from all aspects. Wastewater reuse is considered important, as it will not only serve to increase the water supply available for agriculture, but will also reduce the environmental threat of pollution of the West Bank's and Israel's groundwater resources.

The second order of priority includes investing in a water network in the West Bank. This project may not be economically feasible, but it can make an important contribution toward meeting the millennium development goal of reducing the percentage of people without access to improved water. This project might be carried out with international financial aid. The third priority is seawater desalination projects. Finally, imports by tankers should be used only to provide the amount needed after the previous measures have been exploited. Other methods of importing water and mega projects are more problematic from every aspect discussed here and should not be developed in the near future.

4. CONCLUSIONS

In Israel, it is important to implement water-saving accessories and to initiate a media campaign to save water. In the West Bank, rainwater harvesting and an improved water network are important strategies for sustainable. In general, it is recommended that the plans maintain a diversity of water supply options and technologies, and plan for conservation of water quality and conservation of access to water resources.

Acknowledgment: The paper was presented with the support of NATO Advanced Research Workshop grant.

5. REFERENCES

1. El-Fadel M, Bou-Zeid E. (2001). Climate Change and Water Resources in the Middle East: Vulnerability, Socio-Economic Impacts, and Adaptation (Report 46. 2001). Fondazione Eni Enrico Mattei.
2. Aaron T, Wolf AT. (1995). Hydropolitics along the Jordan River: Scarce Water and its Impact on the Arab–Israeli Conflict. Tokyo · New York · Paris: United Nations University Press. Appendix III: Hydronationalism.
3. Assaf K, al Khatib N, Kally E, Shuval H. (1994). Water in the Israeli–Arab conflict. Palestine-Israel Journal of Politics, Economics & Culture, 3, 11–17.
4. Wolf A, Murakami M. (1995). Techno-political decision making for water resources development. Water Resources Development, 11(2), 147–162.
5. Lein Y. (2001). Not Even a Drop: The Water Crisis in Palestinian Villages Without a Water Network. Btselem website. [online: web]. Cited Oct. 2004.
6. The World Bank. (2003). West Bank and Gaza - Emergency Water Project (Report No. T7597).

7. Palestinian Central Bureau of Statistics. (2004). Survey of Water Resources and its Effects on Households Living Conditions in Palestinian Territory, 2003. Table 1. Percent Distribution of Households in the Palestinian Territory by Water Resources and Region, 2003.
8. The Water Commission. (2002). Transitional Master Plan for Water Sector Development in the Period 2002–2010. State of Israel, Ministry of National Infrastructures, Water Commission Planning Division.
9. Mandil A. (2003). Chronology of a predictable crisis. Water Engineering, 27, 12–16. (In Hebrew).
10. Fisher FM, Askari H. (2001). Optimal Water Management in the Middle East and Other Regions. Finance & Development, 38(3) [online journal].
11. Gleick PH. (1998). Water in crisis: paths to sustainable water use. Ecological Applications, 8(3), 571–579.
12. Haruvy N. (1997). Agricultural reuse of wastewater: nation-wide cost-benefit analysis. Agriculture, Ecosystems and Environment, 66, 113–119.
13. Haruvy N. (1998). Wastewater reuse – regional and economic considerations. Resources, Conservation and Recycling, 23, 57–66.
14. Galil N, Green M, Dorortz K, Zaslavsky D, Lahav U, Friedler E, Smit R. (2002). Wastewater desalination – damage for generations. Water Engineering. 19, p. 40 (In Hebrew).
15. Moore JW. (1994). Defining national property rights to a common property resource: The West Bank aquifers. Research Energy Economics, 16, 373–390.
16. Elmusa SS. (1994). The Israeli-Palestinian water dispute can be resolved. Palestine-Israel Journal of Politics, Economics & Culture, 3, 18–26.
17. Baskin G. (1994). The clash over water: an attempt at demystification. Palestine-Israel Journal of Politics, Economics & Culture, 3, 27–35.
18. The Palestinian Central Bureau of Statistics. (2004). Survey of Water Resources and its Effects on Households Living Conditions in Palestinian Territory, 2003. Table 1. Percent Distribution of Households in the Palestinian Territory by Water Resources and Region, 2003.
19. The Committee on Sustainable Water Supplies for the Middle East. (1999). Water for the Future: The West Bank and Gaza Strip, Israel and Jordan. Washington D.C.: National Academy Press.

WASTEWATER MANAGEMENT IN EGYPT

Hussein I. Abdel-Shafy and Raouf O. Aly
Water Research & Pollution Control Department, National Research Centre, Dokki,
National Centre for Radiation Research & Technology
Nasr City, Cairo, Egypt

Corresponding author: husseinshafy@yahoo.com

Abstract:

The present work aims at highlighting the challenge encountering the environmental management of wastewater in Egypt. In addition, water consumption, industrial, domestic and agricultural pollution, and their adverse impact on water quality is discussed. The weak regularity in compliance and enforcement of the environmental legislation is evaluated. The industrial pollution is at alarming degree.

Keywords: Wastewater, Egypt, industrial discharge, municipal pollution.

1. INTRODUCTION

Population growth, industrialization, and the need for new agricultural areas and water use constitute a growing threat to the very limited water resources. The weight of water intakes could be demonstrated as follows: agriculture, drinking, and industrial are 54.4, 2.9, and 3.9 Bm^3/year, respectively. The various pollution loads led to significant water quality degradation (Table 1)

Table 1. Various pollution loads in Egypt

Parameters	Load (t/d)
BOD	235
COD	423
Oil and grease	168
TDS	296
Heavy metals	1.65

Although agriculture, industrial, and human needs essentially depend on the availability of freshwater, yet the present networks allow eventual mixing up of almost all agricultural and agrochemical drains as well as industrial and domestic wastes. Actually, the burden is too heavy to be tackled by the present water treatment plants. Accordingly, the agricultural network, irrigation, and river systems

M.K. Zaidi (ed.), Wastewater Reuse–Risk Assessment, Decision-Making and Environmental Security, 375–382.

do heavily suffer from excessive water wastes along their passage up to northern lakes and seacoast. Irrigated agriculture is by far the largest consumer of water, almost 84%. Further intensification of crop schedules and extension of agricultural areas put an increasing demand on the already scarce water resources.

The rapid increase in demand is projected to be met by using water from the drains, estimated to be 16.9 Bm^3/year and reducing the flow from drains to the sea. Even though, water in the drains is currently of poor quality due to pollution from industrial, municipal, and agricultural wastes.

Options such as increased efficiency of irrigation and changes in crop types, and reducing areas under irrigation should be seriously considered. Egypt may also have to rely on groundwater aquifers and on expensive desalinization of seawater.

1.2. Economic Dimension

The linkages between a country's national economy and natural resources and environmental sectors are very important. Land, water, energy, and natural resources are vital input into the agricultural and industrial sectors. The economy also is the major stressor of the environment, for instance, production in the industrial and agricultural sectors contributes to pollution of the air and water and the generation of solid and hazardous wastes. Consumption activities lead to the production of solid waste and create demand for wastewater treatment services.

Suitable development sets new challenges to developing countries. The desire to satisfy basic social need could very well override even basic environmental considerations. Nevertheless, a strong economy can be a key factor in the protection of the environment as profitable business that can better afford to invest in environmental protection. Higher incomes are strongly correlated with a higher demand for environmental quality. Pollution prevention and cleaner production measures deal with pollution at its source. Since pollution is regarded as escaping raw materials, chemicals, and products, any pollution minimization represents an economic gain to the facility management unless the implementation measure is not cost effective.

The Egyptian economy grew rapidly from the mid-1970s due to the "open door" policy that slightly liberalized the economy for encouraging foreign investment and the private sector, and limited the predominant roles of the state and public sector. On the other hand, there are some very large basic industries such as companies of iron, steel, coke, aluminum, textile, etc. These industries are characterized by their very high capital investments, low rate of return, high labor intensity.

Applying the current regulations to these old and basic industries seems impractical whereas closure will create other types of social and political stresses.

1.3. Water Resources and Consumptions

Through the Nile Water Agreement with Sudan in 1959 and the completion of the High Aswan Dam in 1968, a stable 55.5 Bm^3/year was allocated to Egypt (Table 2) [1]. The release from the dam ranges from approximately 800 m^3/s during the winter closure period to nearly 2,760 m^3/s during the Summer months [2]. In the Nile Delta, groundwater resource use accounts for approximately 4.4 Bm^3/year mainly being recharged from the Nile and seepage from irrigated agriculture. Rainfalls play a minor role in Egypt water resources, with average rainfall rates declining from 250 mm^3/year to 100 mm^3/year at the Mediterranean Coast and almost zero in upper Egypt [3].

1.4. Status of Sewerage and Water Supply Systems

Approximately 65% of Egypt's population is connected to drinking water supplies and only 24% to sewerage services [4]. In the rural areas, accommodating about half of the population (35 M persons), 95% of the people have no access to sewer systems or wastewater treatment facilities [3]. The rest relies on individual means of excreta and wastewater disposal such as latrines and septic tanks. However, no well-controlled sludge management program exists in Egypt [5, 6]. In Upper Egypt, according to World Bank, 1993, eight wastewater treatment facilities exist with a total capacity of roughly 120,000 m^3/d, with approximately the same amount is recently constructed.

Table 2. Water resources in Egypt

Sources	10^9 m^3/year	
	2000	2020
Nile River	55.50	57.50
Shallow GW	5.50	7.50
Deep GW	0.80	2.75
Drainage water reuse		
Nile Delta	4.50	8.50
Nile Valley	5.00	5.00
Unofficial	3.00	3.00
Wastewater reuse	0.20	2.00
Rainfall	0.50	1.50
Desalinization	–	0.25
Losses	(3.00)	(2.00)
Total	72.00	86.00

Big cities like Cairo, Alexandria, and Ismailia receive well-treatment facilities. In Greater Cairo, six wastewater treatment plants exist. The smallest receives about 330,000 m^3/d for tertiary treatment. The largest receives 2.4 Mm^3/d for treatment where the sludge is anaerobically digested. The produced biogas covers 40% of the plant energy [6]. In the Delta, there are more than 30 wastewater treatment facilities with a total capacity of almost 400,000 m^3/d, with some 100,000 m^3/d under construction. In Upper Egypt and in Delta, some domestic wastewater receives only primary treatment [4]. More than 85% of the industrial water is discharged to the drainage system. Almost all existing sewage treatment plants for urban sewage water receive large amounts of wastewater due to both household consumption and leakage from the water distribution systems. Individual water consumption in Cairo is about 150 l/d, while in the rural areas it decreases to 60 l/d [3].

1.5. Pollution of River Nile and Associates

River Nile suffers from discharge of untreated sanitary wastewater, industrial discharges, heavy navigation of cruises, and commercial transports. Some pollutants added to the river are carcinogenic, mutagenic, or neurotoxin.

1.6. Agricultural Pollution

With the construction of the High Aswan Dam, silt deposits on the Nile floodplain have very likely decreased tenfold. This decrease resulted in a significant increase in the use of chemical fertilizers; and consequently increased values of nutrients in canals and drains. The amount of pesticides has been, on the other hand, steadily declined to nearly 4,000 tons/year; once reached its maximum value in 1975 with 27,000 tons. About 65,000 tons of some 200 different types of pesticides have been used in the Egyptian environment from 1952 to 1989 [7].

In rural areas, local surface and groundwater contamination have been accounted for leaching of pesticides, nitrates from fertilizers, and bacteria from livestock and human wastes. The dissolved oxygen was found to range from 3.1 to 9.5 mg/l with 87% of the observed values over the norm of 5 mg/l. A number of parameters such as phosphorus, nitrates, organochlorines, and BOD_5 stay below their maximum norms in all cases [7]. In Upper Egypt, approximately 4 Bm^3 of drainage water returns to the Nile every year contaminated, for instance, with TDS up to 4,000 mg/l and *E. coli* up to 10^6 MPN/100 ml. In Lower Egypt, there are four major lakes, Lake Maryut close to Alexandria, Lake Idku, Lake Burullus, and Lake Manzala in the eastern part of Delta. The most polluted lakes are lake Maryut and Lake Manzala [3].

Half of the drainage water discharged into the Mediterranean Sea or the northern lakes is of moderate level that can be, such as the El-Salam canal in Sinai, reused for irrigation after mixing [3]. In most drains in the Delta, toxic concentrations of plant nitrates, around 40–50 mg/l are quite common *E. coli* as high as 10^6 MPN/100 ml, and BOD% of 0–160 mg/l were also reported [7].

Bacteriological conditions in the River Nile are alarming; *E. coli* ranges from 14 up to 12,000 MPN/100 ml. Because of poor wastewater treatment, high concentration of coliform bacteria is found in the Nile and its branches downstream of Cairo. Values of 1 to 10 M MPN/100 ml have been measured in the Rosetta branch. This is far above the standard of 5,000 MPN/100 ml as Law 48 of 1982 permits [7]. The Rosetta and Damietta branches receive the water of a number of agricultural drains. The waste in the drain contains high levels of suspended and dissolved solids, oil, grease, nutrients, pesticides, and organic matters [3].

1.7. Industrial Pollution

Industrial pollution is mainly due to the use of old technologies, maintenance, and repair problems, as well as using of nonbiodegradable substances during industrial processes [4]. The amount of industrial wastewater discharged into the Nile is 550 Mm^3/year divided as shown in Table 3. The assessment of wastewater discharges from different industrial sectors is given in Table 4.

Table 3. Amount of industrial wastewater discharged into the river Nile

Water body	Amount (Mm^3/year)
Total discharged to the Nile	550
Main stream	312
Canals	118
Drains	71
Lakes	49

An important feature of the industrialization in Egypt during 1960s was the construction of the new industries in the metropolitan areas along the Nile, North and South Cairo (Table 5) particularly in Helwan and Shoubra El-Kheima, Kafr El-Zayat, Talkha, and Alexandria. Sources of industrial pollution along the River Nile in Upper Egypt area are mainly agro- and small-private industries, such as sugar mills, hydrogenated oil, and onion drying factories.

Alexandria industries include paper, metal, chemical, textile, plastic, pharmaceutical, oil and soap, petroleum, and food processing. Industry is reported to contribute some 20% of the total wastewater of Alexandria discharged mainly to Lake Maryut and partially to the sewage network. At least 17 factories discharge directly to the lake through pipelines; and 22 factories discharge into nearby drains and then to the lake [7].

The total amount of BOD discharged to the River Nile by industrial plants equals 270 t/d. Food processing is responsible for more than 50% of the BOD load. Ten sugar factories between Aswan and Cairo were responsible for about 490 t/d in 1980, gradually and significantly reduced due to the recovery of molasses at the source [7].

In the river's water and sediment, the presences of high concentrations of Cr and Mn in all sediments were found in most examined samples, however, concentrations of Cu and Pb were found to be low. Chemical industry is responsible for more than 60% of the heavy metal discharges. The average is less than 5 μg/l which is slightly more than the normal background [7]. Measurements in the Rosetta branch show that Cd, Cu, and Zn are above standards. Egyptian standards for Cu and Zn are 1,000 μg/l in receiving water and high levels of Hg are also reported for the Alexandria region.

A number of factories discharge their wastewater directly into Lake Manzala. This leads to high concentration of pollutants such as BOD% up to 2,000 mg/l, very high levels of *E. coli*, oil and grease up to 2 mg/l, Hg of 30–90 μg/l, and Cr of 50–200 mg/l. Increased concentration of metals is also found in the bottom sediment.

Table 4. Amount of wastewater produced from the industrial sectors

Industrial Sector	Amount (Mm^3/yr)
Chemicals	98
Food	227
Textiles	88
Engineering	12
Mining	60
Metal sectors	14
Other	51

1.8. Groundwater Pollution

The most important is the Nile aquifer system. The thickness ranges from 300 m in the south to 800 m in the Delta. The flow direction is, generally, south–north. The main source of recharge is percolation from agriculture and infiltration from irrigation and drainage canals. Due to the interaction of groundwater with surface waters, pollution of aquifers is closely related to adjacent polluted surface waters [2]. Pollution is more severe on the edges and desert fringes of the Nile Valley and in the shallow portions of the aquifers underneath urban areas, domestic sewage. Contamination such as *E. coli* of >100 MPN/100 ml, and organophosphorus and carbamate pesticides were also observed at shallow depths in regions with highly vulnerable groundwater. Meanwhile, organochlorine pesticides were found at greater depths. Nitrate, occasionally 70–100 mg/l is expected to increase with time in regions with intensive fertilizers application. High salinity was observed and ascribed to seawater intrusion and local return flow from irrigation. Natural iron and manganese were also noted.

.

Table 5. Classification of the industrial factories in Cairo

Type of activity	Number of factories
Chemicals	23
Textiles and spinning	27
Steel and galvanization	7
Food processing	32
Engineering	29
Mining and refraction	9
Bakeries	>350
Marble and tile	>120

2. ENVIRONMENTAL MANAGEMENT LEGISLATION

Substantial management laws and ministerial decrees were enacted during the era of 1970s and the Egyptian Environmental Affairs Agency (EEAA) was established as the coordinating body for environmental policy in Egypt. The EEAA broad authority regulates air pollution, control hazardous substances and waste management, control discharges to marine waters and the use of the Environmental Impact Assessment (EIA) process in the context of licensing new expansion, or rehabilitation projects, based upon the following main principles:

1. Type of activities performed 2. extent of natural resources exploitation, 3. location. 4. Type of energy used to operate. Three categories according to severity of possible environmental impact were identified as follows: (1) white list projects (for establishments with minor environmental impact), (2) gray list projects (for establishments which may result in substantial environmental impacts), (3) black list projects (for establishments which require full EIA due to their potential impact).

Most of the regulatory process is based on pollution discharge sanctions. Point polluters are held responsible for keeping pollution discharges below a standard. Yet standards, generally, do not allow for the polluter and the governmental agency to seek a quick agreement on a compliance schedule. In addition, the intensity of pollution should be described not only by the pollutant's concentration but also by its load. The commutative effect of pollutants is not considered. The law also disregards the carrying capacity of regions at the present and as a cumulative impact in the future, which is a requirement for sustainable development.

EIA is conducted for specific developments whereas the environmental impacts result not from a single intervention but from the cluster of establishments concentrated in an area.

3. CONCLUSIONS

The paper reviews the state of water pollution and wastewater management in Egypt, including economic issues; characteristics of agricultural and industrial pollution and their effect on surface water and groundwater; environmental legislations; and suggestions for improving the country's wastewater management system.

4. REFERENCES

1. Dijkman JPM. (1993), "Environmental Action Plan of Egypt" A Working Paper on Water. Directorate of General International Cooperation, Ministry of Foreign Affairs, the Netherlands, pp. 116–127.
2. Harris J, Nasser AG. (1995), Assessment of Water Quality in Egypt, National Water Quality Conservation Unit, PRIDE.
3. Abdel-Shafy HI, Aly RO. (2002), "Water Issue in Egypt Resources, Pollution and Protection Endeavors", Central European J. Occupational and Environmental Medicine, 8 (1), pp. 3–21.
4. Abdel-Wahaab R. (1995), Wastewater Treatment and Reuse; Environmental Health and Safety Consideration, Inter. J. Environ. , Health Research, 5, pp. 2–13.
5. Abdel-Shafy HI. (1996), "Environmental Transformation of Bioenergy via the Anaerobic Digestion", "Environment Xenobiotic", M. Richardson (Ed), Taylor & Francis Ltd, London, pp. 95–119.
6. Abdel-Sabour MF, Abdel-Shafy HI, Mohamed AR. (2005), "Plant Yield Production and Heavy Metals Accumulation as Affected by Sewage Sludge Application on Desert Soil", Sustainable Water Management, 1, pp. 27–31.
7. Abdel-Shafy HI, 1996, Environmental Transformation of Bio-energy, Environ. Xenobiotics. M. Richradson (Ed), Taylor and Francis, London, pp. 95–119.

SESSION 10. DECISION-MAKING AND RISK ASSESSMENT– INTERNATIONAL CONFLICTS

HEALTH ASSESSMENT OF WASTEWATER REUSE IN JORDAN

Mu'taz Al-Alawi
Water and Environmental Research Center, Mu'tah University
Karak 61110 Jordan

Corresponding author: alawi1979@yahoo.com

Abstract:

Irrigation with low quality of treated wastewater in the Jordan Valley has been practiced for several decades. The present study was designed to determine the health effects of this practice on the transmission of two protozoan infections, giardiasis and amoebiasis. Results of this study showed that at the exposed group 73% of the subjects were infected by Giardia and/or *Entamoeba histolytica*, and only 24% in the control zone. This excess of parasitic infestation may be related to the low quality of wastewater reuse in agriculture.

Keywords: Wastewater, risk assessment, pathogenic organisms, Jordan.

1. INTRODUCTION

Jordan is about 90,000 km in area. The Jordanian climate is arid to semiarid. Rainfall ranges between 50 mm in the desert region to about 600 mm in the eastern mountains adjacent to the Jordan Valley. The total rain fall in Jordan is estimated at 8.5 billion m^3 of which about 85% is lost to evaporation with the remainder flowing into valley and partially infiltrating into deep aquifer.

Recently, the problem of water shortage in Jordan has been exacerbated as a result of high natural population growth, influxes of refugees and returnees to the country. In 1997, the Government of Jordan's Water Strategy, that "wastewater should not be managed as 'waste.' It should be collected and treated to standard that allows its reuse in unrestricted agriculture and other non domestic purposes, including groundwater recharge." More than 70 million m^3 of reclaimed water, around 10% of the total national water supply, is used either directly or indirectly in Jordan each year [1]. The categories of use are (a) planned direct use within or adjacent to wastewater treatment plant, (b) unplanned use in the wadi (a dry-stream bed or the valley in which such a stream bed is located), and (c) indirect use after mixing with natural surface water supplies and freshwater supplies downstream, primarily in parts of the Jordan Valley. The treated wastewater is mostly discharged to

M.K. Zaidi (ed.), *Wastewater Reuse–Risk Assessment, Decision-Making and Environmental Security*, 385–392.

watercourses 13% of the total water used for irrigation. The irrigated area using direct treated sewage is estimated at 630 ha while 3,000 are irrigated by indirect irrigation using treated effluent in the Jordan Valley [2].

The aim of this study was to compare the prevalence amoebiasis and giardiasis among children groups coming from two populations which are similar in social living standards, in general. The first group lives in the spreading zone of middle and southern Jordan valley where low quality of treated wastewater is used for agriculture purposes, while the second group (nonexposed group) lives in a control zone. This epidemiological study would therefore allow describing the risk of infections attributable to wastewater reuse in agriculture.

2. MATERIAL AND METHODS

2.1. Populations Investigated

An epidemiological investigation was undertaken during the summer period on two children populations (n = 647), comprising both males and females from 2–14 years old (Table 1). The first population constituted the exposed group (n = 340 children) living in a peri-urban area where diluted treated wastewater was used for agricultural irrigation purposes. The second population which acted as a control group (n = 307 children) were derived from a peri-urban area with similar social living standards. Within this population, well water was used for irrigation (control zone). Data on individual exposure and on potential confounding factors (gender, age, water supply, sanitation, hygiene, level of education, family size, father's occupation, etc.) were collected using a standard questionnaire. During visits to both areas, children (boys and girls) were randomly selected from families.

Table 1. Distribution of investigated children by age and sex

Groups studied	Exposed group		Control group	
Age (years)	Male	Female	Male	Female
2–8	102	84	79	86
9–14	81	73	69	73
Total/sex	183	157	148	159
Total/area	340		307	
Total	647			

Treated wastewater used for agricultural irrigation in the central and southern Jordan Valley originates from Jordan's largest treatment plant at Al-Samra, which is located about 80 km east of the Jordan Valley. The plant at Al-Samra treats the domestic wastewater from Amman and Zarqa. Treated effluent from the Al-Samra wastewater treatment plant is discharged to Wadi Zarqa. The wadi flows into the King Talal Reservoir (KTR), picking up whatever surface runoff occurs in the Amman-Zarqa catchment. The water in the reservoir, blended with water from the King Abdullah Canal, when available, is used for irrigation in the Jordan Valley [3]. The treatment plant at Al-Samra has been operational since 1985, and its design capacity is 68,000 m^3/d, while the inflow in 2005 was 140.763 m^3/d. This has led to deterioration in effluent quality [4].

2.2. Collection and Examining of Faecal Samples

Stools samples were collected in standard 50 ml plastic container labelled with individual ID-number. The formalin-ether sedimentation method was used in Parasep (DiaSys Europe Ltd., Wokingham, UK) faecal parasite concentrator tubes to determine protozoan pathogens prevalence. The tubes were centrifuged at 1,000 rpm for 10 min and then fat and format water was drained off. Three slides of sediment were analyzed in a saline solution with the optic microscope at ×400 and ×1,000. Cysts are identified by iodine staining. Four stool samples were studied for each child. The Chi-Squared test was used to compare different prevalence values.

3. RESULTS AND DISCUSSION

3.1. Prevalence of Giardiasis and Amoebiasis

There are several protozoan pathogens which have been isolated from wastewater and recycled water sources [5]. The most common detected are *Entamoeba histolytica, Giardia intestinalis* (formerly known as *Giardia lamblia*), and Cryptosporidium parvum [6]. The prevalence expresses the proportion of people affected by a disease or infection at one particular time. Analysis of the different faecal samples showed that global rate of infection much higher in the exposed group than the control one, reaching 73% (247/340) and 24% (73/307), respectively (p<0.001).

3.1.1. Giardiasis

Giardiasis is a diarrheal illness caused by a one-celled, microscopic parasite, Giardia intestinalis (also known as *Giardia lamblia*). Giardia lives in the intestine of infected humans or animals. Millions of germs can be released in a bowel movement from an infected human or animal. You can become infected after accidentally swallowing the parasite. Giardia may be found in soil, food, water, or surfaces that have been contaminated with the feces from infected humans or animals. Giardia and Cryptosporidium are ubiquitous in fresh and estuarine waters and have been detected in numerous countries around the globe [7–11]. As little as 10 or fewer Giardia cysts are sufficient to cause illness [12].

In this epidemiological study, we observed a higher prevalence (Figure 1) in the wastewater spreading area of Jordan valley (exposed group) which is 41.24% (140/340) against 16.54% (51/307) in the control zone (nonexposed group), ($p<0.001$). These levels seem to be comparable to some data reported in the literature. According to WHO [13], the prevalence of giardiasis goes from less than 1% to more than 50%, it depends on demographic characteristics and the dominating transmission mode. Farthing [14] reported that the prevalence of Giardia varies between 2% and 5% in the industrialized world and up to 20–30% in the developing world.

3.1.2. Amoebiasis

Amoebiasis is an intestinal illness caused by the parasite, *Entamoeba histolytica*. The parasite multiplies in the intestine and produces cysts which are passed from the body in the stool (bowel movement). A person becomes infected with Entamoeba histolytica by swallowing the cyst stage of the parasite. This can occur by eating food which has been prepared by an infected individual who has poor personal hygiene, drinking water contaminated by sewage.

According to our study results, prevalence of the amoebiasis among children of the nonexposed group was 7.27% (22/307) against 31.35% (107/340) for children of spreading fields (exposed group) (Figure 1), what shows the net abundance of this infection in children of the spreading area compared to those of the witness group ($p<0.001$).

3.2. Distribution by Sex and Age

In order to determine the most exposed children to protozoan infection in the in middle and southern Jordan valley, we studied the variation of infections rates according to their sex and age.

The 340 examined children in the spreading fields comprise 183 males and 157 females. Males were infected with a level of 33.15% (61/183) by *Entamoeba histolytica*, against 28.82% (45/157) of females. For giardiasis, the respective prevalences for the two sexes were 46.53% (85/183) and 36.43% (57/157) (Figure 2). Thus, males seem to be more infected than females, however no statistically significant difference was noticed between the two sexes ($p>0.05$).

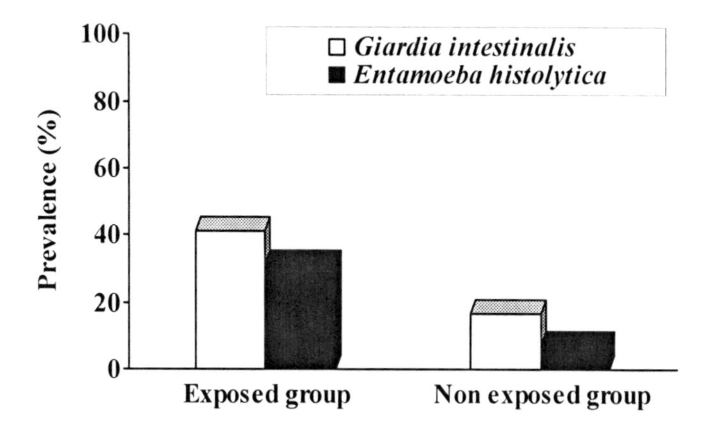

Figure 1. Prevalences of giardiasis and amoebiasis in the exposed and non-exposed groups

The examined children were subdivided in two age classes, 2–8 years (186) and 9–14 years (154). Among the children of the spreading area, we observed that the prevalence of the intestinal protozoa varies according to age (Figure 3). For giardiasis, this prevalence reached 49.14% (91/186) among children of 2–8 years, against 32.26% (50/154) for children of 9–14 years, the difference is statistically significant ($p<0.001$). For amoebisis the respective prevalences in the two age groups were 35.62% (66/186) and 25.27% (39/154) and there is no difference statistically significant ($p>0.05$). The variation of the infestation rate according to age would be attributed to the imprudence of the little children and to their non hygienic practices with contamination sources [15]. Indeed, during our visits to the spreading area we noted that children at low age spend the majority of their time playing in irrigated fields without any precaution.

3.3. Risk of infection attributable to wastewater reuse

Even though wastewater reuse has been considered as a potential route for pathogen dissemination, the role it plays in transmitting parasite infections requires more detailed epidemiological investigation. The current study revealed that children living in the wastewater spreading fields of Jordan Valley were more likely to have high incidence of protozoan infections than a control group of the same socioeconomic characteristics.

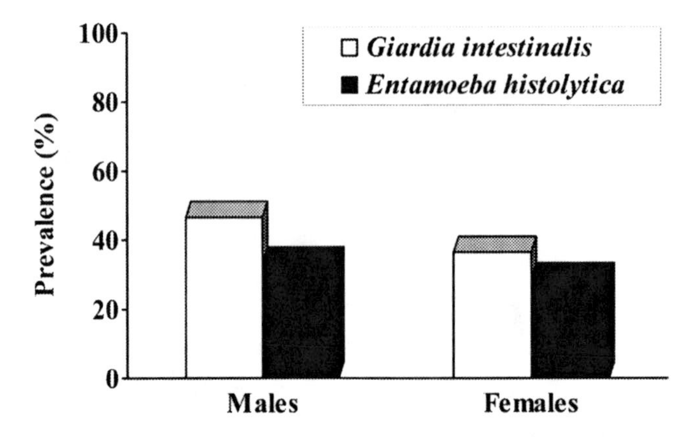

Figure 2. Variation of parasites prevalences according to children's sex in the spreading area

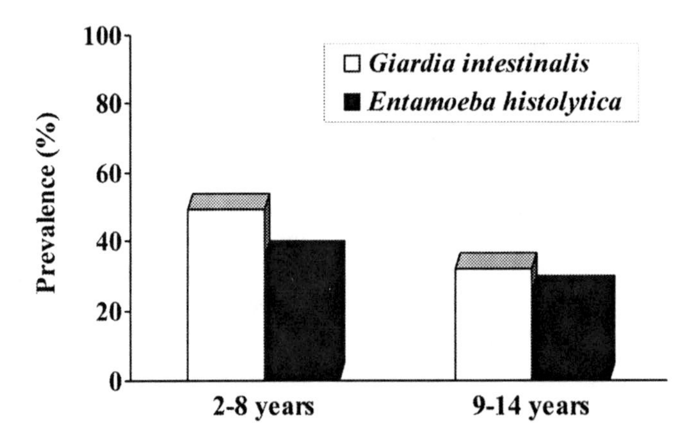

Figure 3. Variation of infections prevalences according to age in the spreading area

Therefore, the global risk of the two infections attributable to wastewater reuse was about 48.78% (24.7 and 24.08% for giadiasis and amoebiasis, respectively). This may be related to the participation of children in their relatives tasks in rustic works, which tends to put them in direct contact with contamination sources represented by fields and cultures irrigated with low quality of treated wastewater loaded of pathogenic agents, and the consumption of vegetables irrigated also with this low quality of treated wastewater may explain this excess of infection. Thus, the excess of parasite infestations among children of the spreading zone would be attributed to the utilisation low quality of treated wastewater for irrigation in this zone.

4. CONCLUSION

The current study has assessment the health hazards related to wastewater reuse in the transmission of two protozoan infections. It becomes evident that this practice leads to high infection rates among the children of the Jordan valley compared to a control group. The prevention of these risks can be achieved by an integrated set of measures which may include crop restriction, appropriate wastewater application techniques and human exposure control. Also, increase the capacity of Al-Samra treatment plant in order to produce wastewater free from pathogens.

Acknowledgment: NATO provided financial support to present the paper at the ARW – Wastewater Reuse – Risk Assessment, Decision making and Environmental Security held in Istanbul, Turkey, during October 12–16, 2006.

5. REFERENCES

1. McCornick P. (2001). Plan for Managing Water Reuse in the Amman-Zarqa Basin and Jordan Valley. Water Reuse Component Working Paper, Water Policy Support Project, Ministry of Water and Irrigation, Jordan, 70 pp.
2. FAO. (2001). Strengthening Capacity for the Reuse of Treated Wastewater in Irrigation. Draft Project Document for Syria. FAO Regional Office for the Near East, Cairo, Egypt.
3. McCornick PG, Taha S, El-Nasser H. (2002). Planning for Reclaimed Water in the Amman-Zarqa Basin and Jordan Valley. American Society of Civil Engineering-Environmental and Water Resources Institute, Conference and Symposium on Droughts and Floods, Roanoke, Virginia,10 pp.
4. Water Authority of Jordan. (2005). Internal Working Report. Amman, Jordan
5. Gennaccaro AL, McLaughlin MR, Quintero-Betancourt W, Huffman DE, Rose JB. (2003). Infectious Cryptosporidium parvum Oocysts in Final Reclaimed Effluent. Appl. Environ. Microbiol. 69, 4983–4984.
6. Toze S. (1997). Microbial Pathogens in Wastewater. CSIRO Land and Water Technical Report 1/97.
7. Ferguson CM, Coote BG, Ashbolt NJ, Stevenson IM. (1996). Relationships Between Indicators, Pathogens and Water Quality in an Estuarine System. Water Res. 30, 2045–2054.

8. Ho BSW, Tam TY, Hutton P, Yam WC. (1995). Detection and Enumeration of Giardia cysts in River Waters of Hong Kong by Flocculationpercoll/Sucrose Gradient-Immunofluorescence Method. Water. Sci. Tech. 31, 431–434.
9. Kfir R, Hilner C, du Preez M, Bateman B. (1995). Studies on the Prevalence of Giardia cysts and Cryptosporidium Oocysts. Water Sci. Tech. 31, 435–438.
10. Ongerth JE, Hunter GD, DeWalle FB. (1995). Watershed Use and Giardia Cyst Presence. Water Res. 29, 1295–1299.
11. Wallis PM, Erlandsen SL, Isaac-Renton JL, Olson ME, Robertson WJ, Van Keulen H. (1996). Prevalence of Giardia cysts and Cryptosporidium Oocysts and Characterization of Giardia spp. Ap. Env. Microb. 62, 2789–2797.
12. Brooks GF, Butel JS, Ornston LN. (1991). Medical Microbiology. 9th Edition. New York. 488 pp.
13. WHO. (1987). Prevention and control of intestinal parasitic infections. Report of a WHO Expert Committee. WHO Technical Report No. 749.
14. Farthing MJG. (1993). Diarrhoeal Disease: Current Concepts-Future Challenges, Pathogenis of giardiasis. Trans. R. Soc. Trop. Med. Hyg. 3, 17–21.
15. Feachem RG, Bradly DT, Mara DD, 1983, Sanitation and Disease: Health Aspects of Excreta and Wastewater Management. Chichester, UK, John Wiley & Sons, London.

WASTEWATER RISK ASSESSMENT OF POLYESTER CABINS IN EGYPT

Mohamed A. Amasha and Tarek M. Abou Elmaaty
Faculty of Specific Education and Department of Textile Printing, Dyeing & Finishing, Faculty of Applied Arts, Mansoura University, 34512 Egypt

Corresponding author: mw_amasha@yahoo.com

Abstract:

The technological revolution has brought a great progress in furniture industry which considered as one of the most vital resources in Egypt economy. The government has established a great number of polyester cabins (in Damietta governorate) which used in furniture painting and varnishing. The wastewater is transferred by the same route to the water purification station which according to our research has no proper means to eliminate the polyester hazard. In this work we conducted a survey among the cabin workers to assist their knowledge about the wastewater hazard. Also, we have suggested a strategy to overcome this problem.

Keywords: Wastewater, risk assessment, polyester cabins.

1. INTRODUCTION

Damietta Governorate comprises 25% of the enterprises operating in the furniture manufacturing, as well as 25% of the workers who work in this field. This industry is affected by the technological revolution. Now computer is used in mixing, blending, and stabilizing the paints of the furniture. Some manufacturers use modern paints which contain very dangerous materials as polyester [1–3], polyurethane [3]. These chemical materials are carcinogenic [1, 2] and have harmful emissions come during the process of spraying. When polyester mixes with the air, it makes explosive nitrocellulose mixture which causes extensive damage to the respiratory system of the human being. It is a solid material in the form of filaments or powder and it is flammable.

It also causes explosion and forms toxic gases, as nitrogen oxides, hydrogen cyanide, and carbon monoxide. These gases are very dangerous when they mix together. Also all of the previous materials should be treated by solvents that are carcinogenic or semicarcinogen as acetone–toluene–carbon tetrachloride–benzene [1].

M.K. Zaidi (ed.), Wastewater Reuse–Risk Assessment, Decision-Making and Environmental Security, 393–401.

Because of the critical social conditions for the workers in this field, they use locally designed cabins for the process of spraying. The cabin is a room composed of four corners with a total area of 54" × 62", in which there is an outlet watercourse and a ventilating fan lined with local filter which is altered from time to time after finishing the spray process. The government has established a new type of cabin that reduces environmental pollution, called electronic cabin.

An electronic cabin can be schematized into three types. The first is called spray booth in which the ventilation system is especially designed to manage the processes of furniture spray in order to handle and determine the emitted vapors and splashes forwarding them securely to a group of filters for disposal. The second is called water spray booth—the most commonly used spray cabin, and it's equipped with "a water carpet" to handle dusts and spray vapors for disposal using special drainage system.

The third cabin is called dry spray booth—filled with filters to absorb harmful dusts and vapors, and allows air to pass through certain direction. Dusts and vapors are set for disposal via filter before they are released into the atmosphere. The second type of cabins uses water or watercourse. It eliminates the emitted vapors via condensing them inside the cabin where they are mixed with water and then disposed off.

According to the reports of physicians and environmental experts these cabins have affected the life of thousands of citizens, especially children, pregnant women, and the elderly [4, 5].

2. DISCUSSION AND RESULTS

In this work, we have noticed that there is a great deal of water loss in some cabins in Al-Shou'raa village due to the ongoing use of spray. The quantity of water lost during spray process in one of the cabins that use water carpets and a watercourse at its bottom has been calculated using Bernoulli equation [6]. The quantity of water was found to be 11 l^3.

It is known that chemically contaminated water through dumping sprays which contain harmful compounds such as polyethylene, nitrocellulose, and polyester represents a main source for water contamination and results in grave health hazards. Water pollution has greatly prompted to enact law No. 48 in 1994 [7], which stipulates "It's banned to discharge any liquid, solid, gas wastes from any establishment including drainage plants into water supplies without being granted a license. The law imposes a $400 fine on breaching the law. Al-Shou'raa village is located" in Dameitta Governorate with an area of 550, population census of 40,080, and it has 481 workshops all of which are located inside the populated area. The village is home to 29 electric cabins used in furniture spraying. Personal interviews were conducted with the owners of the electronic cabins that particularly use water carpets during spraying process.

Moreover, to achieve the goals of the study, the researcher prepared the questionnaire. The questionnaire falls into three pivots: the first is concerned with water consumption inside the cabin and it includes 12 elements; the second pivot deals with using electronic cabins and it involves 9 elements; the third pivot tackles "the government role in spreading awareness" and it implies 7 elements.

To achieve the aims of the research, we tested a random population inside electronic cabins, more particularly inside the second type. The selected population is 30 spray workers representing 15 cabins elected at random. The following Table 1 shows the number of workers and their education levels.

Table 1. Relation between frequencies and percentages of workers

Educational level	Number	%
Illiterate	26	86.67
Intermediate education	3	10.0
Institute graduate	1	3.33
High education	0	0

The following table shows the sample individuals' responses towards the 1st pivot that deals with "the consumptions of water inside the cabin".

The Table 2 shows that the value of X^2 for the responses of the sample individuals around the first pivot deals with water consumption inside cabins is bigger than the value of X^2 which hit 3.84 at a significant rate of 0.05 and freedom degree. This denotes that the distinction between the observed frequency and the expected ones of the answers around that pivot is statistically significant and, therefore, it's not subjected to coincidence. The study demonstrates the importance of using water inside cabins to absorb the splashes emitted during spraying process in order not to affect the workers' health and to reduce environment pollution. This process consumes a great deal of water per day assessed to be 11 l^3. Water is also used in cleansing the operating tools during spraying process. Furthermore, water is so often used in spraying in front of cabins so as to alleviate the effects of dusts because of the in appropriateness of the surroundings of spray booths. This result in an increase in water loss that can not be assessed but the existence of such a loss should be admitted.

Table 2. Frequenies, percentages, and the value of X^2 as well as the significance level of the sample individuals' response toward the 1st pivot concerned with "the consumption of water inside the cabin"

SN	Phrases	Agree		Disagree		X^2
		f	%	f	%	
1	Is the consumption of water inside the cabin necessary?	29	96.67	1	3.33	26.13
2	Is the consumption of water inside the cabin is limited to spraying process only?	26	86.67	4	13.33	16.13
3	Do you use water for other purposes inside the cabin?	23	76.67	7	23.33	8.53
4	Do you know that seeping organic dissolvent used in spraying into water affects health?	5	16.67	25	83.33	13.33
5	Do you use deposit sinks to separate organic dissolvent from water?	30	100	0	0	30
6	Do you know that the used water affects kidney because it includes dissolvent used in spraying as gasoline, distillates and petroleum?	27	90	3	10	19.2
7	Can you assess the quantity of water you use per day?	28	93.33	2	6.67	22.53
8	Is the used water dumped into special drainage?	4	13.33	26	86.67	16.13
9	Do you know that quantity of lost water can be retreated and reused?	0	0	30	100	30
10	Is there special place to discharge water emitting from vapor condensation inside the cabin?	4	13.33	26	86.67	16.13
11	Have you ever received any education about the danger of contaminated water?	27	9	3	10	19.2

The results pointed out that the water used during spraying process is not drained into specialized drainage channels but with house drainage. Hence, organic dissolvent used in spraying are drained into house drainage that's directed to the wells of water plants for treatment in order to be reused, the matter that affects public health badly.

Moreover, the results indicated that workers don't know anything about the risks of using organic dissolvent used during spraying process on public health. In addition, the results indicated that special filters are not used in those cabins to refine water before disposal in drainage channels, and that those workers don't know that the quantity of lost water can be retreated and reused.

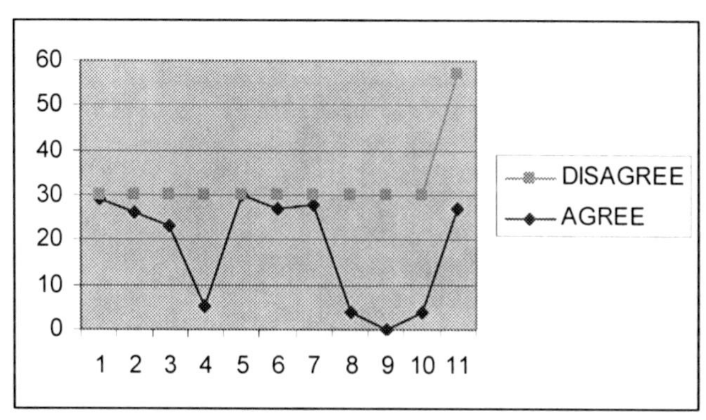

Figure 1. Responses toward the 1st pivot concerned with "the consumption of water inside the cabin"

This reveals that they know nothing about this culture and its importance to Egypt's vein of life, i.e., the River Nile. The following table shows the response of the sample individuals around the 2nd pivot in the questionnaire dealing with "using electronic cabins"

The Table 3 shows that the value of X^2 for all the responses of the sample individuals around the 2nd pivot of the questionnaire that deals with "Using electronic cabins" is bigger than the value of X^2 which amounts 3.84 at a significant rate of 0.05. This denotes that the distinction between the observed occurrences and the expected ones of the responses concerned with that pivot is statistically significant and, therefore, it rules out coincidence factor.

The results pointed out that using electronic cabins in spraying process is more secure and it reduces environmental harms/effects and the importance of using water carpets inside cabins and making a stream of water at the bottom to help absorb rising splashes and vapors emitted from spraying process. Moreover, the results, indicated that spray workers have no knowledge about the scientific composition of the material used in spraying process and they don't know the harmful effects of these materials, particularly carbon tetrachloride, carbon hydrides and ethanol when mixed with water and then dumped into the house drainage. The workers, additionally, asserted that they don't use filters to get rid of these dangerous substances before discharging. The workers also emphasized using water during spraying process to absorb the emitted exhaust which is composed of dusts like powder that affects spray workers badly.

Table 3. The repetitions, percentages, the value of X^2 and the statistical, significance level of the sample individuals' responses towards the 2nd pivot that tackles "using electronic cabins"

SN	Phrases	Agree		Disagree		X^2
		f	%	f	%	
1	Do you think that spraying inside electronic booths is useful?	25	83.33	5	16.67	21.66
2	Is the usage of water carpets during spraying process is preferable?	27	90	3	10	24.6
3	Do you know the risks of chemical compounds used during spraying process?	6	20	24	86.67	20.4
4	Do you have any information about the scientific method of synthesizing the material used in spraying process?	3	10	27	90	24.6
5	Are you aware of the hazards of dissolving carbon tetrachloride, carbon hydrides, and ethanol into water?	0	0	30	100	30
6	Do you know that the leaked toxic vapors outside the booth are mixed with water used in spraying area?	0	0	30	100	30
7	Do you use filters to eliminate vapors?	29	96.67	1	3.33	28.06
8	Do you use any type of filter to purify water inside the cabin before disposal?	30	100	0	0	30
9	Are the booths fitted with special drainage as drain tanks?	30	100	0	0	30

On addressing another question to the sample individuals around whether they how tanks to store the water used in spraying process and if not where they dump this water, they answered that they discharge it in nearby house drainage. The following table shows the sample individuals' responses on the third pivot of the questionnaire that deals with" the provision of education and the government role.

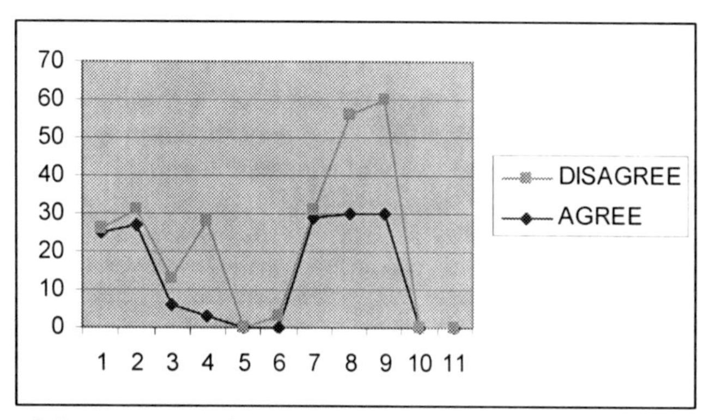

Figure 2. Responses towards the 2nd pivot that tackles "using electronic cabins"

Table 4 shows that the value of X^2 for all the responses of the sample individuals around the third pivot that tackles "the provision of education and the government role," is bigger than the value of X^2 which hit 3.84 at a significant rate of 0.05. This denotes that the distinction between the observed occurrences and the expected ones of the responses concerned with that pivot is statistically significant and therefore, it shrugs out the role of chance. The results illustrated that despite the education and training to spray workers received through the government and other concerned associations, they are not aware of the risks of seeping chemicals into water as a result of spraying process and then, dumping it into house drainage.

The workers also didn't demonstrate any willingness to move into other places out of the populated zones because of their attachment to the places of their work. The Dammiettan workers are always inclined to work in a workshop in their houses as well as their economic conditions hinders them to move out. Asking them about the environmentally friendly paints which use water as being recommended by several conferences and symposiums to convince them to use these paints in furniture spraying instead of chemicals, they maintained that paints proved to be in efficient and yield poor results compared to that of chemicals. It's worthy of note that this was the workers' viewpoint. In addition, the workers highlight their ignorance of issuance of a law banning water misuse or discharging contaminated water into irrigation courses or nearby drainage channels. They also pointed out that they are not ready to attend conferences and symposiums because this wouldn't be useful for them and that they don't want to squander a wage of a day as these symposiums don't make up for that wage.

Asking them about the alternatives and whether they want to get the education via video tapes, leaflets, or computer programs written on CDs, 85% of the individuals maintained their desire to get their education through CDs because there is a plenty of internet cafes in the village. As for video tapes, they don't own video sets to run the tapes and most of them are not good at reading or writing to read leaflets. Consequently, designing a CD based-educational program and providing it to those workers may be the best way of educating them.

Table 4. Repetitions, percentages, and the value of X^2 to the sample individuals' responses on the third pivot which tackles "the provision of education and the government role"

SN	Phrases	Agree		Disagree		X^2
		f	%	f	%	
1	Are you aware of the dangers of water contaminated with chemicals during spraying process?	2	6.67	28	93.33	26.26
2	Have you ever been acquainted with the risks of chemically contaminated water?	29	96.67	1	3.33	28.06
3	Are you ready to move into other places that are well equipped for spraying?	0	0	30	100	30
4	Are you ready to switch into a more secure spraying process in the same place?	16	53.33	14	46.67	15.06
5	Are you aware of environmentally friendly paints which use water as a solvent instead of chemicals used in spraying?	22	73.33	8	26.67	18.26
6	Do you know that there is a law that criminalizes dumping any wastes into house drainage?	9	30	21	70	17.4
7	Are you keen on broadening your awareness through symposiums, leaflets and conferences, computer software?	4	13.33	26	86.67	23.06

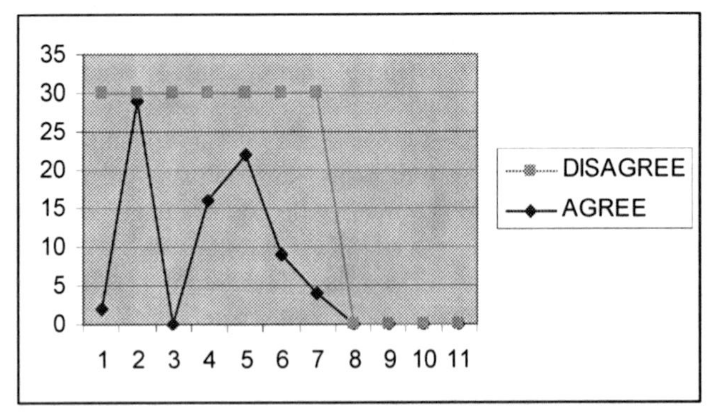

Figure 3. The provision of education and the government role

Therefore, the authors would recommend designing an educational software addresses to spray workers to overcome this problem.

3. CONCLUSION

1. There is a great deal of lost water inside the electronic cabins used in furniture spraying. The quantity of water used in one cabin in 11 m^3/d a day and there is lack of independent drainage route for these cabins to get ride of the used water or to save it to be reused after treatment.
2. Lack of filters equipped to refine the resulting water from spraying process before disposal into drainage, the matter that poses threats on public health in case of retreated.
3. Workers are not aware of the risks of the using chemicals such as carbon tetrachloride, carbon hydrides, and ethanol are mixed with water.
4. Those workers have never received any education about the hazards of chemically contaminated water and they are not aware of the issuance of a law that criminalizes dumping contaminated water into water supplies or house drainage channels.

Acknowledgment: The authors thank NATO-ARW for financial support to present this paper at the NATO ARW.

4. REFERENCES

1. Center for Disease Control Report, National Report on Human Exposure to Environmental Chemicals (2001).
2. Elton S, Fire and materials, 22, 19–23 (1998).
3. Selikoff IJ, Hammond EC, Am J Public Health Nations Health, September, 58(9): 1658–1666, (1968).
4. Ecology center Plastic Task Force Report, Berkeley, CA, (1996).
5. National Environmental Protection Council, (1998a), National EP Measure for the National Pollutant Inventory.
6. Bruce RM et al., Fundamentals of Fluid Mechanics, J. Wiley, NY, USA, 332–335, (1994).
7. Egypt Environmental Protection Agency, Pollution Prevention and Toxics, Facts Sheet (1994).

REMOVAL OF URANIUM FROM WASTEWATER IN TURKEY

Gul Asiye Aycik
Mugla University, Chemistry Department, 48000, Mugla, Turkey

Corresponding author: gulasiye@mu.edu.tr

Abstract:

Adsorbents are thought to be the most effective method for recovering the low concentrations of uranium from the wastewater. Here we will synthesize the interpenetrating polymer networks (IPN) based on acrylonitrile (AN) and polyethylene glycol (PEG) and using its adsorptive ability for removing uranium from aqueous systems. It was found that the polymeric adsorbent has very high-adsorption ability for uranium and can be used in adsorption studies to remove uranium from wastewater. It is to remove uranium and other radioactive elements in their source before mingling to the groundwater. The coal power plants release ashes and they are deposited on the surroundings. Leaching of ash samples to remove toxic, heavy, and radioactive elements before their mingling to the groundwater is studied.

Keywords: Uranium, radioactivity, groundwater, polymers, adsorbents.

1. INTRODUCTION

Uranium is the heaviest element occurring in nature in weighable amounts. Despite its high, atomic number, it is a ubiquitous element and it is by no means a rare element. Its naturally occurring isotopes have masses of U-238, U-235, and U-234. All isotopes are radioactive.

The average U concentration in the earth's crust is about 3 ppm. Concentrations of U in geological materials are highest in continental-type rocks. Uranium forms more than 160 mineral species and accounts for 5% of all known minerals [1].

U in nature does not occur in elemental form. It appears as U(VI) in most natural waters as either a cation or complexed with carbonate (uranyl carbonate). As uranyl hydroxide, the molecule is a cation, and as uranyl carbonate, it is an anion.

M.K. Zaidi (ed.), Wastewater Reuse–Risk Assessment, Decision-Making and Environmental Security, 403–414.

1.1. Radioactive Elements in Coal and Fly Ash

Coal is an extremely important fuel in most parts of the world and will remain so. As population increases worldwide, coal combustion continues to be the dominant fuel source for electricity. Some 23% of primary energy needs are met by coal and 39% of electricity is generated from coal. About 70% of world steel production depends on coal feedstock. Coal is the world's most abundant and widely distributed fossil fuel source.

Coal is largely composed of organic matter, but it is the inorganic matter in coal-minerals and trace elements that have been cited as possible causes of health, environmental, and technological problems associated with the use of coal. It is one of the most impure form fuels. Its impurities range from trace quantities of many metals, including uranium and thorium, to much larger quantities of silicon, aluminum, iron, calcium, magnesium, titanium, sodium, potassium and to still larger quantities of impurities such as sulfur.

Fly ash generated during the combustion of coal for energy production is an industrial by-product and is recognized as an environmental pollutant. Because of the environmental problem of fly ash, there are much efforts on the utilization of fly ash throughout the world.

Releases from coal combustion contain naturally occurring radioactive materials—mainly, uranium (U), thorium (Th), and their numerous decay products, including radium (Ra), and radon (Rn). Although these elements are less chemically toxic than other coal constituents such as arsenic, selenium, or mercury, questions have been raised concerning possible risk from radiation. Since uranium has alpha-emitting radionuclides, the removal of it is of great interest. Radon is one of the important daughters of uranium that is known to cause cancer when inhaled or consumed over long periods of time. Radon is produced by the natural radioactive decay of radium (from uranium) in the ground. Some rocks, such as granite, limestone, and sandstone, contain high concentrations of radium and produce significant amounts of radon. Groundwater found in these rock formations may contain elevated concentrations of uranium and its daughter, radon.

Assessment of the radiation exposure from coal burning is critically dependent on the concentration of radioactive elements in coal and in the fly ash that remains after combustion. Concentrations of uranium in coal range from less than 1 part per million (ppm) in some samples to around 10 ppm in others. During coal combustion most of the uranium, thorium, and their decay products are released from the original coal matrix and are distributed between the gas phase and solid combustion products.

Less volatile elements such as thorium, uranium, and the majority of their decay products present in feed coal are almost entirely retained in the solid combustion waste. As the average ash yield of coal burned is approximately 20 wt %, therefore, the concentration of most radioactive elements in solid combustion wastes will be approximately five times the concentration in the original coal because the uranium content is not decreased as the volume of coal is reduced [2, 3]. Uranium has been discharged into the environment by coal and especially by deposited fly ash, causing serious soil and water pollution. Various physicochemical and biological methods for removal of uranyl ions have been studied.

Uranium is found in both the mineral and organic fractions of coal. Some uranium may be added slowly over geologic times because organic matter can extract dissolved uranium from ground water. In fly ash, the uranium is more concentrated in the finer-sized particles. If during coal combustion some uranium is concentrated on surface as a condensate, then this surface-bound uranium is potentially more susceptible to leaching. Coal-fired power plants and mine leaching and cement plants contribute these elements to our water resources.

1.2. Health and Environmental Impact of Radioactive Elements Associated with Coal Utilization

The vast majority of coal and the majority of fly ash are in associated radioactivity, compared to common soils or rocks. Radioactive elements from coal and fly ash may come in contact with the general public when they are dispersed in air and water or are included in commercial products that contain fly ash. Approximately three-fourths of the annual production of fly ash is destined for disposal in engineered surface impoundments and landfills, or in abandoned mines and quarries. The primary environmental concern associated with these disposal sites is the potential for groundwater contamination. When fly ash is not properly disposed they may come into contact with the terrestrial, aquatic, and atmospheric environment and if leached, these elements cause the contamination of subsurface water. Thus rain and other drainage provide a pathway for potentially radioactive elements to enter the food chain and human life cycle from these disposal sites. Depending on the chemical/physicochemical features of fly ash, the leachability of uranium may vary.

Existing coal-fired plants of 1000 MW capacity annually burn about 4 million tons of coal. Using these data, the releases of radioactive materials per typical plant can be calculated for any year. Assuming coal contains uranium concentrations of 1.3 ppm, each typical plant released 5.2 tons of uranium that year. Releases from the combustion of coal in power plants from worldwide combustion of about 2,800 million tons of coal; more than 150 million tons of fly ash and 3,800 tons of uranium are produced annually. At least half of this amount is disposed of by landfill, thus contributing to environmental pollution due to leaching of its constituents. Disposal of huge amounts of fly ash in landfills and surface impoundments or its reuse in construction materials is of environmental concern.

All studies of potential health hazards associated with the release of radioactive elements from coal combustion conclude that the perturbation of natural background dose levels is almost negligible. However, because the half-life of radioactive uranium is practically infinite in terms of human lifetimes, the accumulation of these species in the biosphere is directly proportional to the length of time that a quantity of coal is burned.

Uranium is known to be detrimental to the living beings if it is taken by digestion and/or by inhalation systems. Uranium is an important contaminant of surface and groundwater because of emissions from it and its daughters. It has been observed that increased concentrations of uranium in the aquatic environment can be considered to be an environmental concern.

Two different approaches can be proposed for the removal of U as ecologically: The first one is the leaching of radioactive elements from disposed fly ash. The other one is the removal of uranium from contaminated water.

1.3. Leaching of Radioactive Elements from Disposed Fly Ash

Acid leaching is an alternative to the conventional stabilization of ash by carbonization or polymer induced coagulation/flocculation [4].

Previous studies of radioelement mobility in the environment, and in particular, in the vicinity of uranium mines and mills, provide a basis for predicting which chemical conditions are likely to influence leachability of uranium, barium (a chemical analog for radium), and thorium from fly ash. For example, leachability of radioactive elements is critically influenced by the pH that results from reaction of water with fly ash. Extremes of either acidity ($pH<4$) or alkalinity ($pH>8$) can enhance solubility of radioactive elements. Acidic solutions attack a variety of mineral phases that are found in fly ash. However, neutralization of acid solutions by subsequent reaction with natural rock or soil promotes precipitation or sorption of many dissolved elements including uranium, thorium, and their decay products. Highly alkaline solutions promote dissolution of the glassy components of fly ash that are an identified host of uranium; this can, in particular, increase uranium solubility as uranium carbonate species. Fortunately, most leachates of fly ash are rich in dissolved sulfate, and this minimizes the solubility of barium (and radium), which form highly insoluble sulfates.

The location and form of radioactive elements in fly ash determine the availability of elements for leaching during ash utilization or disposal. Existing measurements of uranium distribution in fly ash particles indicate a uniform distribution of uranium throughout the glassy particles. The apparent absence of abundant, surface-bound, relatively available uranium suggests that the rate of release of uranium is dominantly controlled by the relatively slow dissolution of host ash particles.

The leachability of radioactive elements from fly ash has been studied for three types of coal ashes in Turkey: Mugla-Yatagan-Eskihisar, Kahramanmaras-Elbistan-Kislakoy, and Ankara-Beypazari

1.4. Removal of Uranium from Contaminated Water

Water contamination resulted from coal and ash is a major concern. Discharging large amounts of fly ash, can affect the physical and chemical properties of water sources. Radionuclides are one of the important pollutants in surface and groundwater contaminants. Groundwater, especially bedrock water, may contain great amounts of natural radionuclides, derived mainly from the U-238 series. Elevated levels of natural radionuclides in groundwaters are mainly associated with uranium and thorium rich soil and rock minerals, or with uranium, thorium, and radium deposits. The surveys made indicate that high concentrations of radon and other radionuclides usually occur in water from wells drilled in bedrock.

Elevated levels of natural radionuclides in drinking water are accompanied with potential health risks for the population by increasing the radiation dose. Therefore, water should be purified before using it. Various processes based on different principles can be applied to remove the radioactivity from water. Aeration is a method that is usually applied to remove radon (Rn-222) from drinking water. Aeration should be used if the concentration of radon is high, whereas the granular activated carbon filtration can be used when the radon concentration of water is moderately low.

Mainly used processes of wastewater purification such as adsorption, ion-exchange, precipitation, oxidation, ultrafiltration, reverse osmosis, and electrodialysis seem to be more attractive methods.

One of the most widely used techniques involves the process of adsorption in the separation and isolation of elements. The greatest advantage of adsorption is the possibility to separate all substances from a very large volume of solution. Adsorption is a physicochemical wastewater treatment process, which has gained prominence as a means of producing quality treated effluents from effluents, which contain low concentration of dissolved chemical contaminants. Adsorption of source-active molecules at the solid–liquid interface is an important phenomenon in water-purification industry. An ideal heavy metal adsorbent should have a very strong affinity for the target metal, binding it irreversibly under ambient conditions, and simultaneously possessing the ability to release the metal from the structure under different conditions such that the adsorbent can be regenerated for further cleanup.

The process using adsorbents is thought to be the most effective and convenient technique for removing of dissolved uranium from water because of the high selectivity, the ease for handling, simplicity, efficiency, the safety to the environment, low cost in adsorption, and so on.

In recent years there has been considerable interest in the use of different kind of adsorbents for the sorption of uranium from aqueous solutions [5, 6].

Our study focuses on the removal of uranium by adsorbents containing amidoximated IPN.

2. EXPERIMENTAL

2.1. Leaching of Radioactive Elements from Disposed Fly Ash

Dry coal or ash was mixed (solid:liquid 1:10, by weight) with deionized water (18 MΩ cm) and the slurry stirred overnight (>12 h). The pH was reduced to 1.0 by the addition of 1 M H_2SO_4 (VWW, PA grade) and maintained at 1.0 by the subsequent additions of 1 M H_2SO_4 [7]. Leaching was terminated by the separation of effluent and solid residue after 30 min (Munktell, OOR filter paper) and filtration was completed within 30–40 s the volume of the effluent was measured and a sample was taken for gamma spectrometric and ICP-AES/MS/SMS analysis. The solid residue was dried and a sample from solid residue was also taken for gamma spectrometric analysis to compare the results of radioisotopic analysis [7].

The experiment was then repeated for a new stopping time, with a fresh sample from the homogenized batch. We report analyses of the aqueous phase after 0.5, 1, 2, 4, 8, 16, and 32 h of leaching.

2.2 Removal of Uranium from Contaminated Water

This study is based on the adsorption of uranium by IPN. IPN of polyethylene glycol (PEG) and acrylonitrile (AN), which are synthesized by using ^{60}Co gamma radiation as a source of polymerization of AN and crosslinking of PEG. However, during irradiation PEG–AN graft type cross-links and the crosslinking of AN to itself also takes place to some extent. These IPN are amidoximated and used for uranium adsorption from aqueous media [6].

3. RESULTS AND DISCUSSION

3.1. Leaching of Radioactive Elements from Disposed Fly Ash

Table 1. Specific activity and concentration of elements in untreated coal and asphalt ashes

Sample	Nuclide	Specific activity, Bq/kg	Element	Gamma spect. ppm	ICP, ppm
Yatagan	^{238}U	397	U	30	47.3
	^{232}Th	9.67	Th	2.4	1.8
	^{40}K	10.7	K	3450	5010
	^{226}Ra	17.9	–	–	–
Tuncbile	^{238}U	311	U	24	23.9
	^{232}Th	4.23	Th	1.1	3.4
	^{40}K	44.1	K	14200	10800
	^{226}Ra	10.7	–	–	–
Silopi	^{238}U	1035	U	80	71
	^{232}Th	0.72	Th	0.23	0.48
	^{40}K	70.5	K	24350	26400
	^{226}Ra	33.1	–	–	–

3.2. Ashes Contaminated by Ra-226

The specific activities of untreated ash from Yatagan, Silopi, and Tuncbilek. Gamma intensities for the liquid phase were estimated from ICP data and gamma spectra gave stable total activities, i.e., for the solid plus liquid phases, for all three samples. Concentration of elements in untreated coal and asphaltit ashes by ICP and gamma spectrometric system are shown in Table1.

Figure 1 shows the Ra-226 specific activity during acid leaching of lignite and asphaltite–ash. We notice that the activity in the coal is low and set by the original U-238 concentration, since little material is moved, save for possible volcanic activity and the dissolution and precipitation of minerals by underground water.

The parallel behavior of Calcium (Ca) and Radium (Ra) is not universal. Ca is leached to near unity from Tuncbilek ash [8], but Ra is not. The Ca concentration is unusually low in this ash, which means that Ca solubility, although small, is sufficient to dissolve the element. Radium sulfate solubility is 0.00002 g/l at 25°C and 0.00005 g/l at 45°C. The limited solubility has lead to suggestions to coprecipitate Ca and Ra for disposal/metal recovery [9].

Conclusions regarding radium leaching are not easily drawn from data of the solid phase/residue alone. It appears from Figure 1 that some, possibly as much as 50%, of the radium atoms are dissolved from Yatagan and Tuncbilek lignite–ash, but little from Silopi asphaltite ash. Ra association to Ca components in these ashes can not be firmly concluded, but we note that radium dissolution from Tuncbilek lignite–ash resembles aluminum leaching, suggesting the possibility that Ra may be present as a dopant in an oxide with aluminum as a major element. Comparisons with the leaching profiles for Yatagan lignite–ash reveal a similarity between Ra, As, Mg, Mn, and P [10].

3.3. Ashes Contaminated by U-238

Data for lignite-and asphaltite–ash (Figure 2) shows very rapid leaching of uranium during the first 30 min, followed by slow leaching during a longer period. The uranium activity during leaching (<0 h) agreed well with the measured specific activity of Tuncbilek untreated ash (311 Bq/kg), but deviated slightly (234 Bq/kg) at longer times. The total specific activity of Yatagan during leaching was nearly constant (567 Bq/kg), but higher than the measured value for the untreated ash (397 Bq/kg). The situation was similar for Silopi with a stable calculated total specific activity (1846 Bq/kg) vs. a lower measured activity for the untreated ash (1035 Bq/kg). It is concluded that the deviations come from batch inhomogeneities rather than adverse loss of material.

The rapid initial phase and the absence of leaching after 8–16 h suggest surface condensation, a separate uranium phase or association with calcium containing phases [10]. Surface condensation is a fly-ash phenomenon and less likely for the present ashing procedure. The measured specific activity shows a reduction from 39 Bq/kg (untreated ash) to 75 Bq/kg (32 h, Figure 2). 30 min treatment gave 142 Bq/kg. The equivalent values for Tuncbilek were: untreated ash, 311 Bq/kg; solid phase after 30 min leaching, 198 Bq/kg; and solid phase after 32 h, 76 Bq/kg. Silopi showed an unprecedented specific activity of 1035 Bq/kg for the untreated ash. Acid leaching effectively reduced this to 414 Bq/kg (30 min) and 262 Bq/kg (32 h).

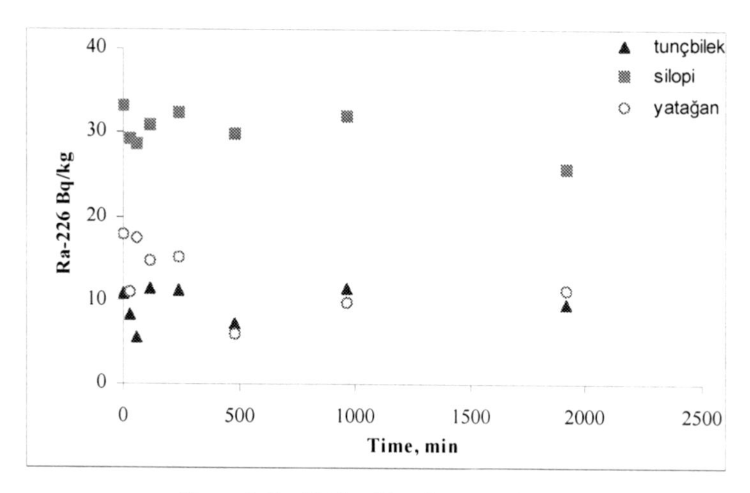

Figure 1. Ra-226 leaching from coal fly ash

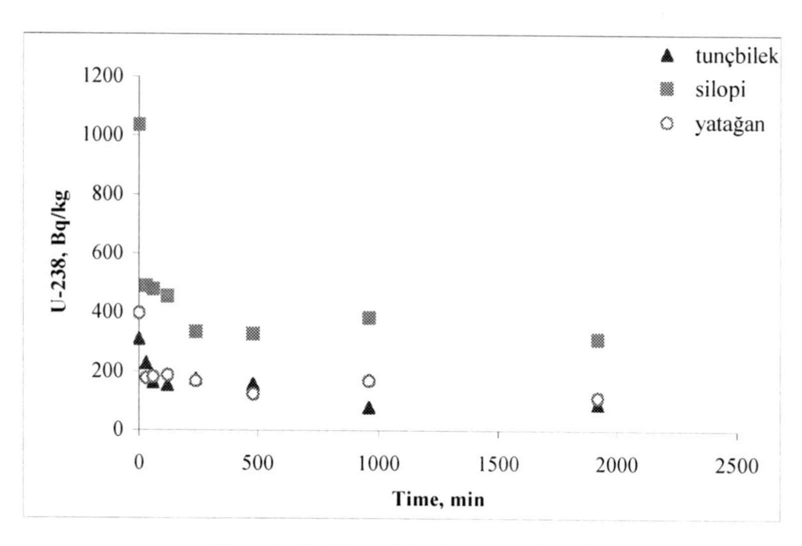

Figure 2. U-238 leaching from coal fly ash

The predominant valence states of uranium in nature are U(VI) and U(IV). Briefly, U(VI) is water soluble, but U(IV) is not [11]. Ashing at modest temperatures boosts uranium dissolution in sulfuric acid [8], probably due to the oxidation of U(IV). Moreover, that $U(SO_4)_2.4–9\ H_2O$ is soluble to more than 100 g/l and $UO_2SO_4.3H_2O$ to around 200 g/l. $2(UO_2S0_4).7H_2O$ has low solubility in water. High temperature combustion may promote uranium doped oxides, which must be dissolved to extract uranium during acid leaching.

Acidity is also critical. UO^{2+} (uranyl) is the stable form in solution at pH 1.0 [11, 12] and U will not leach at high pH. At pH 11 very low concentrations of uranium were measured in a lake adjacent to an ash depository, while radium concentration was an order of magnitude higher [13]. It is obvious that pH as well as oxidation potential must be considered.

3.4. Removal of Uranium from Contaminated Water

The removal of U, Ra-226, and other natural radionuclides from drinking water depends on their speciation. In order to find effective removal methods for these nuclides the knowledge on their speciation in ground water should be known. When selecting methods for removal of the radionuclides from ground waters it must be taken into consideration that these radionuclides exist mainly bound in particles in water. Thus, it can be assumed that the oxidation state of uranium has no significant role in removing uranium from drinking water.

The results obtained in adsorption by IPN demonstrate that the uranium potential of IPN for the removal of uranyl ions from aqua solutions reached about 550 mg U/g IPN within 500 min of contact time, Figure 3. Since amidoxime polymers displays a significant affinity for a variety of metal ions present in waters more research should be conducted to assess these co-products along with uranium from aqua media [14].

It was found that natural radionuclides, especially uranium and Ra-226 from drinking water were best removed by ion exchange techniques (reductions from 50 to 99%) when both anion and cation resins were applied. Strong basic anion resins for the removal of uranium and strong acidic cation resins for radium removal performed best. The efficiency for Pb-210 and Po-210 varied a lot, since the main proportion of these nuclides is supposed to be particle-bound in natural waters, and therefore no ion exchange process in the real sense, but adsorption to the resins is responsible for their reduction. Hydroxylapatite was found to have a good capability to remove in average from 95.6 to 99.8% of uranium, Ra-226, Pb-210 and Po-210. Aeration methods were highly effective in removing radon from water at water works. Removal efficiencies of more than 98% can be achieved, for example, with diffused bubble and packed tower aerators. Most aeration facilities can be constructed to achieve radon removal efficiencies of more than 95% or even more than 99% [15].

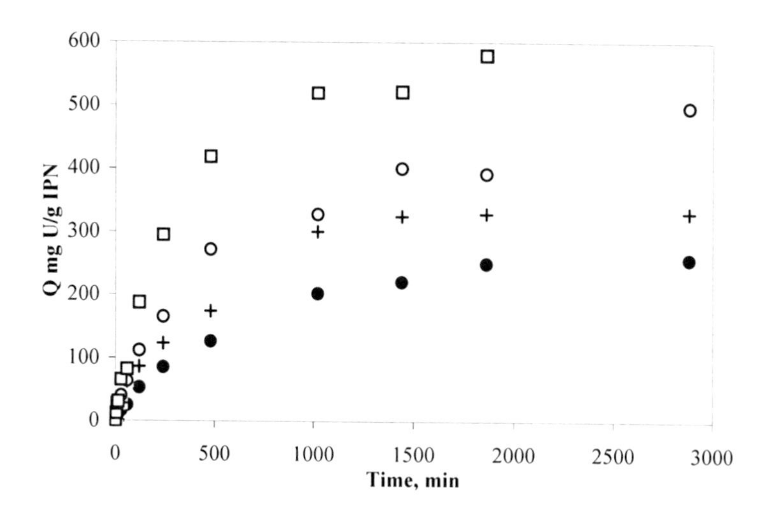

Figure 3. Uranium adsorption per gram of IPN with the adsorption time at temperatures of 290 K (•), 298 K (+), 308 K (o), and 318 K (□)

4. CONCLUSION

Using adsorbents is thought to be the most effective method for recovering the low concentrations of uranium in the aqueous media because of their fast and selective uptake of uranium, a sufficient adsorption capacity and high physical and chemical stability against the media. It was found that the polymeric adsorbent has a very high adsorption ability for uranium (~600 mg U/g IPN) and can be used in adsorption studies to remove uranium from aqueous system. Another point is to remove uranium from wastewate and other radioactive elements in their source before mingling to the groundwater.

Acknowledgment: NATO provided finacial support to the author to present this paper in the Advanced Research Workshop in Istanbul, Turkey during October 12–16, 2006.

5. REFERENCES

1. Kalin M, Wheeler WN, Meinrath GJ. (2005). Envi. Radioactivity. 78, 151–177.
2. Aycik GA, Ercan AJ. (1997). Environ. Radiact., 35, 23–35.
3. Aycik GA, Ercan A. (1998), Applied Spectroscopy Reviews 33, 237–251.
4. Querol X, Umana JC, Alastuey A, Ayora C, Lopez-Soler A, Plana F. (2001). Fuel 80 801–813.
5. Kilislioglu A, Bilgin B. (2003). Applied Radiation and Isotopes 58, 155–160.
6. Guler, H, Sahiner N, Aycik, GA, Guven, O. (1997). J. Applied Polymer Science 66 66 (1997) 2475–2480.
7. Aycik GA, Paul M, Sandstrom A Paul J. (2003). IV. International Conference on Nuclear and Radioation Physics, 15–17 September 2003, Institute of Nuclear Physics, Almaty, Kazakshtan.
8. Seferinoglu M, Paul M, Sandström Å, Köker A, Toprak S, Paul J. (2003). Fuel 82, 1721–1734.
9. Million JB, Sartain JB, Gonzalez RX, Carrier WD. (1994). J.Environmental Quality 23, 671–676.
10. Paul M, Seferinoglu M, Aycik GA, Sandstrom A, Smith M, Paul J. (2006). Acid leaching of ash and coal: Time dependence and trace element occurrences, Int. J. Mineral Process, 79, 27–41.
11. Murphy WM, Shock EL. (1999). In Uranium: Mineralogy, Geochemistry and the Environment, Reviews in Mineralogy, Eds. P.C. Burns and R. Finch, Mineralogical Society of America, 38, 221–253.
12. Suzuki Y, Banfield JF. (1999). In Uranium: Mineralogy, Geochemistry and the Environment, Reviews in Mineralogy, Eds. P.C. Burns and R. Finch, Mineralogical Society of America, 38, 393–432.
13. Hubert S, Barthelot K, Fourest B, Lagarde G, Dacheux N, Baglan N, http://www.cea.fr/conferences/atalante 2000/p5–13.pdf
14. Aycik GA, Gurellier R, Sarikaya Y. (2006). J.Radioanal. Nucl. Chem. (2006) in press.
15. Treatment Techniques for Removing Natural Radionuclides from Drinking Water (TENAWA), (1999). Nuclear fission safety programme of the European Union Report, Contract No: FI4P-CT96-0054.

WASTEWATER MANAGEMENT AND ITS SUSTAINABLE USE IN TURKEY

Murat Türkeş and Zahide Acar
Çanakkale Onsekiz Mart University, Faculty of Arts and Sciences, Department of Geography, 17020, Çanakkale, Turkey

Corresponding author: zahideacar@hotmail.com

Abstract:

Mankind has utilized water sources and wetlands since their earlier times. As mankind's dependency on water increased, water became a strategic commodity. The process of industrialization has increased the human pressure on the resources, and consumption has become unsustainable. The usable freshwater resources in the world have been significantly declining. The available potential of freshwater resources on the planet Earth is decreasing very rapidly as a direct consequence of unnatural processes including human-induced climate change and globalization as-well-as the lack of proper management strategies. The climate change is likely to create significant droughts in many parts of the World.

Keywords: Water scarcity, climate change, precipitation, drought and desertification, sustainable usage.

1. INTRODUCTION

Water is a vital resource for sustainable development. Although the world's population increased three times in the twentieth century, the consumption of water resources increased six times [1]. This overconsumption is not sustainable in any way, and does not provide equal opportunity for its beneficiaries.

In the future decades, many large cities of the Northern Hemisphere that are located in the North America, the Europe, and the Middle East will be under the stress of high population. Particularly the water consumption levels of these large metropolitan cities are predicted to increase significantly. According to a Food and Agriculture Organization (FAO) report, 41% of the world population has already experienced water scarcity in the year 1995 [2]. This ratio is projected to increase to 49% in the year 2025. In additions, the world population is projected to increase at a much faster rate at the regions that already suffer from serious water scarcity (Table 1).

M.K. Zaidi (ed.), Wastewater Reuse–Risk Assessment, Decision-Making and Environmental Security, 415–420.

Table 1. The comparison of world's present and projected water resources between the years of 1995 and 2025 considering the population [2]

Status	Water resources (m^3/capita)	Year			
		1995		2025	
		Population (million)	Population (%)	Population (million)	Population (%)
Water scarcity	<500 500–1,000	1,077,587	19 10	1,783,624	259
Water stress	1,000–1,700	669	12	1,077	15
Water availability	>1,700	3,091	55	3,494	48
Unclassified	–	241	4	296	4
Total		5,665	100	7,274	100

1.1. Sustainable Use of Water

The increased speed of industrialization that occurred after the industrialization revolution as well as the changing patterns of production and consumption have created extra burden on the environment. Increase in global and regional average surface mean temperatures started after the nineteenth century became more profound after the year 1980, and numerous record high temperatures were observed in many parts of the world. The most sophisticated climate models predicted temperature increases of 1.4–5.8°C between the years of 1990 and 2100 in global average surface mean temperatures [3]. As in many countries of the subtropical and mid- and high-latitude regions of the Northern Hemisphere, surface air temperatures in Turkey have increased significantly particularly in the warm period of the year [4]. Increased temperature patterns dominated particularly in spring and summer minimum (nighttime) temperatures. Along with these significant changes, diurnal temperature ranges of Turkey have also indicated statistically significant and spatially coherent decreasing trends in Turkey [5].

Observed changes in the global climate system have created a decrease in the total areas of some lakes, snow covers, sea and land ices, mountain glacials and icecaps over the poles, an increase in the average sea levels and serious deterioration in the ecological system. It is also very likely that the global climate change would be responsible for the abnormal occurrences of floods and droughts as well as the increased frequencies of epidemic diseases in many parts of the world [6]. According to the various reports published by the United Nations, the water potential of countries changes with respect to their levels of development as well as the climate region that they are located. In this regard, many industrialized countries utilize a majority of their water resources for the purpose of industrial production, whereas developing and less-developed countries located in the monsoon Asia, such as India and Bangladesh use a large portion of their water resources for agricultural purposes. The distributions of water use in the world as a function of different sectors are presented in Table 2.

Table 2. The sectorial percent distribution of clean and fresh water resources with respect to development levels of countries [7]

Sector	World	Developed countries	Developing countries	Less-developed countries	Europe	Turkey
Agriculture	67–70	39	52	86	33	72
Industry	22–23	46	38	7	51	12
Human consump-tion	8–10	15	10	7	16	16

Today, a common problem of all countries experiencing water scarcity is that they are using a large portion of their water resources for agricultural purposes. The water used in agriculture is easily polluted upon the application of various fertilizers and chemicals. Unconscious irrigation practices create salinity problems in soils and are responsible for decreases in yields. Furthermore, incorrect implementations of water management plans together with uncontrolled discharges of wastewaters amplify the water quality and quantity problems and deplete the available fresh-water resources in many countries.

1.2. General Condition of World's Water Resources

Variability over a wide range of scales in space and time is the most obvious feature of the global pattern of the hydrological cycle and of its component parts which determine water resources. But space and time deal very unfairly with certain parts of the world: while some regions and periods experience water scarcity, others are replete with water resources [1].

The freshwater which is used by humans is little amount in the world. The total amount of water in the world is around 1.4 billion km^3. Of this total, about 1.3 billion km^3 (97.5%) is saline water and only 0.035 billion km^3 (2.5%) is considered to be freshwater [1].

The average per capita water use in the world is about 800 m^3. It must be noted that a population of 1.4 billion that corresponds to about 20% of the world population lives without adequate water resources. Moreover, 2.3 billion of the world's population does not have good quality water [8].

Some average water consumption values are presented in Table 3. It should be mentioned that an uneven distribution is observed when the surface areas and the populations of some countries in these continents are compared with their corresponding water consumption values. It could also be seen from the figures that Turkey has a water consumption value that is much less than the world's average. Turkey is not a water-rich country.

Table 3. Average water consumption rates [9]

Region/country	Water per capita (m^3/year)
Turkey	1,735
Asia	3,000
Western Europe	5,000
Africa	7,000
South America	23,000
World	7,600

1.3. Water Resources and Wastewater Management in Turkey

Reuse of wastewater is an integral part of the contemporary water management is one of the well-known plans of action to cope the problem of water security and water shortage in the world. National actions for the reuse of wastewater have been commonly increased since the 1980s, because the global environmental issues related with the water have been getting worst. Some applications dealing with the reuse of wastewater also exist in Turkey even though those have not been enough and widely applied yet.

2. RESULTS AND DISCUSSION

Observed increases in the global average surface mean temperatures after the year 1980 and observed decreases in the precipitations totals of some tropical and subtropical regions such as in the Sahel and the Mediterranean basin since 1960s and 1970s, respectively, are likely to intensify the demand for water.

Water resources will be under more stress in many countries in the near future. It is also expected for the future that drier conditions will prevail in Turkey, which is characterized mainly with the macro Mediterranean climate type. Consequently, it is very likely that drought and desertification threats will occur much frequently in countries like Turkey due to the negative impacts of projected global climate change, in addition to the existing drier than the long-term average precipitation conditions over much of Turkey persisted since early 1970s [10–13]. There has been a general tendency from humid conditions of around 1960s towards dry subhumid climatic conditions in the aridity index values of many stations of Turkey [14]. For instance, in some stations of Aegean Region, there has been a significant change from humid conditions to dry subhumid or semiarid climatic conditions. The management of water resources should be more effective and sustainable.

3. CONCLUSIONS

The management of water resources should be governed more effective and sustainable. Reuse of the wastewater in irrigation is one of the sustainable and environmentally sound ways of decreasing effect of the water scarcity and of combating the global problem. Resolving the water governance must be a key priority, and any response will need to emphasize an integrated approach to effective water resource management and governance.

Acknowledgment: We express our thanks to NATO ARW organizing committee and the NATO Science Program for financial support for participation at the NATO ARW held in Istanbul, Turkey during October 12–16, 2006.

4. REFERENCES

1. UNWWDR, (2003). Water for People, Water for Life. United Nations World Water Development Report (2003), UNESCO, Paris.
2. FAO, (2002). Crops and Drops Making the Best Use of Water for Agriculture, United Nations Food and Agriculture Organization, Rome.
3. IPCC, (2001a). Climate Change 2001: The Scientific Basic-Contribution of Working Group I to the Third Assessment Report of the Intergovernmental Panel on Climate Change (IPCC), Eds.: J. T. Houghton et al., University Press, Cambridge.
4. Türkeş M, Sümer UM, Demir İ. (2002). Re-evaluation of trends and changes in mean, maximum and minimum temperatures of Turkey for the period 1929–1999. International J. Climatology 22: 947–977.
5. Türkeş M, Sümer UM. (2004). Spatial and temporal patterns of trends and variability in diurnal temperature ranges of Turkey. Theoretical and Applied Climatology 77: 195–227. 1. ÇOB, 2005, Ankara. (In Turkish).
6. IPCC, (2001b). Climate Change 2001: Impacts, Adaptation and Vulnerability – Contribution of Working Group II to the Third Assessment Report of the Intergovernmental Panel on Climate Change (IPCC), Eds.: J. J. McCarthy et al., University Press, Cambridge.
7. ÇOB, (2005). Turkey National Action Program to Combat Desertification. Republic of Turkey Ministry of Environment and Forestry (ÇOB), Ankara.

8. WSSD, (2002). United Nations Economic and Social Council's Preparatory Document for WSSD 2002 Implementing Agenda 21.
9. DPT, (2001). Sekizinci Beş Yıllık Kalkınma Planı, İçme Suyu, Kanalizasyon, Arıtma Sistemleri ve Katı Atık Denetimi Özel İhtisas Komisyonu Raporu (2001–2005). Republic of Turkey Prime Ministry State Planning Organization, Ankara. 10. Türkeş M. (1996). Spatial and temporal analysis of annual rainfall variations in Turkey. International J. Climatology 16: 1057–1076.
11. Türkeş M. (1998). Influence of geopotential heights, cyclone frequency and Southern Oscillation on rainfall. Climatology 18: 649–680.
12. Türkeş M, Erlat E. (2003). Precipitation changes and variability in Turkey linked to the North Atlantic Oscillation. Climatology 23: 1771–1796.
13. Türkeş M, Erlat E. (2005). Climatological responses of winter precipitation in Turkey to variability of the North Atlantic Oscillation during the period 1930-2001. Theoretical and Applied Climatology 81: 45–69.
14. Türkeş M. (1999). Vulnerability of Turkey to desertification with respect to precipitation and aridity conditions. Turkish Envir. Science 23: 363–380.

WASTEWATER TREATMENT USING NEW CHELATING GRAFTED MEMBRANE

M. Abdel Geleel
National Centre for Nuclear Safety and Radiation Control,
Atomic Energy Authority,
3 Ahmed El-Zomor St., Nasr City, 11762 Cairo, Egypt

Corresponding author: mageleel2000@yahoo.com

Abstract:

Water reuse plays an important role in water resource, wastewater, and ecosystem management in many countries. Reclaimed water reduces discharges to surface waters, recharges groundwater, and postpones costly investment for development of new water sources and supplies. Water reuse involves taking domestic wastewater, giving it a high degree of treatment, and using the resulting high-quality reclaimed water for a new, beneficial purpose. This study concerning onto the treatment of wastewater by using new chelating grafted membrane. Application of the prepared membrane in the treatment of wastewater was tested.

Keywords: Wastewater, grafting, environment, lead, strontium, heavy metals.

1. INTRODUCTION

A number of methods are available for the removal of metal ions from aqueous solutions. These are ion exchange, solvent extraction, reverse osmosis, precipitation, and adsorption [1–3]. Adsorption is an effective and convenient technique in the separation and isolation of elements from aqueous solutions. The greatest advantage of adsorption is the possibility to separate small amounts of substances from a very large volume of solution [4]. The adsorption of trace amounts of various ions on inorganic and organic adsorbents from aqueous solution has been studied by numerous scientists [5–13].

Ion-exchange membranes play an important role in modern technology, especially in the separation and purification of materials. Grafting of polymers with a mixture of monomers is important since different types of chains containing various functional groups can be introduced into the polymer structure. Copolymer of acrylic and vinyl monomers grafted onto low density polyethylene were used for the extraction of several heavy metal ions [14–18].

M.K. Zaidi (ed.), Wastewater Reuse–Risk Assessment, Decision-Making and Environmental Security, 421–432.

Lead is a poisonous metal that can damage nervous connections (especially in young children) and cause blood and brain disorders. Long-term exposure to lead or its salts (especially soluble salts or the strong oxidant PbO_2) can cause nephropathy, and colic-like abdominal pains.

The concern about the role of lead in mental retardation in children has brought about widespread reduction in its use (lead exposure has been linked to schizophrenia). Paint containing lead has been withdrawn from sale in industrialized countries, though many older houses may still contain substantial lead in their old paint. The most common liquid waste is aqueous in which the waste materials are either dissolved in water or in a liquid which is mainly composed of water. Strontium waste is one of these wastes that are produced from different applications. This type of waste should be treated before disposed of by dispersal into the sewage system (if low-level). The aim of this work is

1. To study the ability of the grafted and chemically treated polypropylene membranes to adsorb lead(II) and strontium(II) ions from its aqueous wastes.
2. To elaborate the parameters affecting the adsorption processes.
3. To evaluate the adsorptive capacity and to describe the sorption mechanisms responsible for the sorption process.

2. EXPERIMENTAL

Polypropylene (PP) of thickness 30 μm was used. Treatment process of the grafted PP was carried out by using 5% aqueous KOH solution, 3 mole HCl, and 1:1 alcoholic solution of methyl alcohol containing 5% hydroxylamine. The process was carried out by boiling the solution in a reflux for 16 h. After treatment, the films immersed in bi-distilled water for 24 h to remove excess reagents, then rinsed with bi-distilled water, and dried at 40–50°C in oven for 24 h.

All reagents were of AR grade chemicals. Stock solutions of the test reagents were prepared by dissolving lead and strontium nitrate in double distilled water. The pH of the test solution was adjusted using reagent grade dilute sulfuric acid and sodium hydroxide. The batch technique was used to study the effect of different parameters on the equilibrium and rate of uptake of lead and strontium by grafted and chemically treated polypropylene membranes. The general procedure used is as follows:

One gram of polypropylene membrane with definite degree of grafting (DG = 170) was added to a known volume (20 ml) of metal feed solution in a 50 ml bottle was shaken, at constant temperature using a thermostated shaker bath, GEL-1083 model. The remaining metal salt in its feed solution was determined by an atomic absorption spectrophotometer (Perkin–Elmer Model 2380) using lamps for Pb and Sr. Merck atomic absorption standard solution was used for the calibration process. The pH was measured using digital pH meter model 609.

2.1. Adsorption Isotherm

Two important physiochemical aspects for the evaluation of the adsorption process as a unit operation are the equilibrium of the adsorption and the kinetics. Equilibrium studies gives the capacity of the adsorbent [19]. The equilibrium relationship between adsorbent and adsorbate are described by adsorption isotherms, usually the ratio between the quantity adsorbed and that remaining in solution at a fixed temperature at equilibrium. There are two types of adsorption isotherms: Langmuir and Freundlich.

2.2. Langmuir Isotherm

The Langmuir adsorption isotherm is often used for adsorption of a solute from a liquid solution. The Langmuir adsorption isotherm is perhaps the best known of all isotherms describing adsorption and is often expressed [20] as

$$Q_e = X_m KC_e /(1 + KC_e) \tag{1}$$

Where Q_e is the adsorption density at the equilibrium solute concentration C_e, C_e is the concentration of adsorbate in solution (ppm), X_m is the maximum adsorption capacity corresponding to complete monolayer coverage (ppm of solute adsorbed per gram of adsorbent), K is the Langmuir constant related to energy of adsorption. The above equation can be rearranged to the following linear form:

$$1/Q_e = 1/X_m + 1/X_mK.\ 1/C_e \tag{2}$$

The linear form can be used for linearization of experimental data by plotting $1/Q_e$ against $1/C_e$. The Langmuir constants X_m and K can be evaluated from the slope and intercept of linear equation.

2.3. Freundlich Isotherm

The Freundlich isotherm is the earliest known relationship describing the adsorption equation and expressed as

$$Q_e = K_f\, C_e\, 1/n \tag{3}$$

Where Q_e is the adsorption density (ppm adsorbate per gram of adsorbent), C_e is the concentration of adsorbate in solution (ppm), K_f and n are the empirical constants dependent on several environmental factors, and n is greater than one. This equation is conveniently used in the linear form by taking the logarithm of both sides:

$$\ln Q_e = \ln K_f + 1/n \ln C_e \tag{4}$$

A plot of ln C_e against ln Q_e yielding a straight line indicates the confirmation of the Freundlich isotherm for adsorption. The constants can be determined from the slope and the intercept of the linear relation.

3. RESULTS AND DISCUSSION

The grafted and chemically treated polypropylene membranes, were studied in the removal of Pb(II) and Sr(II) from aqueous solution under different experimental conditions such as pH, contact time, ion concentration, and temperature. The experimental results and the relevant observations are discussed in the following sections. To assess the type of sorption of Sr(II) and Pb(II), 1 g of the prepared membranes were added to 100, 1,000 ppm concentrations of the investigated metals and the pH adjusted at 5. The change in pH after 4 h was recorded (Table 1).

From this table we can conclude that the type of sorption of Sr(II) and Pb(II) are mainly ion exchange in the case of treated membranes. For untreated membranes, the uptake is mainly sorption on the surface of PP-*g*(AAc/AN).

3.1. Effect of pH

Removal of the metal ions as a function of pH by using grafted and chemically treated polypropylene membranes is presented in Figures 1 and 2. These studies were carried out at the initial concentration of 150 ppm for Pb(II) and Sr(II). It was observed that the amount of Pb(II) and Sr(II) adsorbion increases with the increase in pH until pH = 5, after this point, lead and strontium adsorption began to decrease. The removal of Pb(II) and Sr(II) are in the following order:
PP-*g*(AAc/AN)-KOH > PP-*g*(AAc/AN)-HCl > PP-*g*(AAc/AN)-NH_2OH > PP-*g*(AAc/AN)

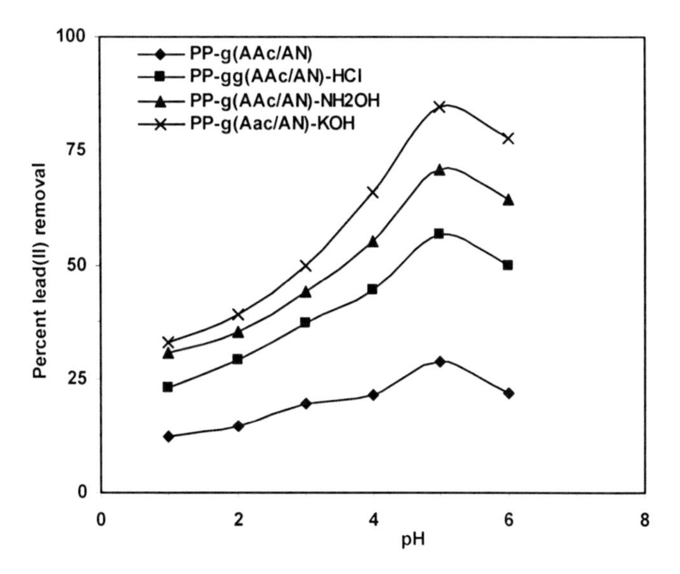

Figure 1. Effect of pH on the removal of lead (II) by using grafted and chemically treated polypropylene membranes (t=4 h, T=50 C, [pb] = 100 ppm, D.G.=170)

The highest Pb(II) and Sr(II) adsorption yield was achieved at pH 5.0 for both grafted and chemically treated polypropylene membranes. Above pH 5.0 however, it is worthy to mention that precipitation occurs. Thus for further studies the optimum pH chosen for Pb(II) and Sr(II) was 5.0 to correlate the removal with the adsorption process.

3.2. Effect of Metal Ion Concentration

The removal of Pb(II) and Sr(II) from its waste solution by using grafted and chemically treated polypropylene membranes as a function of their concentration were studied by varying the metal concentration from 25 to 200 ppm while keeping all other parameters constant. The results are shown in Figures 3 and 4. These figures show that the percentage removal of Pb(II) and Sr(II) increases with increasing metal ion concentration and attains equilibrium at 100 ppm. These results indicate that energetically more favourable active sites become involved with increasing the metal ion concentration [7, 21–23]. On the basis of these results, a 100 ppm concentration of both metal ions was used for all further studies.

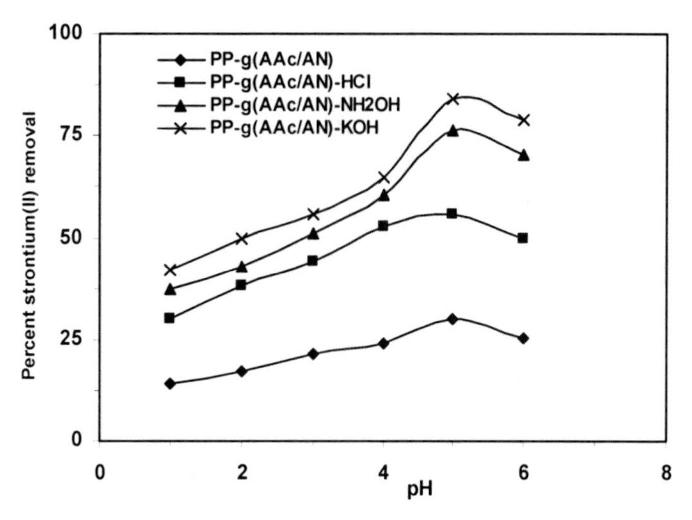

Figure 2. Effect of pH on the removal of lead(II) by using grafted and chemically treated polypropylene membranes (t = 4 h, T = 50°C, [pb] =100 ppm, DG = 170)

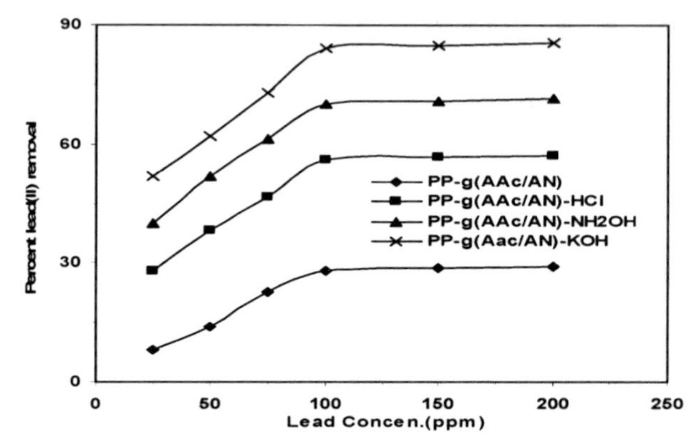

Figure 3. Effect of metal ion concentration on the removal of lead (II) by using grafted and chemically treated membranes (t = 4 h, T = 50°C, pH = 5 DG = 170)

Table 1. Change in pH values after 4 h

PP grafted membranes	Initial pH	pH after ~4 h			
		Pb(II)		Sr(II)	
		100 ppm	1,000 ppm	100 ppm	1,000 ppm
PP-g(AAc/AN)	5	5	5.1	5	4.9–5.1
PP-g(AAc/AN)	5	3.9–4.2	3.5–3.6	4.0–4.3	3.7–3.8
PP-g(AAc/AN)-KOH	5	4.5–4.8	4.3–3.4	4.9–4.7	4.5–4.6
PP-g(AAc/AN)-NH_2OH	5	5.9–6.0	6.1–6.3	5.9–6.2	6.4–6.5

3.3. Effect of Contact Time

From the economical point of view, the time of treatment is an important factor for metal uptake percent. Also, the efficiency of grafted and chemically treated polypropylene membranes can be determined from the time required to adsorb the maximum value of metal by chelation or complexation with its functional groups. Figures 5 and 6 shows the metal uptake as a function of time for Pb(II) and Sr(II) using grafted and chemically treated polypropylene membrane. It can be seen that, the removal percentage increases with time and attains equilibrium after 4 h; this may be due to the diffusion coefficient of these metals through the porous ionic

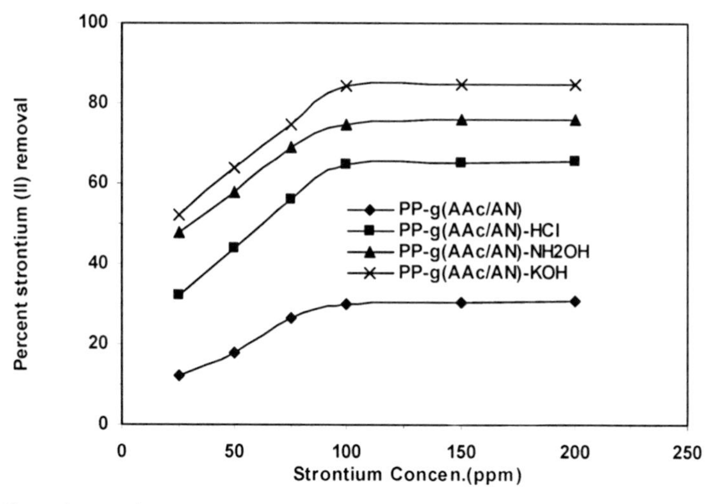

Figure 4. Effect of metal ion concentration on the removal of strontium(II) by using grafted and chemically treated membranes (t = 4 h, T = 50°C, pH = 5, DG = 170)

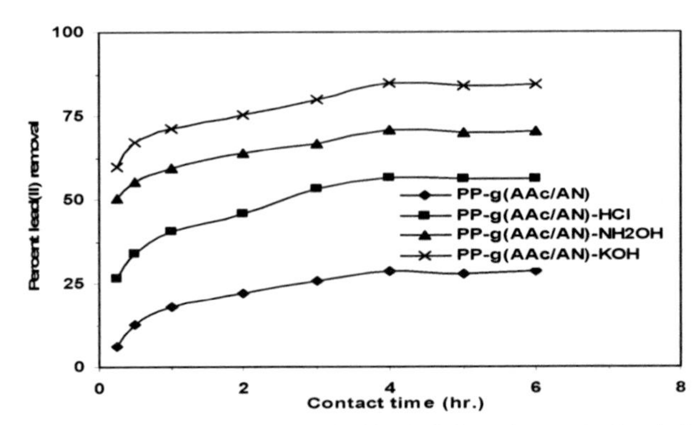

Figure 5. Effect of contact time on the removal of lead(II) by using grafted and chemically treated polypropylene membranes (pH = 5, $T = 50°C$, [Pb] = 100 ppm, DG = 170)

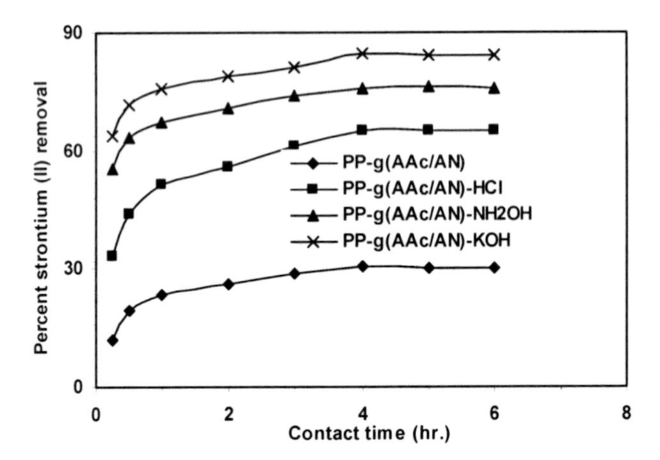

Figure 6. Effect of contact time on the removal of strontium(II) by using grafted and treated polypropylene membranes (pH = 5, $T = 50°C$, [Sr] = 100 ppm, DG = 170)

membrane which is mainly dependent on their polarity, electronic configuration, ionic radii, etc. and also importantly on the nature of interaction with the functional groups of the membranes [24]. It is also observed that the chemically treated polypropylene membranes show high tendency to metal chelation or complexation due to the free penetration of metal ions and the high surface area of the adsorbent materials that having functional groups suitable for adsorption and chelation processes [25]. On the bases of these results a 4 h shaking period was selected for all further studies.

3.4. Effect of Temperature

The effect of temperature on the removal of Pb(II) and Sr(II) was investigated between 30 and 80°C. As shown in Figure 7, the uptake of the metal ions increased with the increase in temperature and attains equilibrium at 50°C thereby indicating the process to be endothermic, at higher temperature no significant increase in the uptake was observed. 50°C was used for all further studies.

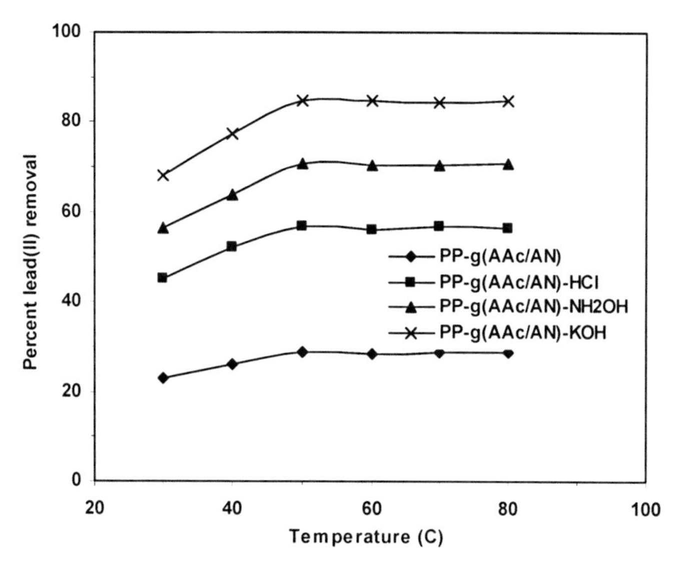

Figure 7. Effect of temperature on the removal of lead(II) by using grafted and chemically treated polypropylene membranes (t = 4 h, pH = 5, [Pb] =100 ppm, DG = 170)

3.5. Adsorption Isotherm

To establish the adsorption isotherms and evaluate the adsorption capacity of grafted and chemically treated polypropylene membranes, individual metal was tested in triplicate. The results described here are the average of three sets of tests.

3.6. Batch Adsorption Test Results for Pb(II) and Sr(II)

The linear equations, the correlation coefficients and isotherm constants, and the fact that the regression coefficients were very close to one, indicating good linearity, confirm that the adsorption of Pb(II) and Sr(II) on grafted and chemically treated polypropylene membranes follows the two theories of adsorption.

Table 2. Linear regression for Langmuir & Freundlich isotherm for Pb(II)

Membrane type	Langmuir data		Langmuir data	
	Linear equ.	Corr. coeff (r^2)	Linear equ.	Corr. coeff (r^2)
PP-g(AAc/AN)	Y=0.088×–0.0032	0.96	Y=3.12×–2.3	0.96
PP-g(AAc/AN)	Y=0.98×–0.0019	0.96	Y=1.85×–0.299	0.98
PP-g(AAc/AN)–KOH	Y=0.14×–0.0018	0.98	Y=2.07×–0.28	0.96
PP-g(AAc/AN)–NH_2OH	Y=0.37×–0.0025	0.95	Y=1.9×–2.1	0.985

Table 3. Linear regression for Langmuir & Freundlich isotherm for Sr(II)

Membrane type	Langmuir data		Langmuir data	
	Linear equ.	Corr. coeff (r^2)	Linear equ.	Corr. coeff (r^2)
PP-g(AAc/AN)	Y=0.45x–0.0039	0.98	Y=1.69x–0.0039	0.98
PP-g(AAc/AN)	Y=0.14x–0.0032	0.96	Y=2.006x–0.58	0.968
PP-g(AAc/AN)-KOH	Y=0.07x–0.0023	0.957	Y=3.12x–2.3	0.96
PP-g(AAc/AN)-NH_2OH	Y=0.072x–0.0017	0.958	Y=2.0x–0.518	0.965

Table 4. Comparison of the isotherm constants for Sr(II)

Membrane type	Langmuir const. *X*m	Freundlich const. *K*f	1/*n*
PP-g(AAc/AN)	546.4	2.67	0.53
PP-g(AAc/AN)	679.2	1.78	0.49
PP-g(AAc/AN)-KOH	763.4	1.53	0.4
PP-g(AAc/AN)-NH_2OH	620	1.68	0.5

Langmuir constant *X*m (maximum adsorption capacity) and the Freundlich constant *K*f were obtained from the linear equations. The values are summarized in Tables 2 and 3. The maximum adsorption capacity of grafted and chemically treated polypropylene membranes for Sr(II) and Pb(II) is highest in case of PP *g*(AAc/AN)-KOH followed by PP-*g*(AAc/AN)-HCl, PP-*g*(AAc/AN)-NH_2OH, and PP-*g*(AAc/AN) are shown in Tables 4 and 5.

Table 5. Comparison of the isotherm constants for Pb(II)

Membrane type	Langmuir const. *X*m	Freundlich const. *Kf*	1/*n*
PP-*g*(AAc/AN)	312.5	9.95	0.32
PP-*g*(AAc/AN)	526.3	1.35	0.54
PP-*g*(AAc/AN)-KOH	555.5	1.33	0.48
PP-*g*(AAc/AN)-NH_2OH	400.0	8.52	0.53

4. CONCLUSIONS

This paper focuses on the treatment of wastewater by using new chelating grafted membrane mainly preparation and characterization of polypropylene membranese. Application of the prepared membrane in the treatment of wastewater was tested. From the results it is evident that the PP membranes have the ability to accumulate relatively large amounts of metals from aqueous solution.

Acknowledgment: I would like to express my deepest gratitude and sincere thanks to the NATO program for security through science and to Prof. Mohammed Zaidi for his kind cooperation and faithful help and for granting financial support to present this paper.

5. REFERENCES

1. Arup KS. 2002, Environmental Separation of Heavy Metals, CRC Press US (7), 105.
2. Pollard S, Fowler T, Sollars CJ, and Perry R. 1992, Sci. Total Environ., (31), 116
3. Lalvanii SB, Wlltowski T, Hubner A, and Mandiich N. 1998, Carbon, (36) 1219.
4. Dal MH and Wu SC. 1975, Sep. Sci., (10), 633.
5. Quek Sy, Wase D, and Forster CF. 1998, Water SA, 24 (3) 251.
6. Ismail IM, El-Sourougy MR, Abdel Moneim N, and Aly HF. 1999, J. Radioanal. Nucl. Chem., 240 (1) 59.
7. Ölmez S, Aytas S, Akyil M, Aslani A, and Aytekin U. 1999, J. Radioanal. Nucl. Chem., (3) 973.
8. Abdel RN and El-Sofany EA. 2003, Arab J. Nucl. Sci. Appl., 36 (3) 35.
9. Abdel RN and Zaki A. 2003, Arab J. Nucl. Sci. Appl., (3) 25.
10. Lasheen YF, El-Sofany EA, Borai EH, and Aly HF. 2003, Arab J. of Nucl. Sci. and Applic., 36 (3) 47.
11. Daneshvar N, Salari D, and Aber S. 2002, J. of Hazardous Materials, (49) 94.
12. Gallatsatou P, Metaxas M, and Kasselour-Rigpoulou V. 2002, J. of Hazardous Materials, (18) 91.
13. Mathialagan T and Vraraghavan T. 2002, J. of Haz. Materials, (291) 94.

14. Hegazy EA, Abd El-Rehim HA, Khalifa NA, Atwa SM, and Shawky HA. 1997, J. Polym. Int., (43) 321.
15. Abd El- Rehim HA, Hegazy EA, and El-Hag Ali A. 1999, J. React Funct. Polym.,14.
16. Hegazy EA, Abd El-Rehim HA, Khalifa NA, and El-Hag AA. 1997, In IAEA TecDoc-1023, Proceedings of the Symposium on Radiation Technology for Conservation of the Environment, Zakopane, Poland, Sept. 8–12 573.
17. Hegazy EA, Abd El-Rehim HA, Ali AMI, Nowier HG, and Aly HF. 1999, J. Nucl. Instrum. Method Phys. Res. B, 393.
18. Hegazy EA, Kamal H, Khalifa NA, and Mahmoud GA. 2001, J. Appl. Polym. Scien., (81) 849.
19. Ho YS, Ph.D. Thesis, 1995, "Adsorption of Heavy Metals from Waste Streams by Natural Materials", University of Birmingham, UK.
20. Casey TJ. 1997, Unit Treatment Processes in Water and Wastewater Engineering, John Wiley and Sons Ltd., England, 113.
21. Nayak D, Lahiri S, Mukhopadhyay A, and Pal R. 2003, J. Radioanal. Nucl. Chem., 256 (3) 535.
22. Radway JAC, Wilde E, Whitaker MJ, and Weisman JC. 2001, J. Appl. Phycol., 13 (451).
23. Low KS, Lee CK, and Leo AC. 1995, J. Bioresour. Technol., 51 (15).
24. Hegazy EA, and Shawky HA. 1997, J. Polym. International, (43) 321.
25. Hegazy EA, El-Arnaouty MB, Abdel GAM, and Taher NH. 2003, Arab J. Nucl. Sci. Appl., 36 (3) 97.

THE WASTEWATER TREATMENT – THE SUSTAINABLE DEVELOPMENT STRATEGY IN LITHUANIA

Gytautas Ignatavicius
University of Vilnius, Lithuania
Gaudenta Sakalauskiene, JSC "Daugela", Zalgirio street
90-505, LT-93053 Vilnius, Lithuania

Corresponding author: gaudenta@daugela.lt

Abstract:

Pollutants discharged directly or indirectly into river basins make harm to the whole water ecosystem. According to these reasons it is important to evaluate the pollution load and its economic value to river basin. According to the Directive 2000/60/EC of the European Parliament and the Council of 23 October 2000, establishing a framework for Community action in the field of water policy it is necessary not only to analyze the specific character of the river basin, impact of human activity, but also to make its economic analysis. These factors could make the main basis in the creation of new programs at protecting and improving water ecosystems. In this article, the pollution load of organic compounds and nutrients (total nitrogen and phosphorus) was estimated in the Lithuanian rivers from point and non point sources.

Keywords: Wastewater, river quality, Water Framework Directive (WFD).

1. INTRODUCTION

The main strategic document in the field of the environmental protection at the national level is the Lithuanian Environmental Strategy, issued and approved by the Lithuanian Republic Parliament in 1996 [1]. Until end of 2015, Lithuania has the intention to implement the main environmental infrastructure investments projects in the mostly sectors of water and waste management, as well as, air protection, in order to ensure requirements for environmental quality.

The following are the priorities of environmental problems in Lithuania that have been recognized by us:

- waste water treatment and reduction of discharges;
- air pollution reduction/stabilization;
- waste management;

M.K. Zaidi (ed.), Wastewater Reuse–Risk Assessment, Decision-Making and Environmental Security, 433–442.

- hazardous wastes management;
- domestic and other non hazardous wastes management;
- protection from physical pollution;
- land use and forest structure optimization;
- prevention of further natural landscape degradation;
- protection of ecologically sensitive and most natural areas;
- rehabilitation of worked out mineral resources quarries;
- rational use of natural resources.

Lithuania has elaborated corresponding action program on implementation of national environmental strategy, covering the water sector, especially. Construction of wastewater treatment facilities and improvement of clear drinking water supply remains the highest priority for investments, particularly for funds from State budget, and loans and subsidies received by the State. It is also necessary to implement measures for the reduction of non point source pollution of ground and surface waters and to develop the necessary water protection laws.

So, water quality protection has been and is still considered one of the priorities of Lithuania's environmental protection, for which major part of all the investments designated for environmental protection was allocated over the last ten years. Nevertheless, recently the need has emerged again for increasing investments into the water sector. This is primarily associated with the implementation of the European Union requirements.Currently, more than 25 directives and decisions regulate the water sector in the European Union.

The main directives determining the quality of surface waters is the Water Framework Directive (WFD) (2000/60/EC) [2] and Directive on the Quality of Fresh Waters Needing Protection or Improvement in Order to Support Fish Life (78/659/EC) [3]; emission limit values are presented in the Directive on Dangerous Substances (76/464/EEC) as well as in its seven daughter directives [4]; pollution from point sources is regulated by the Urban Wastewater Treatment Directive (91/271/EEC) [5].

Before construction or rehabilitation of wastewater treatment plants for towns exceeding the population equivalent of 2000, it is required to apply environmental impact assessment procedures according to both the Lithuanian and EU requirements. During the assessment it is very important to predict and assess the effects on surface water quality that are related to a waste water treatment plant. First of all, it is required that pollution standards in the effluent from a wastewater treatment plant will not exceed standards provided in the Lithuanian legislation. Additionally, it is necessary to ensure that the existing surface water quality will remain at the existing level.

2. BACKGROUND INFORMATION

With a surface of 65,300 km^2 and a population of 3.7 million (of which two-thirds urban) Lithuania is the largest Baltic country in territory of former Soviet Union. The demographic density averages 57 inhabitants per km^2. The capital, Vilnius, has almost 600,000 inhabitants. About 56% of the land is used for agriculture, two-thirds for crops, one-third for livestock. Forests cover 30 % and water bodies 7% of the territory. Lithuania is a low-lying country, peaking at 293.6 m of altitude in Juozapinė. Nine major rivers are longer than 200 km: the Nemunas, the Neris, the Venta, the Šešupė, the Mūša, the Šventoji, the Nevėžis, the Merkys, and the Minija. Apart from the Nevėžis and the Minija, they drain international river basins. The average density of the river network is 1 km/ km^2.

3. POINT AND DIFFUSE SOURCES

3.1. Point Sources

The Urban Wastewater Directive covers 84 Lithuanian cities (Figure 1). The analysis of their sewerage system shows that nitrogen and phosphorus are removed in 15 urban wastewater treatment plants (WWTP). In 61 urban WWTP, biological treatment is applied, while in six towns, effluents are treated mechanically. The wastewater from two towns is discharged without any treatment. As required by the Directive, all 84 cities should have biological treatment facilities and 38 major cities must remove nitrogen and phosphorus from wastewater. Although in the last decade large sums have been invested to build wastewater plants, only 28% of the population is served by biological treatment and 55% by treatment with nitrogen removal. As a result, just 14% of the population is being served by wastewater plants in line with the requirements of EU Directive 91/271/EC.

Figure 2 shows the present status of wastewater treatment in Lithuania. The data for 2003 shows significant increase in volume of effluents undergoing tertiary treatment (planned completion of Vilnius tertiary UWWTP). Analysis of present data shows that 46 agglomerations larger than 2,000 p.e. have to build or upgrade urban waste water treatment plants in order to comply with the requirements of the EU Urban Waste Water Treatment Directive (Table 1).

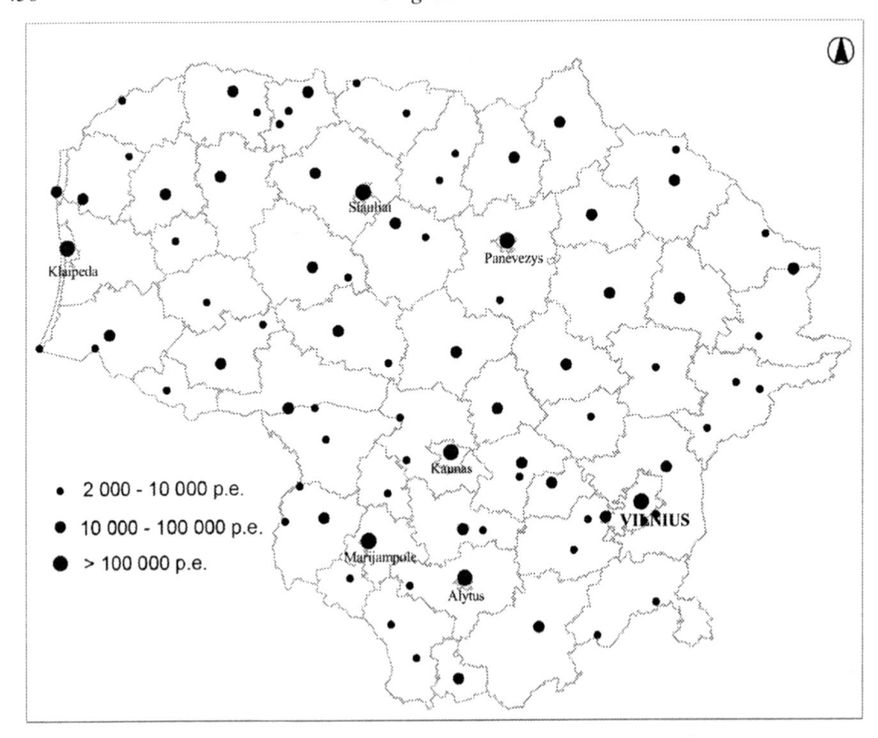

Figure 1. Municipal wastewater treatment plants in Lithuania

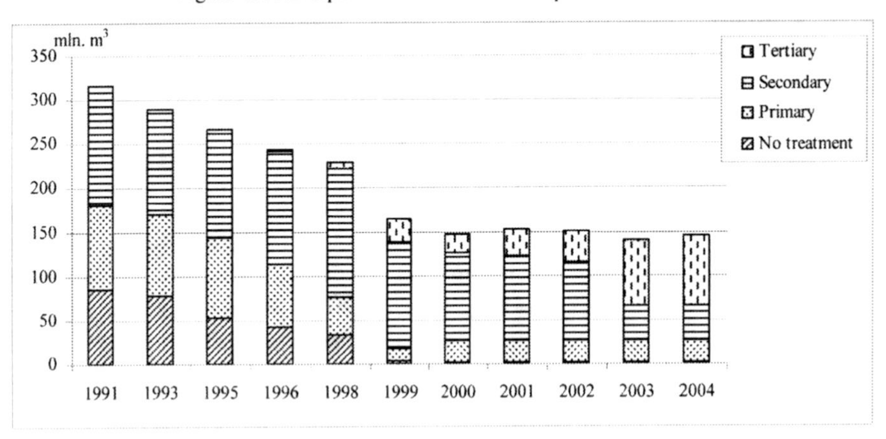

Figure 2. Wastewater treatment in Lithuania in the period 1991–2004

The following data on the organic and nutrient pollution discharged into Lithuanian rivers in 2000–2005 indicate pollution trends over a longer period (Figure 3). The trend toward a decrease in polluting substances (suspend solids, BOD).

Table 1. Summary table of the compliance with the requirements for wastewater treatment (based on data on discharges in 2000)

Size of agglomeration	Number of agglomerations in the size class	Number of agglomeration that comply with requirements of UWWTD			
		BOD_5	SS	N_{tot}	P_{tot}
>100,000 p.e.	7	6	6	1	0
10,000 – 100,000 p.e.	31	26	26	8	5
2,000 – 10,000 p.e.	46	38	36	–	–
Total	84	70	68	9	5

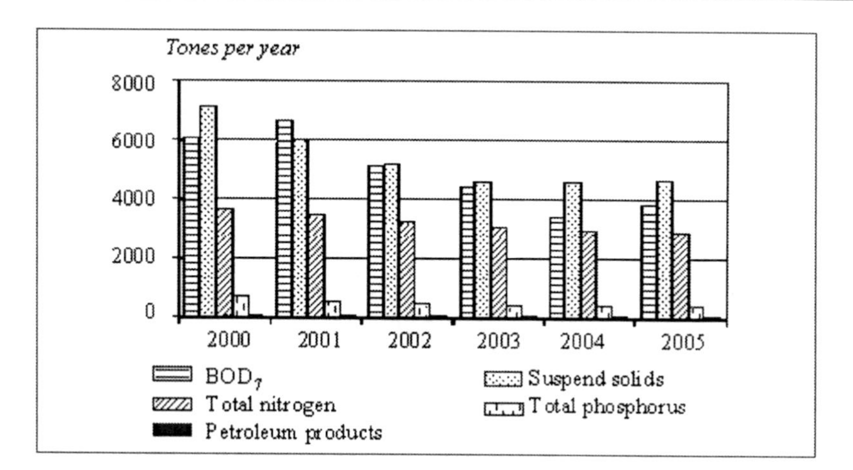

Figure 3. Pollution of Lithuanian rivers by organic compounds and nutrients in year 2000–2005

3.2 Non point Sources

Although point sources are certainly important, many nitrogen and phosphorus compounds enter the streams and rivers by the non point or diffuse way (Figure 4). For example, significant amounts of nitrogen and phosphorus can be brought by urban and agricultural runoff [6]. The main indicators, which characterize the diffuse pollution load, are land use (especially total area of arable land), number of livestock (farming intensity), and usage of fertilizers, plant protection chemicals, and manure. Due to the Lithuanian agriculture 25 thousand tones of nitrogen and 2.2 thousands tones of phosphorus are discharged to the surface waters [7]. The bigger part of this amount flow in to the rivers (≈60%).

Agricultural land use makes 56% of entire territory of the Lithuania. Arable land takes 67% of agricultural land use. Lithuanian agricultural production was significantly reduced in the last decade. Negative impact on the environment decreased as well. The use of fertilizers decreased by 90%. Almost 80% of the Lithuanian agricultural lands were artificially drained. This increased transportation of the nutrients from agricultural fields to the water bodies.

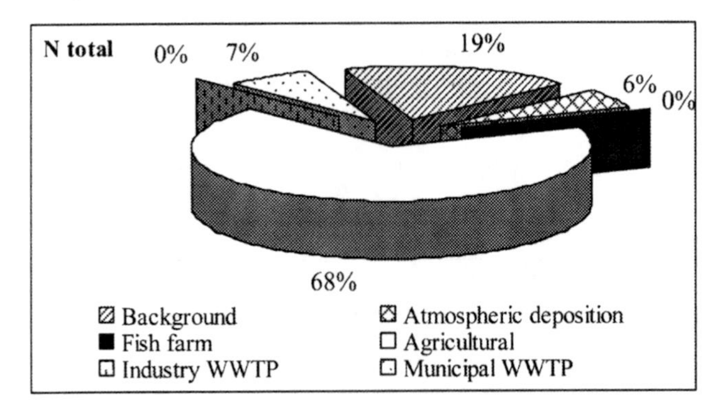

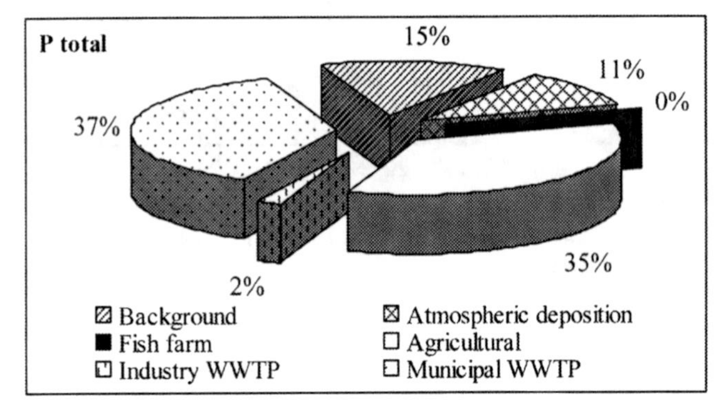

Figure 4. Point and non point sources discharges of total nitrogen and total phosphorus into surface waters in 2000

4. SURFACE WATER QUALITY

The EC-Water Framework Directive (WFD, 2000) has the objective of an integrated catchments-oriented water quality protection for all European waters with the purpose of attaining a good quality status by the year 2015. The water quality evaluation for surface waters shall rely predominantly on biological parameters (such as flora and fauna)—however, aided by hydromorphological (such as flow and substrate conditions) and physical–chemical quality components (such as temperature, oxygen or nutrient conditions)—and on specific pollutants (such as metals or synthetic organic compounds). A good chemical quality status is provided when the environmental quality standards are met for all pollutants or pollutant groups.

The WFD guidance document regarding the analysis of pressures and impacts recommends that the assessment of impacts on river water quality should be based on measurements to the extend possible and that modeling should be supportive only. The purpose of the assessment is of course to verify whether the identified pressures (e.g., point sources) pose a threat to the particular water body. However, water quality measurements provide information about river water quality at the location of monitoring station only and without any link to the upstream pressures. Further, it does not provide a full picture about the actual status of the entire water body.

The GIS models tool like MIKE BASIN, QUAL, and WAMP on the other hand links pressures with the river water quality and at the same time provide a more complete picture of pollutant fate and concentrations within the entire water body (filling of gap between monitoring stations). It is clear that the use of models facilitates the identification of river water quality parameters reaches not meeting water quality objectives.

It is further stated in the same documents that "evaluating the risk of failing objectives should be a straightforward comparison of the state of the water body with threshold values that define the objective," the objective being good ecological status. The challenge is of course to define good ecological status in terms of monitored or modeled parameters. Thus, in order to identify water bodies at risk a criteria for classification of ecological status of individual water bodies was introduced. Point sources were defined as "significant" if the pollution discharge would result in an: increase in biochemical oxygen demand (BOD) in the receiving river by 0.5 mg/l [8]. The pollutant concentration in the river downstream a pollutant outlet (assuming complete mixing) was calculated according to the following formula:

$$C_{river} = \frac{C_{riv\_ups} Q_{riv\_ups} + C_{ps} Q_{ps}}{Q_{riv\_ups} + Q_{ps}}; \tag{1}$$

where

C_{river} – expected pollutant concentration in the river downstream discharger;
C_{riv_ups} – pollutant concentration in the river upstream point source;
Q_{riv_ups} – river discharge upstream point source;
C_{ps} – pollutant concentration in wastewater;
Q_{ps} – waste water discharge.

The analysis revealed 84 point sources may individually pose a potential risk to increase BOD concentrations in a receiving river by more than 0.5 mgO_2/l. However, most of these point sources discharge wastewater into quite small streams. In addition, errors with regard to exact location of many of these point sources are possible and calculation results may be uncertain. So, even though the risk to impair water quality exists, it is not reasonable to include all these point sources into the model individually. Based on the analysis results, 50 point sources were finally identified as being significant. These are mainly WWTP's of medium size settlements and towns discharging municipal wastewater into monitored rivers or their tributaries.

So, Lithuania has 758 rivers, which are longer than 10 km, and 18 longer than 100 km. In 2002, wastewater discharges accounted for 14% of the average annual water flow of all rivers. The greatest amount of pollutants was discharged from the cities of Vilnius, Kaunas, Siauliai, Panevezys, and Alytus (Figure 5).

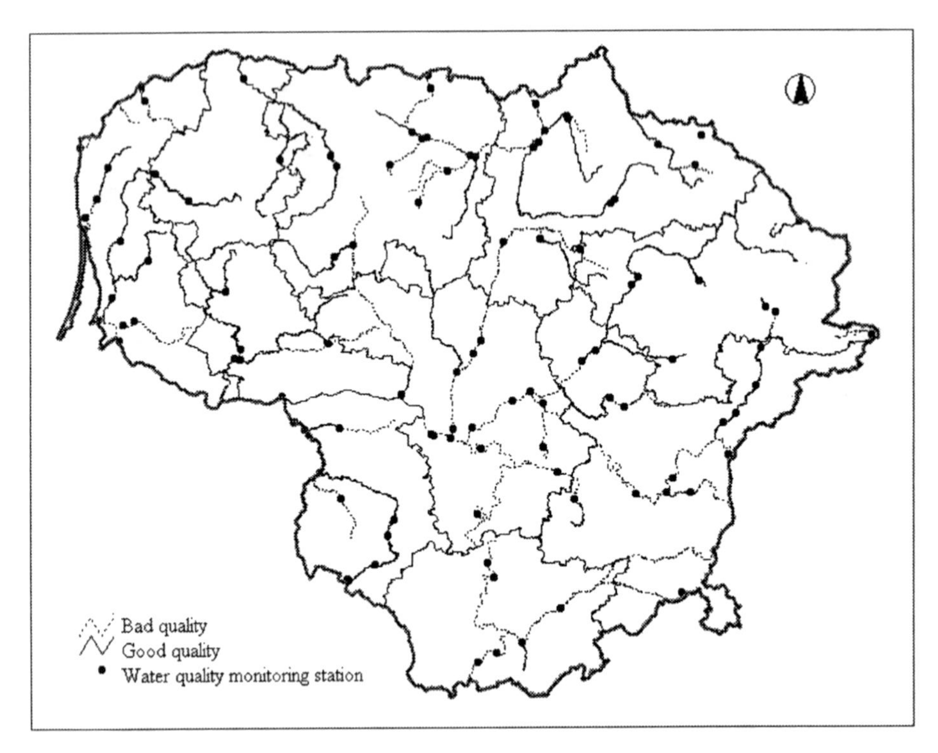

Figure 5. Simple classification of ecological status of Lithuanian rivers based on average BOD_7 concentrations in 2002. Solid line = good ecological status (i.e., $BOD_7<3$ mgO_2/l), dotted line = poor ecological status (i.e., $BOD_7>3$ mgO_2/l)

Under the provisions of the WFD, watercourses have henceforth to be subdivided into five classes of ecological status. The objective there is that all water bodies should achieve good ecological status by 2015. In Lithuania the system and numeric values for high, moderate, poor, and bad water quality in rivers proposed for classifying present status based on monitoring data. This system is based on the general provisions of WFD. Good surface water quality is defined in line with requirements for wish water (Wish water Fish Directive (78/659/EEC)).

Water quality in the Lithuanian rivers is monitored at 234 river's quality monitoring stations-44 surveillance, 174 operations, and 16 references. Most of the monitoring points are located upstream and downstream of the cities.

In order to classify the ecological status of individual water bodies, simple criteria have been adopted. This implies that during the average BOD concentrations should be less than 3 mg/l in order to be classified as good ecological status. With such a relatively simple criteria (considering BOD only) MIKE BASIN can easily provide an overview of water bodies complying or not complying with these conditions.

5. CONCLUSION

Lithuania has 758 rivers, which are longer than 10 km, and 18 longer than 100 km [9]. In 2002, wastewater discharges accounted for 14% of the average annual water flow of all rivers. Water quality protection has been and is still considered one of the priorities of Lithuania's environmental protection, for which major part of all the investments designated for environmental protection was allocated over the last ten years. Water quality in the Lithuanian rivers is monitored at 234 river's quality monitoring stations-44 surveillance, 174 operations, and 16 references. The greatest amount of pollutants was discharged from the largest cities. Lithuania has elaborated corresponding action program on implementation of national environmental strategy, covering the water sector. In order to ensure requirements for environmental quality and with the general intention to implement the main environmental infrastructure for surface water bodies, we plan to achieve a good quality status by the year 2015.

Acknowledgment: NATO ARW advisory committee provided full financial support to present this paper at the NATO ARW–Wastewater reuse–risk assessment, decision-making and environmental security held in Istanbul, Turkey during October 12–16, 2006.

6. REFERENCES

1. Lithuanian Environmental Strategy. (1996). Vilnius.
2. Water Framework Directive, 2000/60/EC. (2000). Office Journal of the European Communities. L327.
3. Council Directive 78/659 of 18 July 1978 on the quality of fresh waters needing protection or improvement in order to support fish life. (1978). Office J.European Communities. L 222/1.
4. Council Directive 76/464 of 4 May 1976 on pollution caused by certain dangerous substances discharged into the aquatic environment of the Community. (1976). Office Journal of the European Communities. L 129/23.
5. Council Directive 91/271 of 21 May 1991 concerning urban waste water treatment. (1991). Offical J.European Communities. L 135/40.
6. The Fourth Baltic Sea Pollution Load Compilation (PLC-4). (2003). Baltic Sea Environmental Proceedings No. 93, 184.
7. Belous O, Zukauskaite A, Ringilaite I. (2003). Nutrient losses in river basin through diffuse pollution form analysis. Environmental research, engineering and management. 1 (23), 12–20.
8. Report “Tools for groundwater and surface water analysis during the implementation of the WFD in Lithuania”. (2004). DHI Water & Environment.
9. Cetkauskaite A, Zarkov D, Stoskus L. (2001). Water-quality control, monitoring and WW treatment in Lithuania. Ambio. 30 (4–5), 297–305.

INDEX

Printed in the United States
93194LV00003B/1/A

9 781402 060267